Linear

Linear Algebra

Richard Kaye

and

Robert Wilson

School of Mathematics and Statistics
The University of Birmingham

OXFORD • NEW YORK • TOKYO
OXFORD UNIVERSITY PRESS

*This book has been printed digitally and produced in a standard specification
in order to ensure its continuing availability*

OXFORD
UNIVERSITY PRESS

Great Clarendon Street, Oxford OX2 6DP

Oxford University Press is a department of the University of Oxford.
It furthers the University's objective of excellence in research, scholarship,
and education by publishing worldwide in

Oxford New York

Auckland Bangkok Buenos Aires Cape Town Chennai
Dar es Salaam Delhi Hong Kong Istanbul Karachi Kolkata
Kuala Lumpur Madrid Melbourne Mexico City Mumbai Nairobi
São Paulo Shanghai Taipei Tokyo Toronto

Oxford is a registered trade mark of Oxford University Press
in the UK and in certain other countries

Published in the United States
by Oxford University Press Inc., New York

ISBN 0-19-850237 0

Preface

This book constitutes a second course in linear algebra, and is based on second-year courses given first by RW and then by RK in Birmingham over the last five years.

The objectives of such a course are as follows. Firstly, the student must learn a whole host of algebraic methods associated with bilinear forms, inner products, eigenvectors, and diagonalization of matrices, and be confident in carrying out calculations with these in all areas of mathematics. Secondly, this course is likely to be one of the first places where the student meets the axiomatic method of abstract algebra, and as such serves as an introduction to abstract algebra in general.

We believe that these two requirements can be successfully married into a single course. By this stage in a student's career, vectors and matrices should be sufficiently familiar that the jump to the full rigour of the axiomatic treatment of vector spaces is not such a great one. Our approach throughout is to show how certain key examples suggest axioms, and then to prove 'structure theorems' showing that the abstract objects satisfying the axioms are isomorphic to one of the 'canonical' examples for finite dimensional vector spaces over $\mathbb{R}$ or $\mathbb{C}$ or other fields. More advanced theorems can then be proved in a 'concrete way' using matrices and column vectors. This approach has the advantages that matrices (with which students are generally quite comfortable) are never very far away, and that proofs coincide very closely with the calculations that the students will be required to do, so algorithms or methods can be obtained from studying proofs (a useful lesson to learn in general). Obviously, we give plenty of examples in the text, and in doing so we are able to point out exactly where the proofs of the corresponding theorems suggest how to do the calculation in question.

We have assumed a certain amount of familiarity with matrices and basic operations on them (addition, multiplication, transpose, determinants, and calculating inverses), and the student should be able to solve simultaneous linear equations of the form $\mathbf{Ax} = \mathbf{b}$, obtaining the full solution set. This much at least will be included in a first-year programme of study, and this material is summarized in Chapter 1. We also assume some basic knowledge of the complex numbers, but we do not assume the student has encountered vector spaces over $\mathbb{C}$ before. Properties of polynomials that are required for understanding of the Cayley–Hamilton Theorem and the minimum polynomial of a linear transformation are set out in a suitable place in Part III.

Typically, a student at this level will have met the concepts of *vector space* over $\mathbb{R}$ and *dimension* before, but full familiarity with these ideas is not necessary since this material is revised fully in Chapter 2. Clearly the extent to which this chapter needs to be revised or expanded upon is at the discretion of the lecturer or the student. We have chosen to include all of the basic material on linear independence, bases and dimension here, even though for most students this will be revision.

Almost all of the time our vector spaces are finite dimensional, but since students at this level will occasionally meet applications (such as Fourier series) which require infinite dimensional vector spaces we have included some material on these too; in most cases we prove results for finite dimensional vector spaces only, indicating afterward whether or not the theorem is also true for infinite dimensional spaces.

We have also included a brief introduction to fields, and this may well be new material to a student at this level. The reasons for including this material are clear: many examples and applications of vector spaces that a typical undergraduate will see will involve fields other than $\mathbb{R}$ or $\mathbb{C}$, and the field axioms provide an important illustration of the axiomatic approach. We do not allow ourselves to dwell on the subject, though, giving for example just a few examples of finite fields rather than a complete classification. In any case the main emphasis of the book is on spaces over $\mathbb{R}$ or $\mathbb{C}$, and this section is optional.

Part II goes on to discuss inner product spaces in general, and also bilinear forms and quadratic forms on real vector spaces, culminating in their full description via the diagonal form given by the Gram–Schmidt orthonormalization (for inner products) and by 'Sylvester's law of inertia' in the more general case of symmetric bilinear forms. In the case of spaces over the complex numbers, conjugate symmetric forms are also considered and the corresponding laws are derived by the same methods. To keep things reasonably straightforward, Part II is concerned mostly with spaces over $\mathbb{R}$ or $\mathbb{C}$.

Part III contains a full discussion of linear transformations from a finite dimensional vector space V to itself, their eigenvalues, eigenvectors, and diagonalization. The algorithms performing these computations are emphasized throughout. Determinants are used as an aid to computations (the characteristic polynomial) but are not required for a full understanding of the theory. The book ends with two chapters that emphasize applications of the material presented in the whole book: one on self-adjoint transformations on inner product spaces and the final chapter on Jordan normal form.

In Part III, our vector spaces are over an arbitrary field F with the only condition on F being that the minimum polynomial of the linear transformation f in question splits over F. The student may prefer to continue to read the text as if F is $\mathbb{R}$ or $\mathbb{C}$.

Of course, there is too much material for a single course here, and it is up to the lecturer to decide on the course content and emphasis. With the material in Part I taken as understood, it would certainly be possible to cover all of the topics here as 'algorithms' or 'methods' in a single course, leaving the brighter students to follow up the sections explaining why some of the more difficult ideas (such as Sylvester's law of inertia, or the Jordan normal form) really do work.

Alternatively, some sections and/or chapters can be omitted altogether without interrupting the flow of the text. For example, Sections 2.6, 4.4, 5.4, 6.2, 7.4 and 7.5 may be omitted and Chapters 13 and 14 are independent of each other, so one or both of these could be omitted (though it should be mentioned that these last two chapters are in some ways the most important of all for applications).

Birmingham   RK
October 1997   RW

Contents

Part I

Matrices and vector spaces

1

Matrices

Column vectors and matrices will appear in many ways in this book. On the one hand, they provide particularly important examples of vector spaces (the subject of the next chapter) and various operations on vector spaces (the subject of Parts II and III). On the other hand, they provide elegant and convenient mathematical tools for performing calculations in all sorts of types of problems, including problems which at first sight have nothing to do with matrices.

You should already have some familiarity with matrices and some of the more common matrix operations. The purpose of this chapter is to review the required material that will be needed later.

This will be done in a somewhat more 'advanced' way than when you perhaps learnt this material. The same basic ideas concerning matrices appear in several different contexts, and can be usefully presented in a rather 'unified' way.

1.1 Matrices

An $n \times m$ *matrix* is an array of numbers of the form

$$\begin{pmatrix} a_{11} & a_{12} & \ldots & a_{1m} \\ a_{21} & a_{22} & \ldots & a_{2m} \\ a_{31} & a_{32} & \ldots & a_{3m} \\ \vdots & \vdots & \ddots & \vdots \\ a_{n1} & a_{n2} & \ldots & a_{nm} \end{pmatrix}.$$

Sometimes, this matrix will be denoted by (a_{ij}), where a_{ij} is a 'general' element of the array and the brackets () indicate that the whole array is to be considered. Note that the first coordinate, i in a_{ij}, refers to the *row* of the matrix, and the second, j, refers to the *column*. Similarly, an $n \times m$ matrix has n rows and m columns. In the special case when $n = m$, i.e. the matrix has the same number of rows as columns, the matrix will be said to be *square*.

We will denote the set of $n \times m$ matrices with entries from $\mathbb{R}$ (the real numbers) by $M_{n,m}(\mathbb{R})$, and $n \times m$ matrices with entries from $\mathbb{C}$ (the complex numbers) by $M_{n,m}(\mathbb{C})$.

On its own, a matrix is nothing more than an array of numbers, but matrices can also *represent* other mathematical objects, as we will see later. Because of this they present a very useful mathematical shorthand. Moreover, matrices can be added, multiplied, transposed, and there are other important operations too.

1.2 Addition and multiplication of matrices

When $\mathbf{A} = (a_{ij})$ and $\mathbf{B} = (b_{ij})$ are both $n \times m$ matrices, then they can be added and the result, $\mathbf{A} + \mathbf{B}$, is the matrix whose entries are the sum of the corresponding entries in $\mathbf{A}$ and $\mathbf{B}$, i.e. the $n \times m$ matrix with (i,j)th entry $a_{ij} + b_{ij}$. If $\mathbf{A}$ and $\mathbf{B}$ are of different sizes, then the sum $\mathbf{A} + \mathbf{B}$ is *not defined*. The $n \times m$ *zero matrix*, $\mathbf{0}_{n,m}$, is the matrix all of whose entries are zero. When the size is clear from context, we omit the n, m from the notation, writing simply $\mathbf{0}$.

Note the following laws for matrix addition.

1. (Associativity.) $(\mathbf{A} + \mathbf{B}) + \mathbf{C} = \mathbf{A} + (\mathbf{B} + \mathbf{C})$.
2. (Commutativity.) $\mathbf{A} + \mathbf{B} = \mathbf{B} + \mathbf{A}$.
3. (Zero.) $\mathbf{0} + \mathbf{A} = \mathbf{A} + \mathbf{0} = \mathbf{A}$.

These apply whenever $\mathbf{A}, \mathbf{B}, \mathbf{C}$ are the correct size for the addition operations here to be defined.

Given an $n \times m$ matrix $\mathbf{A} = (a_{ij})$, and a number λ from $\mathbb{R}$ or $\mathbb{C}$, we define $\lambda\mathbf{A}$ to be the $n \times m$ matrix whose entries are λa_{ij}. The operation of multiplying a number λ by a matrix $\mathbf{A}$ is called *scalar multiplication*. The matrix $-\mathbf{A}$ is defined to be $-1\mathbf{A}$. We have the following laws for scalar multiplication.

4. $\lambda(\mu\mathbf{A}) = (\lambda\mu)\mathbf{A}$.
5. $(\lambda + \mu)\mathbf{A} = \lambda\mathbf{A} + \mu\mathbf{A}$.
6. $0(\mathbf{A}) = \mathbf{0}$.
7. $\mathbf{A} + (-\mathbf{A}) = \mathbf{0}$.

Two matrices $\mathbf{A} = (a_{ij})$ and $\mathbf{B} = (b_{ij})$ can sometimes be multiplied together, but only if their sizes agree in a special way. The rule is as follows: the matrix product $\mathbf{AB}$ only exists when $\mathbf{A}$ is $n \times m$ and $\mathbf{B}$ is $m \times k$ for some n, m, k. When this is the case, $\mathbf{AB}$ is the $n \times k$ matrix with (i,j)th entry

$$\sum_{r=1}^{m} a_{ir}b_{rj} = a_{i1}b_{1j} + a_{i2}b_{2j} + \cdots + a_{im}b_{mj}.$$

The $n \times n$ *identity* matrix $\mathbf{I}_n$ is the matrix (a_{ij}) with diagonal entries $a_{ii} = 1$ and all other entries 0. When n is clear from context, it is just denoted $\mathbf{I}$.

Note that matrix multiplication is not in general commutative, and it is not hard to find examples, such as

$$\begin{pmatrix} 1 & 2 \\ -1 & 0 \end{pmatrix} \begin{pmatrix} 1 & -1 \\ 0 & 1 \end{pmatrix} = \begin{pmatrix} 1 & 1 \\ -1 & 1 \end{pmatrix} \neq \begin{pmatrix} 2 & 2 \\ -1 & 0 \end{pmatrix} = \begin{pmatrix} 1 & -1 \\ 0 & 1 \end{pmatrix} \begin{pmatrix} 1 & 2 \\ -1 & 0 \end{pmatrix}.$$

Even worse, it may be that $\mathbf{AB}$ is defined but $\mathbf{BA}$ is not. However, provided the following matrices are all of the correct size for the expressions to be defined, we have the following laws.

8. (Associativity.) $(\mathbf{AB})\mathbf{C} = \mathbf{A}(\mathbf{BC})$.

9. (Zero.) $\mathbf{0A} = \mathbf{A0} = \mathbf{0}$.
10. (Identity.) $\mathbf{IA} = \mathbf{AI} = \mathbf{A}$.
11. (Distributivity.) $\mathbf{A(B + C)} = \mathbf{AB} + \mathbf{AC}$.
12. (Distributivity.) $\mathbf{(A + B)C} = \mathbf{AC} + \mathbf{BC}$.

Square matrices will be particularly important in this book. We note here that if $\mathbf{A}, \mathbf{B}$ are $n \times n$ matrices, then $\mathbf{A} + \mathbf{B}$, $\mathbf{AB}$ and $\lambda\mathbf{A}, \lambda\mathbf{B}$ are all $n \times n$ matrices; thus all of the twelve laws just given hold for $n \times n$ matrices $\mathbf{A}, \mathbf{B}, \mathbf{C}$.

1.3 The inverse of a matrix

If $\mathbf{A}$ is a square matrix, it may be that there is a matrix $\mathbf{B}$ such that $\mathbf{AB} = \mathbf{I}$. When this happens, $\mathbf{B}$ is called a *right inverse* of $\mathbf{A}$. Similarly, it may be that there is a *left inverse* $\mathbf{C}$ of $\mathbf{A}$, satisfying $\mathbf{CA} = \mathbf{I}$. Not all matrices have inverses in this way, but it is an interesting (and not entirely straightforward) fact that the left and right inverses of a square matrix $\mathbf{A}$, if they exist, are the same.

Fact 1.1 *Let $\mathbf{A}$ be an $n \times n$ matrix and suppose $\mathbf{B}$ is either a left or a right inverse of $\mathbf{A}$. Then $\mathbf{B}$ is a two-sided inverse, i.e. $\mathbf{BA} = \mathbf{AB} = \mathbf{I}$, and this inverse is unique, so if $\mathbf{C}$ satisfies either $\mathbf{CA} = \mathbf{I}$ or $\mathbf{AC} = \mathbf{I}$, then $\mathbf{C} = \mathbf{B}$.*

(See Exercise 1.15 for a way to prove this.)

We denote this unique inverse of a square matrix $\mathbf{A}$ when it exists by $\mathbf{A}^{-1}$. If $\mathbf{A}^{-1}$ exists we say $\mathbf{A}$ is *invertible*.

Proposition 1.2 *If $\mathbf{A}$ and $\mathbf{B}$ are invertible square matrices, then $\mathbf{AB}$ is also invertible and $(\mathbf{AB})^{-1}$ equals $\mathbf{B}^{-1}\mathbf{A}^{-1}$.*

Proof Using the associativity law,

$$(\mathbf{AB})(\mathbf{B}^{-1}\mathbf{A}^{-1}) = \mathbf{A}(\mathbf{B}(\mathbf{B}^{-1}\mathbf{A}^{-1})) = \mathbf{A}((\mathbf{BB}^{-1})\mathbf{A}^{-1})$$
$$= \mathbf{A}(\mathbf{IA}^{-1}) = \mathbf{AA}^{-1} = \mathbf{I}.$$

Hence $\mathbf{B}^{-1}\mathbf{A}^{-1}$ is a right inverse of $\mathbf{AB}$, and therefore is the unique inverse of $\mathbf{AB}$. $\qquad\square$

1.4 The transpose of a matrix

The *transpose* operation (which is notated with a ^{T} sign) converts an $n \times m$ matrix to an $m \times n$ matrix as follows:

$$\begin{pmatrix} a_{11} & a_{12} & \cdots & a_{1m} \\ a_{21} & a_{22} & \cdots & a_{2m} \\ a_{31} & a_{32} & \cdots & a_{3m} \\ \vdots & \vdots & \ddots & \vdots \\ a_{n1} & a_{n2} & \cdots & a_{nm} \end{pmatrix}^T = \begin{pmatrix} a_{11} & a_{21} & a_{31} & \cdots & a_{n1} \\ a_{12} & a_{22} & a_{32} & \cdots & a_{n2} \\ \vdots & \vdots & \vdots & \ddots & \vdots \\ a_{1m} & a_{2m} & a_{3m} & \cdots & a_{nm} \end{pmatrix}.$$

Proposition 1.3 *For $n \times n$ matrices $\mathbf{A}, \mathbf{B}$:*

(a) $(\mathbf{AB})^T = \mathbf{B}^T\mathbf{A}^T$;

(b) *if* $\mathbf{A}$ *is invertible, then so is* $\mathbf{A}^T$, *and* $(\mathbf{A}^T)^{-1} = (\mathbf{A}^{-1})^T$.

Proof (a) A typical entry c_{ji} of $\mathbf{C} = \mathbf{AB}$ is $\sum_{k=1}^n a_{jk}b_{ki}$, and this is the (i,j)th entry of $(\mathbf{AB})^T$. On the other hand, the (i,j)th entry of $\mathbf{B}^T\mathbf{A}^T$ is $\sum_{k=1}^n b_{ki}a_{jk}$, which is precisely c_{ji}.

(b) We use the fact that the inverse $\mathbf{B}^{-1}$ of a matrix $\mathbf{B}$, when it exists, is both a left and right inverse (i.e. $\mathbf{B}^{-1}\mathbf{B} = \mathbf{BB}^{-1} = \mathbf{I}$) and is unique. By part (a) we have $(\mathbf{A}^{-1})^T\mathbf{A}^T = (\mathbf{AA}^{-1})^T = \mathbf{I}^T = \mathbf{I}$ so $(\mathbf{A}^{-1})^T$ is *an* inverse of $\mathbf{A}^T$. By uniqueness $(\mathbf{A}^{-1})^T = (\mathbf{A}^T)^{-1}$. $\qquad\square$

1.5 Row and column operations

An elementary *row operation* is a basic type of operation on matrices; for example, 'add row i to row j' is an elementary row operation. The three kinds of elementary row operations are:

1. $\rho_j := \rho_j + \lambda\rho_i$; 'add λ times row i to row j', for any number λ and any rows i, j;
2. $\rho_i := \lambda\rho_i$; 'multiply row i by λ', for any **nonzero** number λ and any row i;
3. $\text{swap}(\rho_i, \rho_j)$; 'swap rows i and j', for any two rows i, j.

(Note: the Greek letter ρ ('rho') sounds almost the same as the English word 'row'!) Each of these operations corresponds to multiplying on the left by a certain matrix. For example, for row operations on $3 \times n$ matrices,

1. $\rho_3 := \rho_3 + 2\rho_2$ corresponds to left-multiplying by $\begin{pmatrix} 1 & 0 & 0 \\ 0 & 1 & 0 \\ 0 & 2 & 1 \end{pmatrix}$;

2. $\rho_2 := \lambda\rho_2$ corresponds to left-multiplying by $\begin{pmatrix} 1 & 0 & 0 \\ 0 & \lambda & 0 \\ 0 & 0 & 1 \end{pmatrix}$; and

3. $\text{swap}(\rho_1, \rho_2)$ corresponds to left-multiplying by $\begin{pmatrix} 0 & 1 & 0 \\ 1 & 0 & 0 \\ 0 & 0 & 1 \end{pmatrix}$.

A *row operation* is a combination of elementary row operations, performed consecutively, and so, by associativity, is equivalent to multiplication on the left by a product of matrices of the forms above.

Exercise 1.1 Check your understanding by calculating $\mathbf{AB}$ directly for

$$\mathbf{B} = \begin{pmatrix} 1 & 2 & 3 & 4 \\ 2 & -1 & 1 & 0 \\ -2 & 3 & 1 & 1 \end{pmatrix},$$

and for each matrix $\mathbf{A}$ in 1–3 above.

By combining elementary row operations, we can obtain row operations to perform several useful transformations of matrices.

Clearing the first column. Our first basic technique using row operations converts any matrix **A** to another matrix **B** of the form

$$\begin{pmatrix} b_{11} & b_{12} & \ldots & b_{1k} \\ 0 & b_{22} & \ldots & b_{2k} \\ \vdots & \vdots & \ddots & \vdots \\ 0 & b_{n2} & \ldots & b_{nk} \end{pmatrix},$$

so all but possibly the first entry in the first column is zero. We can furthermore arrange that b_{11} is either 1 or 0.

1. If all entries in the first column are already zero there is nothing to do.
2. Else use the swap operation to arrange that the top left entry a_{11} in the matrix is nonzero.
3. Optionally, use $\rho_1 := \lambda\rho_1$ where $\lambda = 1/a_{11}$ to ensure that the top left entry is one.
4. Now use operation $\rho_j := \rho_j + \mu\rho_1$ for each $j \geqslant 2$ and for suitable values of μ to ensure the rest of the first column is zero.

Exercise 1.2 Apply this method to convert the following.

(a) $\begin{pmatrix} 0 & 1 & 1 \\ 1 & 1 & 0 \\ 1 & 0 & 1 \end{pmatrix}$ (b) $\begin{pmatrix} 1 & 0 & 1 & 1 \\ -1 & 0 & 2 & 1 \\ -1 & 1 & 5 & 9 \end{pmatrix}$ (c) $\begin{pmatrix} 0 & 4 & 7 & 10 \\ 2 & 5 & 8 & 11 \\ 3 & 6 & 9 & 12 \end{pmatrix}$.

Echelon form. A matrix is in *echelon form* if it contains no adjacent rows of the form

$$\begin{matrix} 0 & 0 & \ldots & 0 & x_1 & x_2 & \ldots \\ 0 & 0 & \ldots & 0 & y_1 & y_2 & \ldots \end{matrix}$$

with $y_1 \neq 0$ (irrespective of what x_1 is). In other words, for a matrix in echelon form, each row starts with a sequence of zeros, and the number of zeros in this initial sequence increases as you go from one row to the next row beneath it until we get to the very last row, or until all other rows are entirely zero.

Exercise 1.3 Decide which of the following are in echelon form. Give reasons.

(a) $\begin{pmatrix} 0 & 1 & 2 & 3 \\ 0 & 0 & 0 & 1 \\ 0 & 0 & 0 & 0 \end{pmatrix}$ (b) $\begin{pmatrix} 1 & 0 & 1 & 2 & 3 \\ 2 & 0 & 0 & 0 & 1 \\ 0 & 1 & 0 & 0 & 0 \end{pmatrix}$ (c) $\begin{pmatrix} 1 & 1 & 1 & 1 \\ 0 & 2 & 2 & 2 \\ 0 & 0 & 3 & 4 \end{pmatrix}$

(d) $\begin{pmatrix} 1 & 1 & 2 & 3 \\ 0 & 0 & 1 & 1 \\ 0 & 0 & 3 & 3 \end{pmatrix}$ (e) $\begin{pmatrix} 1 & 1 & 1 & 1 & 1 \\ 0 & 2 & 2 & 2 & 2 \\ 0 & 0 & 0 & 3 & 3 \end{pmatrix}$.

Converting to echelon form. Any matrix can be converted to echelon form by row operations. The procedure is as follows.

1. Apply 'clearing the first column' (ignoring any initial columns of zeros) until the matrix is of the form

$$\begin{pmatrix} 0 & \cdots & 0 & x & a_1 & \cdots & a_k \\ 0 & \cdots & 0 & 0 & b_{11} & \cdots & b_{1k} \\ \vdots & & \vdots & \vdots & \vdots & \ddots & \vdots \\ 0 & \cdots & 0 & 0 & b_{m1} & \cdots & b_{mk} \end{pmatrix}.$$

2. Now put the matrix

$$\begin{pmatrix} b_{11} & \cdots & b_{1k} \\ \vdots & \ddots & \vdots \\ b_{m1} & \cdots & b_{mk} \end{pmatrix}$$

into echelon form by 'clearing the first column' of this matrix (again, ignoring any initial column of zeros). Note that row operations on this matrix correspond to row operations on the original. Continue in this way until the whole matrix is in echelon form.

If the 'optional' step in 'clearing the first column' above is applied each time, this method gives echelon form in which the first nonzero entry in each row is 1.

Exercise 1.4 Use row operations to put the following into echelon form.

(a) $\begin{pmatrix} 0 & 1 & 1 \\ 0 & 1 & 0 \\ 1 & 1 & 0 \end{pmatrix}$ (b) $\begin{pmatrix} 1 & 1 & 1 & 1 \\ 2 & 3 & 4 & 5 \\ 3 & 4 & 5 & 6 \end{pmatrix}$ (c) $\begin{pmatrix} 1 & 2 & 3 & 5 \\ 1 & 2 & 3 & 4 \\ 0 & 0 & 1 & 1 \end{pmatrix}$

(d) $\begin{pmatrix} 1 & 1 & 2 & 3 \\ 0 & 0 & 1 & 1 \\ 0 & 0 & 3 & 3 \end{pmatrix}$ (e) $\begin{pmatrix} -1 & 0 & 1 & 2 & 3 \\ 0 & 1 & 2 & 3 & 4 \\ -1 & -2 & -3 & -4 & -5 \end{pmatrix}.$

Rank. If **A** can be converted to the echelon form matrix **B** using row operations, and **B** has exactly k nonzero rows, then the *rank* of **A**, rk **A** (sometimes called the *row-rank*), is k.

Exercise 1.5 Say what the ranks of the matrices in Exercise 1.4 are. For further practice, calculate the ranks of the matrices in Exercises 1.2 and 1.3 also.

It is not at all obvious that this notion of rank is well-defined. In other words, it is not obvious whether or not there can be two different sequences of row operations converting **A** into echelon forms **B** and **C** respectively, where **B** has a different number of nonzero rows than **C**. In fact this is impossible, but we will defer a proper discussion of this until Chapter 8.

Converting to the identity matrix. If a matrix **A** is $n \times k$ (i.e. has n rows and k columns) where $n \leqslant k$, then the leftmost $n \times n$ block of **A** can sometimes be converted to the identity matrix using row operations. In fact, this can always be done if **A** has echelon form

$$\begin{pmatrix} a_{11} & \cdots & & & a_{1n} & a_{1\,n+1} & \cdots & a_{1k} \\ 0 & a_{22} & \cdots & & a_{2n} & a_{2\,n+1} & \cdots & a_{2k} \\ 0 & 0 & a_{33} & \cdots & a_{3n} & a_{3\,n+1} & \cdots & a_{3k} \\ \vdots & & & \ddots & \vdots & \vdots & & \vdots \\ 0 & \cdots & & 0 & a_{nn} & a_{n\,n+1} & \cdots & a_{nk} \end{pmatrix}$$

with each of the diagonals $a_{jj} \neq 0$, or in other words, if **A** has rank n.

1. By performing the row operation $\rho_i := (1/a_{ii})\rho_i$ for $i = 1, \ldots, n$ if necessary, convert the above echelon form to a similar one where the diagonal entries a_{ii} are all 1.

2. Now, starting at column n and then working backwards to column $n - 1$, $n-2$, up to column 1, clear all nondiagonal entries in this column by carrying out row operations as follows:

   (a) for column n, use operations $\rho_i := \rho_i + \mu\rho_n$ for $i = 1, 2, \ldots, n - 1$, and suitable values of μ in each case;

   (b) for column $n - 1$, use operations $\rho_i := \rho_i + \mu\rho_{n-1}$ for $i = 1, 2, \ldots, n - 2$;
   $\ldots$

   (c) and so on.

This gives a form

$$\begin{pmatrix} 1 & 0 & 0 & \cdots & 0 & b_{1\,n+1} & \cdots & b_{1k} \\ 0 & 1 & 0 & \cdots & 0 & b_{2\,n+1} & \cdots & b_{2k} \\ 0 & 0 & 1 & & 0 & b_{3\,n+1} & \cdots & b_{3k} \\ \vdots & & & \ddots & \vdots & \vdots & & \vdots \\ 0 & 0 & \cdots & 0 & 1 & b_{n\,n+1} & \cdots & b_{nk} \end{pmatrix}$$

for the matrix.

Exercise 1.6 Where applicable, apply this method to the matrices in Exercises 1.2, 1.3, and 1.4.

Calculating the inverse. If **A** is an $n \times m$ matrix and **B** is an $n \times k$ matrix, the *augmented matrix* $(\mathbf{A}|\mathbf{B})$ is the $n \times (m + k)$ matrix you get by writing down the entries of **A** and **B** next to one another in the obvious way.

To compute the inverse of an $n \times n$ square matrix **A**, apply row operations to the augmented matrix $(\mathbf{A}|\mathbf{I})$, where **I** is the $n \times n$ identity matrix, to get $(\mathbf{I}|\mathbf{B})$ for some matrix **B**, as described in the last section. This is not always possible, but as we have already seen it will be possible if **A** can be converted to some echelon form with n nonzero rows, i.e. if **A** has rank n. Then $\mathbf{A}^{-1} = \mathbf{B}$.

For example, starting with

$$A = \begin{pmatrix} 1 & 2 & 3 \\ 1 & 1 & -1 \\ -1 & 0 & 1 \end{pmatrix}$$

and following the above procedure we can get row operations converting

$$\begin{pmatrix} 1 & 2 & 3 & 1 & 0 & 0 \\ 1 & 1 & -1 & 0 & 1 & 0 \\ -1 & 0 & 1 & 0 & 0 & 1 \end{pmatrix} \quad \text{to} \quad \begin{pmatrix} 1 & 0 & 0 & 1/4 & -1/2 & -5/2 \\ 0 & 1 & 0 & 0 & 1 & 1 \\ 0 & 0 & 1 & 1/4 & -1/2 & -1/4 \end{pmatrix}.$$

(Try it!) Thus

$$\begin{pmatrix} 1 & 2 & 3 \\ 1 & 1 & -1 \\ -1 & 0 & 1 \end{pmatrix}^{-1} = \begin{pmatrix} 1/4 & -1/2 & -5/2 \\ 0 & 1 & 1 \\ 1/4 & -1/2 & -1/4 \end{pmatrix}.$$

To see why the method works, recall that each elementary row operation corresponds to multiplying on the left by an elementary row operation matrix. So applying several row operations to $(\mathbf{A}|\mathbf{I})$ corresponds to multiplying on the left by a product $\mathbf{R} = \mathbf{R}_1 \mathbf{R}_2 \ldots \mathbf{R}_k$ of row operation matrices. If the result of these operations is $(\mathbf{I}|\mathbf{B})$ then by associativity of matrix multiplication $\mathbf{R}(\mathbf{A}|\mathbf{I}) = (\mathbf{I}|\mathbf{B})$, i.e. $\mathbf{R}\mathbf{A} = \mathbf{I}$ and $\mathbf{R}\mathbf{I} = \mathbf{B}$. In other words, $\mathbf{R} = \mathbf{B}$ and $\mathbf{B}$ is a left inverse of $\mathbf{A}$.

We saw that this method will always find the inverse of an $n \times n$ matrix $\mathbf{A}$ if $\mathbf{A}$ has maximum possible rank n. It turns out $\mathbf{A}^{-1}$ exists if and only if rk $\mathbf{A} = n$, so this method will always succeed. The ideas here can be used to prove Fact 1.1. See Exercise 1.15.

Exercise 1.7 Use this method to calculate the inverses of the following matrices.

(a) $\begin{pmatrix} 1 & 1 & 0 & 0 \\ 0 & 1 & 1 & 0 \\ 0 & 0 & 1 & 1 \\ 0 & 0 & 0 & 1 \end{pmatrix}$ (b) $\begin{pmatrix} 1 & 0 & -1 \\ -1 & 1 & 0 \\ 0 & -1 & 0 \end{pmatrix}$ (c) $\begin{pmatrix} 1 & 2 & 0 \\ 0 & -1 & 2 \\ -1 & 2 & 0 \end{pmatrix}$.

Solving linear equations.   Row operations are commonly used to solve simultaneous linear equations, and we may illustrate the method here with an example. To solve

$$\begin{aligned} x + y + 2z &= -1 \\ -x + z &= -1 \\ -x + y + 4z &= 3, \end{aligned}$$

first put the equations in matrix form

$$\begin{pmatrix} 1 & 1 & 2 \\ -1 & 0 & 1 \\ -1 & 1 & 4 \end{pmatrix} \begin{pmatrix} x \\ y \\ z \end{pmatrix} = \begin{pmatrix} -1 \\ -1 \\ 3 \end{pmatrix}$$

and then put the augmented matrix formed from the matrix on the left with the column vector on the right into echelon form:

$$\left(\begin{array}{ccc|c} 1 & 1 & 2 & -1 \\ -1 & 0 & 1 & -1 \\ -1 & 1 & 4 & 3 \end{array}\right) \rightarrow \left(\begin{array}{ccc|c} 1 & 1 & 2 & -1 \\ 0 & 1 & 3 & -2 \\ 0 & 2 & 6 & -4 \end{array}\right) \rightarrow \left(\begin{array}{ccc|c} 1 & 1 & 2 & -1 \\ 0 & 1 & 3 & -2 \\ 0 & 0 & 0 & 0 \end{array}\right).$$

The full solution can now be read off directly:

$$\begin{aligned} x + y + 2z &= -1 \\ y + 3z &= -2 \\ 0 &= 0, \end{aligned}$$

so z may be anything, $y = -2 - 3z$, and $x = -1 - 2z - y = z + 1$.

This method works for any number of simultaneous equations in any number of unknowns, and always gives the most general solution.

Of course, a system of simultaneous linear equations may not have any solution at all. The following is useful in this regard.

Fact 1.4 *The equation*

$$\mathbf{A} \begin{pmatrix} x_1 \\ x_2 \\ \vdots \\ x_n \end{pmatrix} = \mathbf{b}$$

where $\mathbf{A}$ *is a* $k \times n$ *matrix and* $\mathbf{b}$ *is a* $k \times 1$ *column vector has at least one solution if and only if* $\mathrm{rk}(\mathbf{A}) = \mathrm{rk}(\mathbf{A}|\mathbf{b})$. *It has* **exactly one** *solution if and only if* $\mathrm{rk}(\mathbf{A}) = n$.

Column operations. These are analogous to row operations except that they operate on columns instead of rows. In this book, the elementary column operations are denoted by $\kappa_j := \kappa_j + \lambda\kappa_i$, $\kappa_i := \lambda\kappa_i$ (for $\lambda \neq 0$), and $\mathrm{swap}(\kappa_i, \kappa_j)$, in exact analogy with the notation for the row operations. Here, κ_i is used to denote the ith column of a matrix, just as ρ_i denoted the ith row. Column operations correspond to multiplying on the *right* by special column operation matrices, just as row operations correspond to *left*-multiplication by row operation matrices. In fact, column operations are not used as much as row operations in practice, and they will appear here only occasionally.

1.6 Determinant and trace

The *determinant* and *trace* operations take a square matrix $\mathbf{A}$ and return a *number*, $\det \mathbf{A}$ or $\mathrm{tr}\, \mathbf{A}$.

The trace operation is the simplest to calculate, as it is just the sum of the diagonal entries of the matrix. Thus if $\mathbf{A} = (a_{ij})$ is an $n \times n$ matrix, the trace of $\mathbf{A}$ is defined to be

$$\operatorname{tr} \mathbf{A} = \operatorname{tr} \begin{pmatrix} a_{11} & \cdots & a_{1n} \\ \vdots & \ddots & \vdots \\ a_{n1} & \cdots & a_{nn} \end{pmatrix} = a_{11} + a_{22} + \cdots + a_{nn} = \sum_{i=1}^{n} a_{ii}.$$

The significance of this operation is not at all clear from this definition. However, note that it has the obvious property that for $n \times n$ matrices $\mathbf{A}$ and $\mathbf{B}$,

$$\operatorname{tr}(\mathbf{A} + \mathbf{B}) = \operatorname{tr} \mathbf{A} + \operatorname{tr} \mathbf{B}.$$

One much less obvious property that will play an important role later is that for an invertible $n \times n$ matrix $\mathbf{P}$ and any $n \times n$ matrix $\mathbf{A}$,

$$\operatorname{tr}(\mathbf{P}^{-1}\mathbf{A}\mathbf{P}) = \operatorname{tr} \mathbf{A}.$$

At this stage at least, the determinant of $\mathbf{A}$ will seem just as mysterious. Its definition is an inductive one. For 1×1 matrices $\mathbf{A} = (a_{11})$, we define

$$\det(a_{11}) = a_{11}, \tag{1}$$

i.e. the number which is the only entry of the matrix $\mathbf{A}$. For an $n \times n$ matrix $\mathbf{A} = (a_{ij})$, we denote $\det \mathbf{A}$ by the matrix $\mathbf{A}$ with vertical lines round it, and define $\det \mathbf{A} =$

$$\begin{vmatrix} a_{11} & a_{12} & \cdots & a_{1n} \\ a_{21} & a_{22} & \cdots & a_{1n} \\ \vdots & \vdots & \ddots & \vdots \\ a_{n1} & a_{n2} & \cdots & a_{nn} \end{vmatrix} = a_{11} \begin{vmatrix} a_{22} & \cdots & a_{1n} \\ \vdots & \ddots & \vdots \\ a_{n2} & \cdots & a_{nn} \end{vmatrix} - a_{12} \begin{vmatrix} a_{21} & a_{23} & \cdots & a_{1n} \\ \vdots & \vdots & \ddots & \vdots \\ a_{n1} & a_{n3} & \cdots & a_{nn} \end{vmatrix}$$

$$+ \cdots + (-1)^{n+1} a_{1n} \begin{vmatrix} a_{21} & \cdots & a_{1\,n-1} \\ \vdots & \ddots & \vdots \\ a_{n1} & \cdots & a_{n\,n-1} \end{vmatrix}.$$

(This is sometimes called 'expansion by the first row'.) The rule is to take the entries in the top row in turn, with alternating signs, and multiply them by the determinants formed by deleting the row and column of the entry in question. This gives an expression for the $n \times n$ determinant in terms of $(n-1) \times (n-1)$ determinants, and the $(n-1) \times (n-1)$ determinants are evaluated by the same rule repeatedly until we get down to 1×1 determinants which are evaluated using (1).

For example, applying this to the 2×2 case we have

$$\begin{vmatrix} a & b \\ c & d \end{vmatrix} = ad - bc,$$

which should be memorized. In the 3×3 case we have

$$\begin{vmatrix} a_{11} & a_{12} & a_{13} \\ a_{21} & a_{22} & a_{23} \\ a_{31} & a_{32} & a_{33} \end{vmatrix} = a_{11}(a_{22}a_{33} - a_{23}a_{32}) - a_{12}(a_{21}a_{33} - a_{23}a_{31})$$
$$+ a_{13}(a_{21}a_{32} - a_{22}a_{31}).$$

Determinants often have a physical interpretation as areas or volumes. For example, the determinant

$$\begin{vmatrix} a & b \\ c & d \end{vmatrix} = ad - bc$$

has magnitude equal to the area of the parallelogram with corners given by position vectors

$$\begin{pmatrix} 0 \\ 0 \end{pmatrix}, \begin{pmatrix} a \\ c \end{pmatrix}, \begin{pmatrix} b \\ d \end{pmatrix}, \begin{pmatrix} a+b \\ c+d \end{pmatrix}.$$

The first point of significance of determinants for matrix operations is the following.

Fact 1.5 *For an $n \times n$ matrix* **A**, *the following are equivalent:*

(a) $\mathbf{A}^{-1}$ *exists;*

(b) $\det \mathbf{A} \neq 0$;

(c) $\operatorname{rk} \mathbf{A} = n$.

There are various useful rules to help calculate determinants. Firstly, for some matrices it is particularly easy to compute determinants. We say a matrix $\mathbf{A} = (a_{ij})$ is *upper triangular* if $a_{ij} = 0$ for $i > j$; in other words the nonzero entries are all on or above the principal diagonal of the matrix so that

$$\mathbf{A} = \begin{pmatrix} a_{11} & a_{12} & a_{13} & \dots & a_{1n} \\ 0 & a_{22} & a_{23} & \dots & a_{2n} \\ 0 & 0 & a_{33} & \dots & a_{3n} \\ \vdots & & & \ddots & \\ 0 & \dots & & 0 & a_{nn} \end{pmatrix}.$$

Fact 1.6 *The determinant of an upper triangular matrix* **A** *is equal to the product of its diagonal entries.*

We also have

Fact 1.7 *Let* **A**, **B** *be $n \times n$ matrices and* **I** *the $n \times n$ identity matrix. Then:*

(a) $\det \mathbf{A} = \det(\mathbf{A}^T)$;

(b) $\det(\mathbf{AB}) = \det \mathbf{A} \det \mathbf{B}$;

(c) *if* **A** *is invertible,* $\det(\mathbf{A}^{-1}) = (\det \mathbf{A})^{-1}$;

(d) $\det \mathbf{I} = 1$;

(e) *if* **A** *has a row or column which is entirely zero, then* $\det \mathbf{A} = 0$.

Fact 1.7, part (b), is the central fact concerning determinants, and the key to proving all the results concerning determinants mentioned here. A proof of it is outlined in Exercise 1.20 below. It also suggests an alternative way to calculate determinants: instead of using the definition directly, we can apply row operations to get our matrix in echelon form, and then compute the determinant of this (using Fact 1.6) and also the determinants of the row operation matrices used to get to this form.

In fact, the determinants of the basic row operation matrices are easy to calculate.

Fact 1.8 (a) *If* $\mathbf{S}_{i,j}$ *is the matrix for the row operation* swap(ρ_i, ρ_j) *where* $i \neq j$, *then* $\det \mathbf{S}_{i,j} = -1$.

(b) *If* $\mathbf{T}_{i,\lambda}$ *is the matrix for the row operation* $\rho_i := \lambda \rho_i$ *where* $\lambda \neq 0$, *then* $\det \mathbf{T}_{i,\lambda} = \lambda$.

(c) *If* $\mathbf{A}_{i,j,\lambda}$ *is the matrix for the row operation* $\rho_i := \rho_i + \lambda \rho_j$ *where* $i \neq j$, *then* $\det \mathbf{A}_{i,j,\lambda} = 1$.

From the last two basic facts, all the usual rules for evaluating determinants can be deduced.

Proposition 1.9 *If we swap any two rows in a determinant*

$$\begin{vmatrix} a_{11} & a_{12} & \cdots & a_{1n} \\ a_{21} & a_{22} & \cdots & a_{2n} \\ \vdots & \vdots & \ddots & \vdots \\ a_{n1} & a_{n2} & \cdots & a_{nn} \end{vmatrix}$$

the determinant changes sign (i.e. is multiplied by -1*). In particular, if the matrix* $\mathbf{A} = (a_{ij})$ *has two identical rows then* $\det \mathbf{A} = 0$.

Proof Use Fact 1.7 part (b) and Fact 1.8 part (a). If $\mathbf{A}$ has two identical rows, then swapping them does not change $\mathbf{A}$, so $\det \mathbf{A} = -\det \mathbf{A}$ and hence $\det \mathbf{A} = 0$. □

Proposition 1.10 *Multiplying any single row of a determinant by* λ *multiplies the determinant by* λ.

Proof Fact 1.7 part (b) and Fact 1.8 part (b). □

Be careful here: only one row of the matrix is multiplied by λ to multiply the determinant by λ. This is in contrast with scalar multiplication of matrices, $\lambda \mathbf{A}$, where every row is multiplied by λ. In fact, if $\mathbf{A}$ is an $n \times n$ matrix, $\det(\lambda \mathbf{A}) = \lambda^n \det \mathbf{A}$.

Proposition 1.11 *Applying any row operation* $\rho_i := \rho_i + \lambda \rho_j$ $(i \neq j)$ *to a matrix* $\mathbf{A}$ *leaves* $\det \mathbf{A}$ *unchanged.*

Proof Fact 1.7 part (b) and Fact 1.8 part (c). □

Proposition 1.12 *A determinant can be expanded by any row, provided you remember that the sign associated with an entry* a_{ij} *is* $(-1)^{i+j}$.

Proof This can be derived from the definition, Fact 1.7 part (b) and Fact 1.8 part (a). □

Proposition 1.13 *Any of the above rules for evaluating determinants for rows and row operations applies equally to columns and column operations.*

Proof Fact 1.7 part (a). □

For example, a determinant with two identical *columns* is zero, just as one with two identical *rows*.

1.7 Minors and cofactors

Determinants are also used to transform an $n \times n$ square matrix $\mathbf{A} = (a_{ij})$ to another matrix as follows.

Given

$$\mathbf{A} = \begin{pmatrix} a_{11} & a_{12} & \cdots & a_{1n} \\ a_{21} & a_{22} & \cdots & a_{2n} \\ \vdots & \vdots & \ddots & \vdots \\ a_{n1} & a_{n2} & \cdots & a_{nn} \end{pmatrix},$$

define b_{ij} to be the determinant of the $(n-1) \times (n-1)$ matrix obtained by deleting the ith row and the jth column of $\mathbf{A}$. Then the matrix $\mathbf{B} = (b_{ij})$ is called the *matrix of minors* of $\mathbf{A}$.

Now define $c_{ij} = (-1)^{i+j} b_{ij}$. In other words, c_{ij} is the determinant of the $(n-1) \times (n-1)$ submatrix of $\mathbf{A}$ with the sign correction you would use when evaluating $\det \mathbf{A}$ by the ith row (or by the jth column). The number c_{ij} is called the (i, j)th *cofactor* of $\mathbf{A}$, and the matrix $\mathbf{C} = (c_{ij})$ is called the *matrix of cofactors* of $\mathbf{A}$. The transpose of this, $\mathbf{C}^T$, is called the *adjugate matrix* of $\mathbf{A}$, written $\operatorname{adj} \mathbf{A}$.

This matrix $\mathbf{C}^T = \operatorname{adj} \mathbf{A}$ has the useful property that

$$(\operatorname{adj} \mathbf{A})\mathbf{A} = \mathbf{A}(\operatorname{adj} \mathbf{A}) = \det(\mathbf{A})\mathbf{I}. \tag{2}$$

Thus, in principle at least, determinants give another way of calculating inverses: if $\det \mathbf{A} \neq 0$ then

$$\mathbf{A}^{-1} = \frac{1}{\det \mathbf{A}} \operatorname{adj} \mathbf{A}.$$

In practice, this method is only used for particularly simple matrices, including all 2×2 matrices, where we have

$$\begin{pmatrix} a & b \\ c & d \end{pmatrix}^{-1} = \frac{1}{ad - bc} \begin{pmatrix} d & -b \\ -c & a \end{pmatrix}.$$

In most other cases, it is usually simpler to calculate inverses using row operations.

Exercises

Exercise 1.8 To show you understand the definition of 'echelon form', give an example of an upper triangular matrix which is *not* in echelon form.

Exercise 1.9 Let $\mathbf{A}$ be an $m \times n$ matrix and let e_i be the $n \times 1$ column matrix with ith entry equal to 1 and all other entries 0. Show that $\mathbf{A}e_i$ is equal to the ith column of $\mathbf{A}$.

Exercise 1.10 Using the definitions of the matrix operations directly, verify laws 1–3 for matrix addition, 1–4 for scalar multiplication, and 1–5 for matrix multiplication.

Exercise 1.11 In each case, determine if the system of equations has a solution, and if so give the most general solution.

(a)
$$\begin{aligned} x + 2y + z &= 2 \\ -x + 2y &= -1 \\ 5x - 2y + 2z &= 7 \end{aligned}$$

(b)
$$\begin{aligned} x - y - z &= 1 \\ 2x - y &= 1 \\ 2x + 2z &= 1 \end{aligned}$$

(c)
$$\begin{aligned} x + 2y + z &= 1 \\ x + y + z &= 1 \\ -x + z &= 1 \end{aligned}$$

Exercise 1.12 Do the same as the last exercise for the following.

(a)
$$\begin{aligned} x + 2y - z + w &= 1 \\ 2x + y + z &= 2 \end{aligned}$$

(b)
$$\begin{aligned} x + y - z &= 1 \\ y + z &= 2 \\ x + 2z &= 1 \\ x - y + 5z &= 1 \end{aligned}$$

(c)
$$\begin{aligned} x + 2y + 3z + w &= -1 \\ -x + y + z - w &= 2 \\ x + 5y + 7z + w &= 1 \end{aligned}$$

Exercise 1.13 Calculate the adjugate of each of the matrices of Exercise 1.7 and in each case verify that equation (2) on page 15 holds.

Exercise 1.14 Calculate the adjugate of each of the following matrices $\mathbf{A}$.

(a) $\begin{pmatrix} -1 & 2 & 0 \\ 0 & 1 & 3 \\ 2 & -3 & 3 \end{pmatrix}$ (b) $\begin{pmatrix} 1 & 0 & 2 & -1 \\ -1 & 1 & 0 & 0 \\ 0 & 1 & 2 & -1 \\ 2 & -1 & 1 & 1 \end{pmatrix}$.

Exercise 1.15 This exercise proves Fact 1.1 using the idea of row operations, and is for ambitious students only.

(a) For each $n \times n$ row operation matrix $\mathbf{R}$, show directly that $\mathbf{R}$ has an inverse, i.e. some $\mathbf{S}$ with $\mathbf{SR} = \mathbf{RS} = \mathbf{I}$.

(b) Show that if $\mathbf{R} = \mathbf{R}_1\mathbf{R}_2\ldots\mathbf{R}_k$ is a product of elementary row operation matrices then $\mathbf{R}$ does not have a row that is entirely zero. [Hint: use (a) and associativity of matrix multiplication.]

(c) Suppose $\mathbf{A},\mathbf{B}$ are $n \times n$ matrices with $\mathbf{AB} = \mathbf{I}$. Prove that there is some $\mathbf{C}$ with $\mathbf{CA} = \mathbf{I}$. [Hint: let $\mathbf{R} = \mathbf{R}_1\mathbf{R}_2\ldots\mathbf{R}_k$ be row operation matrices so that $\mathbf{RA}$ is in echelon form. By multiplying on the right by $\mathbf{B}$, show that $\mathbf{A}$ has rank n.]

(d) Considering transposes and part (c), show that if $\mathbf{A},\mathbf{B}$ are square matrices with $\mathbf{BA} = \mathbf{I}$ then there is a matrix $\mathbf{C}$ with $\mathbf{AC} = \mathbf{I}$.

(e) Show that if $\mathbf{CA} = \mathbf{AB} = \mathbf{I}$ then $\mathbf{C} = \mathbf{B}$. [Hint: what is $\mathbf{CAB}$?] Similarly, show that if $\mathbf{CA} = \mathbf{BA} = \mathbf{I}$ then $\mathbf{C} = \mathbf{B}$.

Exercise 1.16 Show that any square matrix $\mathbf{A}$ is the product $\mathbf{R}_1\mathbf{R}_2\ldots\mathbf{R}_k$ of 'generalized' elementary row operation matrices, where 'generalized' means that the λ in each of the row operations $\rho_i := \lambda\rho_i$ and $\rho_i := \rho_i + \lambda\rho_j$ is now allowed to be zero. [Hint: modify the 'converting to the identity matrix' method above to find ordinary elementary row operations $\mathbf{R}_i$ and generalized elementary row operations $\mathbf{S}_j$ such that $\mathbf{R}_1\ldots\mathbf{R}_k\mathbf{A} = \mathbf{S}_1\ldots\mathbf{S}_l\mathbf{I}$.]

Exercise 1.17 Show directly from the definition of determinant that

$$\begin{vmatrix} a_{11}+\lambda b_{11} & \cdots & a_{1n}+\lambda b_{1n} \\ a_{21} & \cdots & a_{2n} \\ \vdots & \ddots & \vdots \\ a_{n1} & \cdots & a_{nn} \end{vmatrix} = \begin{vmatrix} a_{11} & \cdots & a_{1n} \\ a_{21} & \cdots & a_{2n} \\ \vdots & \ddots & \vdots \\ a_{n1} & \cdots & a_{nn} \end{vmatrix} + \lambda \begin{vmatrix} b_{11} & \cdots & b_{1n} \\ a_{21} & \cdots & a_{2n} \\ \vdots & \ddots & \vdots \\ a_{n1} & \cdots & a_{nn} \end{vmatrix}.$$

Deduce that the determinant of a matrix $\mathbf{A}$ whose top two rows are equal is zero.

Exercise 1.18 Show that the matrix of any generalized elementary row operation (in the sense of Exercise 1.16) can always be written as one of $\mathbf{S}_i\mathbf{T}_j\mathbf{R}\mathbf{T}_j\mathbf{S}_i$, $\mathbf{S}_i\mathbf{R}\mathbf{S}_i$, $\mathbf{T}_j\mathbf{R}\mathbf{T}_j$, or $\mathbf{R}$, where $\mathbf{R}$ is the matrix of a generalized row operation acting on the first two rows only, and $\mathbf{S}_i$, $\mathbf{T}_j$ are the matrices of 'swap' operations swap(ρ_1,ρ_i) and swap(ρ_2,ρ_j) respectively.

Exercise 1.19 Prove by induction on n that if $\mathbf{A}$, $\mathbf{B}$ are $n \times n$ matrices with $\mathbf{B}$ obtained from $\mathbf{A}$ by the operation swap(ρ_i,ρ_j), where $1 \leqslant i < j \leqslant n$, then $\det\mathbf{B} = -\det\mathbf{A}$. [Hint: if $i = 1$ and $j = 2$, expand $\det\mathbf{B}$ *twice*, using the definition of determinant. Otherwise, use the induction hypothesis.]

Exercise 1.20 Using Exercises 1.16, 1.17, 1.18, and 1.19, show that $\det(\mathbf{AB}) = \det\mathbf{A}\det\mathbf{B}$ for all $n \times n$ matrices $\mathbf{A},\mathbf{B}$.

Exercise 1.21 Using Exercises 1.20 and 1.16, or otherwise, prove the following for an $n \times n$ matrix $\mathbf{A}$.

(a) $\text{rk}\,\mathbf{A} = n$ if and only if $\det\mathbf{A} \neq 0$.

(b) $\det\mathbf{A}^T = \det\mathbf{A}$.

(c) If $\mathbf{A}$ is upper triangular then $\det\mathbf{A}$ is the product of its diagonal entries.

Exercise 1.22 A *permutation* σ of $\{1, 2, \ldots, n\}$ is a bijection

$$\sigma \colon \{1, 2, \ldots, n\} \to \{1, 2, \ldots, n\}.$$

The set of all such permutations is denoted S_n.

The *matrix* $\mathbf{M}_\sigma$ of the permutation σ is the $n \times n$ matrix with 1 in the $(\sigma(i), i)$th position for each $i = 1, 2, \ldots, n$, and all other entries equal to 0. The *sign* of the permutation σ, $\mathrm{sgn}(\sigma)$, is defined to be $\det \mathbf{M}_\sigma$.

(a) Show that $\mathbf{M}_\sigma \mathbf{e}_i = \mathbf{e}_{\sigma(i)}$ for all i, where $\mathbf{e}_i$ is the $n \times 1$ column matrix of Exercise 1.9. Hence show that $\mathbf{M}_\sigma \mathbf{M}_\pi = \mathbf{M}_{\sigma \circ \pi}$, where $\sigma \circ \pi \in S_n$ is the permutation defined by $\sigma \circ \pi(i) = \sigma(\pi(i))$.

(b) Show that $\mathrm{sgn}(\sigma) \in \{1, -1\}$ for all $\sigma \in S_n$.

(c) Calculate the signs of the permutations $\sigma, \pi \in S_3$ given by

$$\sigma(1) = 2, \ \sigma(2) = 1, \ \sigma(3) = 3, \qquad \pi(1) = 2, \ \pi(2) = 3, \ \pi(3) = 1.$$

(d) Verify the formula

$$\begin{vmatrix} a_{11} & a_{12} & a_{13} \\ a_{21} & a_{22} & a_{23} \\ a_{31} & a_{32} & a_{33} \end{vmatrix} = \begin{aligned}& a_{11}a_{22}a_{33} + a_{12}a_{23}a_{31} + a_{13}a_{21}a_{32} \\ & - a_{12}a_{21}a_{33} - a_{13}a_{22}a_{31} - a_{11}a_{23}a_{32}\end{aligned}$$

$$= \sum_\sigma \mathrm{sgn}(\sigma) a_{1\sigma(1)} a_{2\sigma(2)} a_{3\sigma(3)}$$

where the summation is over all six permutations σ of $\{1, 2, 3\}$.

Exercise 1.23 Show by induction on n that the formula

$$\begin{vmatrix} a_{11} & a_{12} & \cdots & a_{1n} \\ a_{21} & a_{22} & \cdots & a_{2n} \\ \vdots & \vdots & \ddots & \vdots \\ a_{n1} & a_{n2} & \cdots & a_{nn} \end{vmatrix} = \sum_{\sigma \in S_n} \mathrm{sgn}(\sigma) a_{1\sigma(1)} a_{2\sigma(2)} \cdots a_{n\sigma(n)}$$

holds for all $n \in \mathbb{N}$.

2

Vector spaces

In this chapter we will review material concerning vector spaces. Some of this (but possibly not all of it) may already be familiar to you. This chapter is very important, however, because it sets out the style in which we shall study linear algebra, how we shall organize the text, and why it is particularly useful to do so in this way.

2.1   Examples and axioms

We start our study of vector spaces with some examples.

Example 2.1 Let $\mathbb{R}^3$ be the set of all column vectors $\begin{pmatrix} x \\ y \\ z \end{pmatrix}$ with entries x, y, z from the real numbers $\mathbb{R}$. We can add two such vectors by the rule

$$\begin{pmatrix} x_1 \\ y_1 \\ z_1 \end{pmatrix} + \begin{pmatrix} x_2 \\ y_2 \\ z_2 \end{pmatrix} = \begin{pmatrix} x_1 + x_2 \\ y_1 + y_2 \\ z_1 + z_2 \end{pmatrix}.$$

It follows that when we add $\begin{pmatrix} 0 \\ 0 \\ 0 \end{pmatrix}$ to a vector $\mathbf{v}$ we get $\mathbf{v}$ again, just like $0 + r = r + 0 = r$ for all real numbers r. So $\begin{pmatrix} 0 \\ 0 \\ 0 \end{pmatrix}$ behaves just like 0 and is called the *zero vector*, denoted $\mathbf{0}$.

We can also multiply a vector $\mathbf{v}$ with a real number r by the rule

$$r \begin{pmatrix} x \\ y \\ z \end{pmatrix} = \begin{pmatrix} rx \\ ry \\ rz \end{pmatrix}.$$

The operations of addition and multiplication by real numbers r are related. For example, it turns out that $2\mathbf{v}$ is equal to $\mathbf{v} + \mathbf{v}$ for all vectors $\mathbf{v}$. More generally, the distributivity law, $(r + s)\mathbf{v} = (r\mathbf{v}) + (s\mathbf{v})$, holds, since

$$(r+s)\begin{pmatrix} x \\ y \\ z \end{pmatrix} = \begin{pmatrix} (r+s)x \\ (r+s)y \\ (r+s)z \end{pmatrix} = \begin{pmatrix} rx+sx \\ ry+sy \\ rz+sz \end{pmatrix} = \begin{pmatrix} rx \\ ry \\ rz \end{pmatrix} + \begin{pmatrix} sx \\ sy \\ sz \end{pmatrix}.$$

Various other laws like this one hold too as you will see in a moment.

Example 2.2   Now consider $M_{3,3}(\mathbb{R})$, the set of 3×3 matrices with entries from $\mathbb{R}$. As for vectors, matrices can be added by the rule

$$\begin{pmatrix} x_{11} & x_{12} & x_{13} \\ x_{21} & x_{22} & x_{23} \\ x_{31} & x_{32} & x_{33} \end{pmatrix} + \begin{pmatrix} y_{11} & y_{12} & y_{13} \\ y_{21} & y_{22} & y_{23} \\ y_{31} & y_{32} & y_{33} \end{pmatrix} =$$

$$\begin{pmatrix} x_{11}+y_{11} & x_{12}+y_{12} & x_{13}+y_{13} \\ x_{21}+y_{21} & x_{22}+y_{22} & x_{23}+y_{23} \\ x_{31}+y_{31} & x_{32}+y_{32} & x_{33}+y_{33} \end{pmatrix}$$

and a matrix can be multiplied by a real number r by

$$r\begin{pmatrix} x_{11} & x_{12} & x_{13} \\ x_{21} & x_{22} & x_{23} \\ x_{31} & x_{32} & x_{33} \end{pmatrix} = \begin{pmatrix} rx_{11} & rx_{12} & rx_{13} \\ rx_{21} & rx_{22} & rx_{23} \\ rx_{31} & rx_{32} & rx_{33} \end{pmatrix}.$$

Similar laws hold for such matrices. For example, the zero matrix (with all entries equal to 0) can be added to any other matrix without changing it, and the commutative and associative laws of addition, and the distributivity law, hold too, as we saw in Section 1.2.

Example 2.3   Consider now $\mathbb{R}[X]$, the set of *polynomials in X* with coefficients from $\mathbb{R}$. A typical polynomial is

$$f(X) = a_n X^n + \cdots + a_2 X^2 + a_1 X^1 + a_0$$

where the a_i are real numbers, and possibly 0. Two such polynomials can be added by

$$(a_n X^n + \cdots + a_1 X^1 + a_0) + (b_n X^n + \cdots + b_1 X^1 + b_0)$$
$$= (a_n + b_n)X^n + \cdots + (a_1 + b_1)X^1 + (a_0 + b_0),$$

and multiplied by r by

$$r(a_n X^n + \cdots + a_1 X^1 + a_0) = (ra_n)X^n + \cdots + (ra_1)X^1 + (ra_0).$$

Real vector spaces.   As will be clear, these examples (and many others like them) have several features in common. When mathematicians want to concentrate on certain features of several examples (and possibly ignore other features of the examples, such as the rule for multiplying two matrices together, or the very different rule for multiplying two polynomials together) they write down *axioms* for the common features. The three examples above are all examples of *real vector spaces* and the next definition gives the axioms for real vector spaces.

Definition 2.4 *A* **real vector space** *or* **vector space over** $\mathbb{R}$ *is a set V containing a special* **zero vector** $\mathbf{0}$, *together with operations of addition of two vectors, giving* $\mathbf{u} + \mathbf{v}$, *and multiplication of a vector* $\mathbf{v}$ *with a real number* λ, *giving* $\lambda\mathbf{v}$, *satisfying the following laws for all* $\mathbf{u}, \mathbf{v}, \mathbf{w} \in V$ *and* $\lambda, \mu \in \mathbb{R}$.

$$(\mathbf{u} + \mathbf{v}) + \mathbf{w} = \mathbf{u} + (\mathbf{v} + \mathbf{w}) \tag{1}$$

$$\mathbf{u} + \mathbf{v} = \mathbf{v} + \mathbf{u} \tag{2}$$

$$\mathbf{u} + \mathbf{0} = \mathbf{u} \tag{3}$$

$$\mathbf{v} + (-1)\mathbf{v} = \mathbf{0} \tag{4}$$

$$\lambda(\mu\mathbf{v}) = (\lambda\mu)\mathbf{v} \tag{5}$$

$$\lambda(\mathbf{u} + \mathbf{v}) = \lambda\mathbf{u} + \lambda\mathbf{v} \tag{6}$$

$$(\lambda + \mu)\mathbf{u} = \lambda\mathbf{u} + \mu\mathbf{u}. \tag{7}$$

The vector $(-1)\mathbf{v}$ *is defined to be the result of multiplying* -1 *and* $\mathbf{v}$, *but is more usually written* $-\mathbf{v}$; *similarly,* $\mathbf{u} + (-\mathbf{v})$ *is written* $\mathbf{u} - \mathbf{v}$. *The elements of V are called the* **vectors** *of the space. Real numbers are often referred to as the* **scalars** *of the space.*

Some of the axioms above have special names: (1) is called the *associativity* of addition; (2) is the *commutativity* of addition; and (6) and (7) are the *distributivity* laws.

Several advantages of the axiomatic approach are already apparent. We now have a way of dealing with examples like the three mentioned above (and many others like them) in one go, rather than proving theorems for each example individually, so this approach saves time and energy. It also helps understanding the examples, since features which are irrelevant for some problems (matrix multiplication perhaps) are not mentioned in the axioms. Thirdly, it helps considerably in checking that proofs are correct, since for example a proof of a theorem about vector spaces can only use the axioms listed in the definition, and no other properties of any special example the reader or author might have in mind.

Of course, to apply any theorems we might prove about real vector spaces to a given space V, one must first prove that the axioms for real vector spaces are true of V. This is usually straightforward.

Example 2.5 The set of complex numbers $\mathbb{C}$ forms a real vector space with the usual addition $z_1 + z_2$ of complex numbers, and scalar multiplication,

$$\lambda(x + iy) = (\lambda x) + i(\lambda y),$$

where x, y, λ are all real.

Proof Most of the vector space axioms express well-known facts about $\mathbb{C}$. For example, $(z_1 + z_2) + z_3 = z_1 + (z_2 + z_3)$ is just associativity of addition of complex numbers, and $z_1 + z_2 = z_2 + z_1$ is commutativity. The zero vector $\mathbf{0}$ here is just the complex number $0 = 0 + i0$, so $z + 0 = z$ and $z + (-1)z = 0$ are clear. Also $\lambda(\mu z) = (\lambda\mu)z$, $\lambda(z_1 + z_2) = \lambda z_1 + \lambda z_2$, and $(\lambda + \mu)z = \lambda z + \mu z$ are consequences of associativity and distributivity of complex multiplication. $\qquad\square$

Note how a vector space is defined. It is necessary to say what the set V of vectors is, *and also* to say what the rules for addition of vectors and scalar multiplication are too.

The next example is very important, despite its simplicity. In fact, this example describes the simplest kind of vector space of all.

Example 2.6 The *zero space* over $\mathbb{R}$ is the vector space $V = \{0\}$, where (obviously) 0 denotes the zero vector. Addition and scalar multiplication are given by the rules $0 + 0 = 0$ and $\lambda 0 = 0$ for all $\lambda \in \mathbb{R}$. The vector space axioms are all true for this V for a very simple reason: since every meaningful expression involving scalars and the vector 0 gives the vector 0, any two such expressions are equal, and hence equations (1)–(7) are all true in this special zero space.

The next proposition gives the first example of the use of the axioms to prove statements about vector spaces. It shows that many of the laws which you might have expected to be axioms in fact follow from the axioms already given.

Proposition 2.7 *In a real vector space V, (a)* $\mathbf{v} = 1\mathbf{v}$, *(b)* $0\mathbf{v} = \mathbf{0}$, *and (c)* $\lambda\mathbf{0} = \mathbf{0}$ *for all* $\mathbf{v} \in V$ *and all scalars* λ.

Proof (a) Given $\mathbf{v}$, we have

$$
\begin{aligned}
1\mathbf{v} &= (-1)(-1)\mathbf{v} & \text{by (5)} \\
&= (-1)(-1)\mathbf{v} + \mathbf{0} & \text{by (3)} \\
&= (-1)(-1)\mathbf{v} + (\mathbf{v} + (-1)\mathbf{v}) & \text{by (4)} \\
&= (-1)(-1)\mathbf{v} + ((-1)\mathbf{v} + \mathbf{v}) & \text{by (2)} \\
&= ((-1)(-1)\mathbf{v} + (-1)\mathbf{v}) + \mathbf{v} & \text{by (1)} \\
&= \mathbf{0} + \mathbf{v} & \text{by (4) and (2)} \\
&= \mathbf{v} & \text{by (3) and (2).}
\end{aligned}
$$

(b) Again,

$$
\begin{aligned}
0\mathbf{v} &= (1 + (-1))\mathbf{v} & \\
&= 1\mathbf{v} + (-1)\mathbf{v} & \text{by (7)} \\
&= \mathbf{v} + (-1)\mathbf{v} & \text{by (a)} \\
&= \mathbf{0} & \text{by (4).}
\end{aligned}
$$

For (c),

$$
\begin{aligned}
\lambda\mathbf{0} &= \lambda(\mathbf{0} + (-1)\mathbf{0}) & \text{by (4)} \\
&= \lambda\mathbf{0} + \lambda((-1)\mathbf{0}) & \text{by (6)} \\
&= \lambda\mathbf{0} + (-1)(\lambda\mathbf{0}) & \text{by (5)} \\
&= \mathbf{0} & \text{by (4),}
\end{aligned}
$$

as required.   □

Because of the axioms and because of propositions like this last one, we will adopt a slightly more relaxed approach to notation, writing for example $\mathbf{u} + \mathbf{v} + \mathbf{w}$ for the sum of three vectors $\mathbf{u}, \mathbf{v}, \mathbf{w}$ in a vector space V without specifying the order in which they are to be added. (This order does not matter, since by (1) and (2) $(\mathbf{u} + \mathbf{v}) + \mathbf{w}$, $\mathbf{u} + (\mathbf{v} + \mathbf{w})$, $\mathbf{u} + (\mathbf{w} + \mathbf{v})$, $\mathbf{w} + (\mathbf{v} + \mathbf{u})$, $(\mathbf{w} + \mathbf{u}) + \mathbf{v}$, and so on, are all the same.) Similarly, by (5), $(-2\lambda)\mathbf{v} = (-2)(\lambda\mathbf{v}) = -(2\lambda)\mathbf{v}$, etc., and this vector will be denoted more simply by $-2\lambda\mathbf{v}$. We will use the axioms (1)–(7) and Proposition 2.7 all the time, usually without explicit mention.

Complex vector spaces. The definition given above of a vector space over $\mathbb{R}$ can easily be modified to give the notion of a *vector space over the complex numbers* $\mathbb{C}$, by just replacing $\mathbb{R}$ by $\mathbb{C}$ in Definition 2.4 and letting λ and μ range over all complex numbers in the laws (1)–(7) for scalar multiplication. For example, the set of column vectors $\mathbb{C}^3$ of height 3 and with entries from $\mathbb{C}$ can be regarded as a complex vector space, in just the same way as $\mathbb{R}^3$ can be regarded as a real vector space. However, care is needed. For example, the set of complex numbers $\mathbb{C}$ can be regarded as a *real* vector space as we have already seen, or as a *complex* vector space with usual complex-number addition $z_1 + z_2$ and for scalar multiplication the *complex-number multiplication* λz. Similarly, $\mathbb{C}^3$ can be regarded as either a real vector space or a complex vector space. The properties of these spaces (e.g. $\mathbb{C}$ as a real vector space, and $\mathbb{C}$ as a complex vector space) are quite different, as we shall see. The distinction of what scalars you allow is a very important one.

Canonical examples. This book is concerned with vector spaces in general, but we will constantly refer back to the more familiar vector spaces $\mathbb{R}^n$ and $\mathbb{C}^n$ with addition defined by

$$\begin{pmatrix} x_1 \\ y_1 \\ \vdots \\ z_1 \end{pmatrix} + \begin{pmatrix} x_2 \\ y_2 \\ \vdots \\ z_2 \end{pmatrix} = \begin{pmatrix} x_1 + x_2 \\ y_1 + y_2 \\ \vdots \\ z_1 + z_2 \end{pmatrix} \tag{8}$$

and scalar multiplication defined by

$$\lambda \begin{pmatrix} x \\ y \\ \vdots \\ z \end{pmatrix} = \begin{pmatrix} \lambda x \\ \lambda y \\ \vdots \\ \lambda z \end{pmatrix} \tag{9}$$

for scalars λ. Unless otherwise specified, $\mathbb{C}^n$ will be regarded as a complex vector space (so the scalars λ in (9) will be complex numbers, and multiplication, λx etc., is ordinary multiplication of complex numbers). $\mathbb{R}^n$ is always considered as a real vector space. The bold face notation for vectors, as in $\mathbf{v}, \mathbf{w}, \dots$, will be reserved for the vector spaces $\mathbb{R}^n$ and $\mathbb{C}^n$ only, and we will use light face letters $x, y, z, u, v, w, \dots$ for vectors (i.e. elements of V) in the general case. Scalars will be denoted with Greek letters $\lambda, \mu, \nu, \dots$ to distinguish them from vectors.

We will also refer to the spaces $\mathbb{R}^0$ and $\mathbb{C}^0$. If $\mathbb{R}^n$ is the set of column vectors with n entries, $\mathbb{R}^0$ should be the set of column vectors with no entries. There is only one such vector, and it might be written (), just as the empty set is sometimes written {}. The only possible way to define addition and scalar multiplication is by defining $() + () = ()$ and $\lambda() = ()$. In other words $\mathbb{R}^0$ is just the zero space over $\mathbb{R}$, as in Example 2.6, where () is thought of as 0. Similarly, $\mathbb{C}^0$ is the *complex* zero vector space—it too has a single vector () thought of as 0, but this time the scalars are the complex numbers.

The main disadvantage with column notation for vectors in $\mathbb{R}^n$ and $\mathbb{C}^n$ is the amount of paper it requires. For this reason, you may see people use so-called *row vectors*, but changing between row and column vectors is sometimes confusing. In this book, we will use column vectors throughout, but will sometimes notate a column vector as a row with a transpose sign ^{T} as a reminder. Thus we often write the vector

$$\begin{pmatrix} x_1 \\ x_2 \\ x_3 \end{pmatrix}$$

as $(x_1, x_2, x_3)^T$.

In our examples of vector spaces over the real numbers $\mathbb{R}$ (or the complex numbers $\mathbb{C}$), $\mathbb{R}$ (or $\mathbb{C}$) is called the *field of scalars*. To simplify terminology, we shall often talk about a *vector space over F*, or an *F-vector space*, where F is $\mathbb{R}$ or $\mathbb{C}$. The optional Section 2.6 below takes this terminology a bit further and gives axioms and further examples of fields.

2.2   Subspaces

Almost invariably in pure mathematics, whenever you see a definition for an 'object' of some kind, you will also have a definition of a 'subobject'.

Definition 2.8 *Given a vector space V over $\mathbb{R}$ or $\mathbb{C}$, a **subspace of V** is a subset $W \subseteq V$ which contains the zero vector of V and is closed under the operations of addition and scalar multiplication. That is, for each $u, v \in W$ and each scalar λ, each of $u + v$, λu (and λv) must be in W.*

Since -1 is a scalar, every subspace W contains $-v$ for each v in W, so W is also closed under subtraction of vectors, $v - u$, too. You should be able to check easily that $\{0\}$ and V itself are both subspaces of the vector space V.

The following lemma gives a 'minimal' condition for a subset W of V to be a subspace of V.

Lemma 2.9 *Let $W \subseteq V$ be nonempty, where V is a vector space over $\mathbb{R}$ or $\mathbb{C}$. Then W is a subspace of V if and only if $v + \lambda w \in W$ for each $v, w \in W$ and each scalar λ.*

Proof   Given $v, w \in W$ and $\lambda \in F$, we have $0 = v + (-1)v \in W$, $v + w = v + 1w \in W$, and $\lambda w = 0 + \lambda w \in W$. This proves the 'if' part. For the 'only if' part, suppose W is a subspace of V, $v, w \in W$, and λ is a scalar. Then $\lambda w \in W$ so $v + \lambda w \in W$. □

The laws (1)–(7) clearly hold in any subspace, so a subspace W of a vector space V is a vector space in its own right.

Subspaces will turn out to be very important indeed. We would like some way of specifying a subspace accurately and succinctly. For example, suppose we knew the vectors $a_1, \ldots, a_n$ were in a subspace W of V. Does this determine W? Or, if not, is there some 'special' or 'best' subspace of V containing $a_1, \ldots, a_n$?

With this in mind we make the tentative definition,

If V is a vector space over $F = \mathbb{R}$ or $\mathbb{C}$, and $A \subseteq V$ is any subset of vectors from V, then the *smallest* subspace W of V that contains A is called the subspace *spanned* by A.

The problem with this is that it is not immediately obvious that this subspace W exists for all A. Certainly, V itself is a subspace of V containing A, so *some* subspace of V containing A exists, but why is there a *smallest* subspace containing A?

However, one can rescue this idea as follows. If $a_1, a_2, \ldots, a_n \in W$ and W is a subspace of V then it follows from the fact that W is closed under addition and scalar multiplication that

$$\lambda_1 a_1 + \lambda_2 a_2 + \cdots + \lambda_n a_n \in W \tag{10}$$

for any scalars $\lambda_1, \lambda_2, \ldots, \lambda_n$. (An expression like (10) is called a *linear combination* of the vectors $a_1, \ldots, a_n$.) What is more, by the associativity and distributivity laws in the vector space,

$$(\lambda_1 a_1 + \lambda_2 a_2 + \cdots + \lambda_n a_n) + (\mu_1 a_1 + \mu_2 a_2 + \cdots + \mu_n a_n)$$
$$= (\lambda_1 + \mu_1)a_1 + (\lambda_2 + \mu_2)a_2 + \cdots + (\lambda_n + \mu_n)a_n$$

and

$$\mu(\lambda_1 a_1 + \lambda_2 a_2 + \cdots + \lambda_n a_n) = (\mu\lambda_1)a_1 + (\mu\lambda_2)a_2 + \cdots + (\mu\lambda_n)a_n,$$

so the sum of any two linear combinations of $a_1, \ldots, a_n$ or the scalar product of a linear combination of $a_1, \ldots, a_n$ is again a linear combination. In other words the set of linear combinations of $a_1, \ldots, a_n$ is a subspace of V, so we make the following definition.

Definition 2.10 *Given a vector space V over $F = \mathbb{R}$ or $\mathbb{C}$, and given a subset $A = \{a_1, a_2, \ldots, a_n\}$ of V,*

$$W = \{\lambda_1 a_1 + \lambda_2 a_2 + \cdots + \lambda_n a_n : \lambda_1, \ldots, \lambda_n \in F\}$$

*is the subspace of V **spanned by** A. The elements of W*

$$\lambda_1 a_1 + \lambda_2 a_2 + \cdots + \lambda_n a_n$$

*are called **linear combinations of vectors from** A. This subspace W is denoted* span A *or* span$(a_1, a_2, \ldots, a_n)$.

We note again that span A is a subspace of V, and any subspace containing all the vectors from A must contain every linear combination of vectors from A, i.e. must contain each element of W, so W is indeed the smallest subspace containing A. In particular, the zero vector 0 is always a linear combination of $a_1, a_2, \ldots, a_n$ since

$$0 = 0a_1 + 0a_2 + \cdots + 0a_n.$$

Example 2.11 Let $V = \mathbb{R}^3$ with the usual addition and scalar multiplication, and consider vectors $\mathbf{a} = (1, 2, 0)^T$ and $\mathbf{b} = (0, 1, -1)^T$. Then a typical linear combination of $\mathbf{a}, \mathbf{b}$ is

$$\lambda \mathbf{a} + \mu \mathbf{b} = \begin{pmatrix} \lambda \\ 2\lambda \\ 0 \end{pmatrix} + \begin{pmatrix} 0 \\ \mu \\ -\mu \end{pmatrix} = \begin{pmatrix} \lambda \\ 2\lambda + \mu \\ -\mu \end{pmatrix}.$$

If we write this as $(x, y, z)^T$ we easily see that $2x - y - z = 0$ since $x = \lambda$, $y = 2\lambda + \mu$, and $z = -\mu$. So every vector $(x, y, z)^T$ in span$(\mathbf{a}, \mathbf{b})$ satisfies $2x - y - z = 0$.

On the other hand, given a vector $(x, y, z)^T$ such that $2x - y - z = 0$, we have

$$\begin{pmatrix} x \\ y \\ z \end{pmatrix} = \begin{pmatrix} x \\ 2x - z \\ z \end{pmatrix} = x \begin{pmatrix} 1 \\ 2 \\ 0 \end{pmatrix} - z \begin{pmatrix} 0 \\ 1 \\ -1 \end{pmatrix} = x\mathbf{a} + (-z)\mathbf{b}$$

so $(x, y, z)^T$ is in span$(\mathbf{a}, \mathbf{b})$. Thus we have proved that

$$\text{span}(\mathbf{a}, \mathbf{b}) = \left\{ \begin{pmatrix} x \\ y \\ z \end{pmatrix} : 2x - y - z = 0 \right\}.$$

If A is the empty set, we define span A to be $\{0\}$. This is just a convention; if you like, you can think of 0 as the sum of an empty sequence of terms of the form $\lambda_i a_i$, but if this doesn't appeal, just learn the convention.

Example 2.12 If $V = \mathbb{R}^4$ with the usual addition and scalar multiplication, and

$$W = \left\{ \begin{pmatrix} x \\ y \\ z \\ w \end{pmatrix} : \begin{array}{c} 3x + y = 0 \\ x + y + z = w \end{array} \right\}$$

we can easily check that W is a subspace of V: if $x, y, z, w, x', y', z', w' \in \mathbb{R}$ with $3x+y = 0$, $x+y+z = w$, $3x'+y' = 0$, $x'+y'+z' = w'$, then $3(x+\lambda x')+(y+\lambda y') = 0$ and $(x+\lambda x')+(y+\lambda y')+(z+\lambda z') = (w+\lambda w')$, so W is a subspace by Lemma 2.9.

For any vector $(x, y, z, w)^T \in W$, $y = -3x$ and $w = x + y + z = z - 2x$ so every vector of W is of the form $(x, -3x, z, z - 2x)^T$ for some $x, z \in \mathbb{R}$. In particular the vectors $\mathbf{a} = (1, -3, 0, -2)^T$ and $\mathbf{b} = (0, 0, 1, 1)^T$ are in W, as you can check. In fact span$(\mathbf{a}, \mathbf{b}) = W$ since

$$
\begin{pmatrix} x \\ -3x \\ z \\ z - 2x \end{pmatrix} = x \begin{pmatrix} 1 \\ -3 \\ 0 \\ -2 \end{pmatrix} + z \begin{pmatrix} 0 \\ 0 \\ 1 \\ 1 \end{pmatrix}.
$$

It is sometimes useful to be able to define span A when A is infinite. In this case we do not have any way to form an infinite sum of terms of the form $\lambda_i a_i$, so instead we are guided by our principle that span A should be the smallest vector subspace of V containing A.

Definition 2.13 *If A is an infinite subset of V, where V is a vector space over a field F, we define* span A, *the **subspace spanned by** A, to be the set of all linear combinations of finite subsets of A.*

Thus span A is the union of subspaces of V of the form span B where $B \subseteq A$ is finite. In symbols,

$$
\text{span } A = \bigcup_{\substack{B \subseteq A \\ B \text{ finite}}} \text{span } B
$$
$$
= \{\lambda_1 a_1 + \cdots + \lambda_n a_n : n \in \mathbb{N}, \lambda_1, \ldots, \lambda_n \in F, a_1, \ldots, a_n \in A\}.
$$

You can check that this definition makes span A into a subspace of V, the smallest subspace of V to contain A. This is so important it is worth noting as a separate proposition.

Proposition 2.14 *If V is a vector space over $\mathbb{R}$ or $\mathbb{C}$, $B \subseteq V$, and $a_1, a_2, \ldots, a_k$ are vectors in* span B *then* span$(a_1, a_2, \ldots, a_k) \subseteq$ span B.

If A is infinite, a *linear combination* of vectors from A is just an element of span A; that is, a linear combination of a finite number of elements of A. We repeat that, in general, there is no way to combine infinitely many elements into a single linear combination.

2.3 Linear independence

Suppose that $A = \{a_1, a_2, \ldots, a_n\} \subseteq V$ where V is a vector space over $\mathbb{R}$ or $\mathbb{C}$. We have seen that the zero vector is always a linear combination of vectors from A, and we can ask if the expression

$$
0 = 0a_1 + 0a_2 + \cdots + 0a_n
$$

for 0 is unique. The set of vectors A is said to be *linearly independent* if the expression above is the only linear combination of vectors from A that gives 0, and A is *linearly dependent* otherwise.

Generalizing slightly to include the case when A may be infinite we have,

Definition 2.15 *A set $A \subseteq V$ of vectors in a vector space V over $F = \mathbb{R}$ or $\mathbb{C}$ is **linearly dependent** if there is $n \in \mathbb{N}$, vectors $a_1, a_2, \ldots, a_n \in A$, and scalars $\lambda_1, \lambda_2, \ldots, \lambda_n$ not all zero such that*

$$\lambda_1 a_1 + \lambda_2 a_2 + \cdots + \lambda_n a_n = 0.$$

*Otherwise, A is **linearly independent**.*

So a finite set $A = \{a_1, a_2, \ldots, a_n\}$ is linearly independent if and only if for all scalars $\lambda_1, \lambda_2, \ldots, \lambda_n \in F$

$$\lambda_1 a_1 + \lambda_2 a_2 + \cdots + \lambda_n a_n = 0 \text{ implies } \lambda_1 = \lambda_2 = \cdots = \lambda_n = 0.$$

Also, if A is infinite, it is linearly independent if and only if every finite subset of A is linearly independent. By convention, the empty set containing no vectors is linearly independent.

Example 2.16 In the real vector space $\mathbb{R}^3$ the vectors $\mathbf{a} = (1, 2, 0)^T$, $\mathbf{b} = (1, 1, 1)^T$, and $\mathbf{c} = (0, 0, 1)^T$ form a linearly independent set, for if

$$\lambda \mathbf{a} + \mu \mathbf{b} + \nu \mathbf{c} = \mathbf{0}$$

for some scalars λ, μ, ν then the following system of equations is satisfied:

$$
\begin{array}{rrrcl}
\lambda & + & \mu & & = & 0 \\
2\lambda & + & \mu & & = & 0 \\
& & \mu & + \nu & = & 0.
\end{array}
$$

But this system has $\lambda = \mu = \nu = 0$ as its only solution.

On the other hand, the vectors $\mathbf{a} = (1, 2, 0)^T$, $\mathbf{b} = (1, 1, 1)^T$, and $\mathbf{d} = (1, -1, 3)^T$ form a linearly dependent set since

$$2\mathbf{a} - 3\mathbf{b} + \mathbf{d} = \mathbf{0}.$$

Example 2.17 Let V be $\mathbb{C}$ considered as a *real* vector space with addition

$$(x_1 + iy_1) + (x_2 + iy_2) = (x_1 + x_2) + i(y_1 + y_2) \tag{11}$$

and scalar multiplication

$$\lambda(x + iy) = (\lambda x) + i(\lambda y). \tag{12}$$

Then $\{1, i\}$ is linearly independent, since if λ, μ are real numbers with

$$0 = \lambda \cdot 1 + \mu \cdot i = \lambda + i\mu$$

then the real part, λ, and the imaginary part, μ, of $\lambda + i\mu$ are both zero.

Now consider $V = \mathbb{C}$ as a complex vector space with operations as in (11) and (12) except now λ may be a complex number. This time $\{1, i\}$ is linearly dependent, since

$$1 \cdot 1 + i \cdot i = 0$$

so for $\lambda = 1$ and $\mu = i$

$$\lambda \cdot 1 + \mu \cdot i = 0.$$

In $\mathbb{R}^3$, a one-element set $\{\mathbf{a}\}$ is linearly independent just in case $\mathbf{a} \neq \mathbf{0}$. (Note in particular that $\{\mathbf{0}\}$ is linearly *dependent* since $\mathbf{0} = 1\mathbf{0}$, and the scalar 1 used here is nonzero.) Also, $\{\mathbf{a}, \mathbf{b}\}$ is linearly independent if and only if $\mathbf{a}$ and $\mathbf{b}$ do not lie on a single line through $\mathbf{0}$, and $\{\mathbf{a}, \mathbf{b}, \mathbf{c}\}$ is linearly independent if and only if $\mathbf{a}$, $\mathbf{b}$, and $\mathbf{c}$ do not lie on a single plane.

We started talking about linear independence via the uniqueness of the linear combination $0 = 0a_1 + 0a_2 + \cdots + 0a_n$ for the zero vector. However, if A is linearly independent and v is in the subspace spanned by A then the linear combination for v is also unique, as the following useful proposition shows.

Proposition 2.18 *Suppose $A = \{a_1, \ldots, a_n\} \subseteq V$ is linearly independent, where V is a vector space over $\mathbb{R}$ or $\mathbb{C}$. Suppose also that $v \in V$ and there are scalars $\lambda_1, \ldots, \lambda_n$ and $\mu_1, \ldots, \mu_n$ such that*

$$v = \lambda_1 a_1 + \lambda_2 a_2 + \cdots + \lambda_n a_n$$

and

$$v = \mu_1 a_1 + \mu_2 a_2 + \cdots + \mu_n a_n.$$

Then $\lambda_1 = \mu_1$, $\lambda_2 = \mu_2$, $\ldots$, $\lambda_n = \mu_n$.

Proof We have

$$\lambda_1 a_1 + \lambda_2 a_2 + \cdots + \lambda_n a_n = v = \mu_1 a_1 + \mu_2 a_2 + \cdots + \mu_n a_n$$

so

$$(\lambda_1 a_1 + \lambda_2 a_2 + \cdots + \lambda_n a_n) - (\mu_1 a_1 + \mu_2 a_2 + \cdots + \mu_n a_n) = 0$$

giving

$$(\lambda_1 - \mu_1)a_1 + (\lambda_2 - \mu_2)a_2 + \cdots + (\lambda_n - \mu_n)a_n = 0,$$

and hence

$$(\lambda_1 - \mu_1) = (\lambda_2 - \mu_2) = \cdots = (\lambda_n - \mu_n) = 0$$

since $\{a_1, a_2, \ldots, a_n\}$ is linearly independent. So $\lambda_1 = \mu_1$, $\lambda_2 = \mu_2$, $\ldots$, $\lambda_n = \mu_n$ as required. $\square$

2.4 Bases

Definition 2.19 *A **basis** of a vector space V is a linearly independent set $B \subseteq V$ which spans V.*

Example 2.20 The real vector space $\mathbb{R}^3$ has basis $\{e_1, e_2, e_3\}$, where $e_1 = (1, 0, 0)^T$, $e_2 = (0, 1, 0)^T$, and $e_3 = (0, 0, 1)^T$. To prove this you need to check the set is linearly independent and spans the vector space in question. For the first, if

$$\lambda_1 e_1 + \lambda_2 e_2 + \lambda_3 e_3 = 0$$

then

$$(\lambda_1, \lambda_2, \lambda_3)^T = (0,0,0)^T$$

so $\lambda_1 = \lambda_2 = \lambda_3 = 0$, as required. For the second, an arbitrary vector in $\mathbb{R}^3$ is $(x, y, z)^T$ where $x, y, z \in \mathbb{R}$. But

$$(x, y, z)^T = x e_1 + y e_2 + z e_3$$

so $(x, y, z)^T$ is a linear combination of e_1, e_2, e_3.

Similarly, $\mathbb{R}^n$ has basis $\{e_1, e_2, \ldots, e_n\}$, where e_i is the $n \times 1$ column vector with ith entry equal to 1 and all other entries zero. This basis is used so often that it is called the *usual basis* or *standard basis* of $\mathbb{R}^n$.

The following theorem is particularly important.

Theorem 2.21 *Let V be a vector space over $\mathbb{R}$ or $\mathbb{C}$, and let $B \subseteq V$ be linearly independent. Then there is a basis B' of V with $B \subseteq B'$.*

Proof Although the theorem is true generally, we give a proof in the case when B is finite and there are $a_1, a_2, \ldots, a_k \in V$ such that $V = \mathrm{span}(a_1, a_2, \ldots, a_k)$.

Suppose $B = \{b_1, b_2, \ldots, b_n\}$ is given and is linearly independent. Then:

either $a_i \in \mathrm{span}\, B$ for all i, so $V = \mathrm{span}(a_1, a_2, \ldots, a_k) \subseteq \mathrm{span}\, B$ and hence span $B = V$ so B itself is a basis and we may take $B' = B$;

or some a_i is not in span B. By reordering $a_1, a_2, \ldots, a_k$ if necessary we may assume that $a_1 \notin \mathrm{span}\, B$. We show that $B \cup \{a_1\}$ is linearly independent. If

$$\lambda_1 b_1 + \cdots + \lambda_n b_n + \lambda_{n+1} a_1 = 0$$

then $\lambda_{n+1} = 0$, for else

$$a_1 = -\lambda_{n+1}^{-1}(\lambda_1 b_1 + \cdots + \lambda_n b_n) \in \mathrm{span}\, B.$$

Therefore

$$\lambda_1 b_1 + \cdots + \lambda_n b_n = 0$$

giving $\lambda_1 = \lambda_2 = \cdots = \lambda_n = 0$ by the linear independence of B.

Continuing this process at most k times we obtain a linearly independent set $B' \subseteq B \cup \{a_1, \ldots, a_k\}$ for which $a_i \in \mathrm{span}\, B'$ for all i, i.e. B' spans V and hence is a basis. $\square$

Note that the above argument proves the following fact in the particular case when A, B are finite.

Theorem 2.22 *Suppose span $A = V$ and $B \subseteq V$ is linearly independent. Then there is a basis B' of V with $B \subseteq B' \subseteq A \cup B$.*

Again, this theorem is true generally even in the infinite case, but requires more sophisticated set theory to prove.

Example 2.23 As for most proofs in this book, the proof of Theorem 2.21 also provides a method for calculating a suitable basis. For example, suppose V is the real vector space $\mathbb{R}^4$ and $\mathbf{a} = (1,1,0,0)^T$, $\mathbf{b} = (1,1,1,1)^T$. Then the set $B = \{\mathbf{a}, \mathbf{b}\}$ is linearly independent so can be extended to a basis. To find such a basis, start with the usual basis vectors

$$\mathbf{e}_1 = (1,0,0,0)^T, \ \mathbf{e}_2 = (0,1,0,0)^T, \ \mathbf{e}_3 = (0,0,1,0)^T, \ \mathbf{e}_4 = (0,0,0,1)^T$$

of $\mathbb{R}^4$. It is easy to check that $\mathbf{e}_1 \notin \text{span}(\mathbf{a}, \mathbf{b})$, so $\{\mathbf{a}, \mathbf{b}, \mathbf{e}_1\}$ is linearly independent by the argument in the proof of the theorem. We can therefore add $\mathbf{e}_1$ to the basis we are constructing.

We now look at $\mathbf{e}_2$, and this time find that $\mathbf{e}_2 \in \text{span}(\mathbf{a}, \mathbf{b}, \mathbf{e}_1)$, since $\mathbf{e}_2 = \mathbf{a} - \mathbf{e}_1$. However, $\{\mathbf{a}, \mathbf{b}, \mathbf{e}_1, \mathbf{e}_3\}$ is linearly independent, as you can check, so we adjoin $\mathbf{e}_3$ to our basis. This gives a basis $B' = \{\mathbf{a}, \mathbf{b}, \mathbf{e}_1, \mathbf{e}_3\}$ of $\mathbb{R}^4$, for $\mathbf{e}_4 = -\mathbf{a} + \mathbf{b} - \mathbf{e}_3$ and therefore B' spans V since span B' contains the spanning set $\{\mathbf{e}_1, \mathbf{e}_2, \mathbf{e}_3, \mathbf{e}_4\}$.

Note that the basis B' extending B is by no means unique. For example, $\{\mathbf{a}, \mathbf{b}, \mathbf{e}_2, \mathbf{e}_4\}$ is another basis of $\mathbb{R}^4$, as is $\{\mathbf{a}, \mathbf{b}, (1,-1,0,0)^T, (0,0,1,-1)^T\}$.

Bases are used to define the notion of the *dimension* of a vector space V. The key to getting this to work is the following simple lemma.

Lemma 2.24 (The exchange lemma) *Suppose $a_1, a_2, \ldots, a_n, b$ are vectors in a vector space V, and suppose that*

$$b \in \text{span}(a_1, \ldots, a_{n-1}, a_n)$$

but

$$b \notin \text{span}(a_1, \ldots, a_{n-1}).$$

Then $a_n \in \text{span}(a_1, \ldots, a_{n-1}, b)$. If, in addition, $\{a_1, \ldots, a_{n-1}, a_n\}$ is linearly independent, then so is $\{a_1, \ldots, a_{n-1}, b\}$.

Proof Since $b \in \text{span}(a_1, \ldots, a_{n-1}, a_n)$, there are scalars λ_i such that

$$b = \lambda_1 a_1 + \cdots + \lambda_{n-1} a_{n-1} + \lambda_n a_n.$$

Now, if $\lambda_n = 0$ then

$$b = \lambda_1 a_1 + \cdots + \lambda_{n-1} a_{n-1}$$

so $b \in \text{span}(a_1, \ldots, a_{n-1})$, which is false. So $\lambda_n \neq 0$ and

$$
\begin{aligned}
a_n &= \lambda_n^{-1}(b - \lambda_1 a_1 + \cdots + \lambda_{n-1} a_{n-1}) \\
&= \lambda_n^{-1} b - \lambda_n^{-1} \lambda_1 a_1 - \cdots - \lambda_n^{-1} \lambda_{n-1} a_{n-1}.
\end{aligned}
$$

Hence $a_n \in \text{span}(a_1, \ldots, a_{n-1}, b)$, as required.

For the additional part, we are given that $\{a_1, \dots, a_{n-1}, a_n\}$ is linearly independent; suppose scalars μ_i are given with

$$\mu_1 a_1 + \cdots + \mu_{n-1} a_{n-1} + \mu_n b = 0.$$

Substituting $b = \lambda_1 a_1 + \cdots + \lambda_{n-1} a_{n-1} + \lambda_n a_n$ into this we get

$$(\mu_1 + \mu_n \lambda_1) a_1 + \cdots + (\mu_{n-1} + \mu_n \lambda_n) a_{n-1} + \mu_n \lambda_n a_n = 0.$$

Now $\{a_1, \dots, a_{n-1}, a_n\}$ is linearly independent so all these coefficients are zero. In particular, $\lambda_n \mu_n = 0$ so $\mu_n = 0$ since $\lambda_n \neq 0$. But this gives

$$\mu_1 a_1 + \cdots + \mu_{n-1} a_{n-1} = 0$$

so $\mu_1 = \mu_2 = \cdots = \mu_{n-1} = 0$ by the linear independence of $\{a_1, a_2, \dots, a_n\}$, as required. $\qquad\qquad\qquad\qquad\qquad\qquad\qquad\qquad\qquad\qquad\qquad\qquad\qquad\qquad\square$

Theorem 2.25 *Suppose A, B are both bases of a vector space V over $\mathbb{R}$ or $\mathbb{C}$. Then A, B have the same number of elements.*

Again, the theorem is true generally, but we will prove it here in the special case when one of the two sets A, B is finite.

Proof Suppose A has at least as many elements as B, and B is finite. List all the elements of B as $b_1, b_2, \dots, b_n$, and let $a_1, a_2, \dots, a_n$ be distinct elements of A. Our task is to show that this in fact lists *all* the elements of A.

Now $a_1 \in V = \mathrm{span}(b_1, b_2, \dots, b_n)$ so

$$a_1 = \lambda_1 b_1 + \lambda_2 b_2 + \cdots + \lambda_n b_n$$

for some scalars λ_i. Certainly not all the λ_i are zero, for otherwise $a_1 = 0 \in A$ so A would not be linearly independent. By reordering the b_i if necessary we may assume $\lambda_1 \neq 0$. Then $a_1 \notin \mathrm{span}(b_2, \dots, b_n)$, for else there would be scalars μ_i with

$$a_1 = 0 b_1 + \mu_2 b_2 + \cdots + \mu_n b_n = \lambda_1 b_1 + \lambda_2 b_2 + \cdots + \lambda_n b_n$$

and $0 \neq \lambda_1$, contradicting the uniqueness of the coefficients in linear combinations of linearly independent sets (Proposition 2.18). So by the exchange lemma, $a_1, b_2, b_3, \dots, b_n$ is a basis of V.

Now consider a_2. Again,

$$a_2 = \lambda_1 a_1 + \lambda_2 b_2 + \cdots + \lambda_n b_n$$

for some scalars λ_i. Not all of $\lambda_2, \lambda_3, \dots, \lambda_n$ are zero, else

$$\lambda_1 a_1 - a_2 = 0$$

contradicting the linear independence of A. By reordering if necessary, we may assume $\lambda_2 \neq 0$. So $a_2 \in \mathrm{span}(a_1, b_2, b_3, \dots, b_n)$. But $a_2 \notin \mathrm{span}(a_1, b_3, \dots, b_n)$, for else

$$\lambda_1 a_1 + \lambda_2 b_2 + \lambda_3 b_3 + \cdots + \lambda_n b_n = \mu_1 a_1 + 0 b_2 + \mu_3 b_3 + \cdots + \mu_n b_n$$

for some scalars μ_i, with $\lambda_2 \neq 0$ contradicting Proposition 2.18. Therefore, by the exchange lemma $a_1, a_2, b_3, \dots, b_n$ is a basis of V.

Continuing in this way, we eventually get that $a_1, a_2, \ldots, a_n$ is a basis of V. Now if $A \neq \{a_1, a_2, \ldots, a_n\}$ take $a \in A$ not equal to any a_i. Since $\{a_1, a_2, \ldots, a_n\}$ spans V there are scalars ν_i with

$$a = \nu_1 a_1 + \nu_2 a_2 + \cdots + \nu_n a_n$$

so $\{a_1, a_2, \ldots, a_n, a\}$ is not linearly independent, a contradiction. Therefore $A = \{a_1, a_2, \ldots, a_n\}$, and A and B have the same number of elements. $\square$

Definition 2.26 *The number of elements of a basis of V (which depends only on V, and not on the choice of basis) is called the **dimension** of V. The dimension of V is denoted* $\dim V$.

The usual examples turn out to have the dimension you would expect. For example, $\mathbb{R}^3$ has dimension 3 since $e_1 = (1, 0, 0)^T$, $e_2 = (0, 1, 0)^T$, $e_3 = (0, 0, 1)^T$ forms a basis of size 3. Similarly, $\mathbb{R}^n$ has dimension n. The *complex* vector space $\mathbb{C}^n$ also has dimension n, since the usual basis $\{e_1, e_2, \ldots, e_n\}$ is a basis for $\mathbb{C}^n$ too (but see also Example 2.17 and Exercise 2.5 for the dimension of $\mathbb{C}^n$ as a *real* vector space).

It is wise not to forget the case of dimension 0. This is when a vector space is spanned by the empty linearly independent set, $\varnothing$. But what vectors are a linear combination of vectors in $\varnothing$? The zero vector (by convention) is one such, and in fact it is the only one. So a vector space V of dimension 0 is the zero space, i.e. $V = \{0\}$.

Corollary 2.27 *If V is a vector space over $\mathbb{R}$ or $\mathbb{C}$ and $U \subseteq V$ is a subspace of V then $\dim U \leqslant \dim V$. If, additionally, $\dim V$ is finite and $U \neq V$ then $\dim U < \dim V$.*

Proof Let $B \subseteq V$ be a basis of U. Then by Theorem 2.21 B extends to a basis $B' \supseteq B$ of V. Clearly as $B \subseteq B'$, B' has at least as many elements as B.

If $\dim V$ is finite and $U \neq V$, then B' is finite and hence B is also finite, so U has finite dimension. But $U = \text{span}\, B \neq V = \text{span}\, B'$, so $B' \neq B$ and hence B' has strictly more elements than B. $\square$

The second part of this very useful corollary can be stated in an alternative form as follows. Note too that the finiteness assumption is essential here (unlike some of the results here which were proved only in the finite case but are nevertheless true in the infinite case too). See Exercise 2.10.

Corollary 2.28 *Suppose that V is a vector space over $\mathbb{R}$ or $\mathbb{C}$, $\dim V$ is finite, and $U \subseteq V$ is a subspace of V with $\dim U = \dim V$. Then $U = V$.*

Often, a vector space V has finite dimension, in which case all bases of V are finite, but this may not be the case. In this book we are mostly concerned with finite dimensional vector spaces, but occasionally infinite dimensional spaces are required.

Example 2.29 Let V be the set of all functions $f, g, \ldots$ from the natural numbers $\mathbb{N} = \{0, 1, 2, 3, \ldots\}$ to $\mathbb{R}$, with addition $f + g$ defined by

$$(f + g)(n) = f(n) + g(n) \qquad \text{all } n \in \mathbb{N}$$

and scalar multiplication λf by

$$(\lambda f)(n) = \lambda \cdot f(n) \qquad \text{all } n \in \mathbb{N}.$$

Then V is a real vector space with infinite dimension.

Proof   We leave the verification of the vector space axioms as a straightforward exercise.

To show that V has infinite dimension, let $e_i \in V$ be the function defined by $e_i(n) = 0$ for $i \neq n$ and $e_i(i) = 1$. We show that $\{e_i : i \in \mathbb{N}\}$ is linearly independent, and hence by Theorem 2.21 can be extended to a (necessarily infinite) basis. Let

$$f = \lambda_0 e_0 + \lambda_1 e_1 + \cdots + \lambda_n e_n$$

be an arbitrary linear combination of vectors from V, and suppose $f = 0$. We must show $\lambda_0 = \lambda_1 = \cdots = \lambda_n = 0$. The vector $f \in V$ is of course a function $\mathbb{N} \to \mathbb{R}$, and checking the definitions of $+$ and scalar multiplication we see that

$$f(i) = \begin{cases} \lambda_i & \text{if } i \leqslant n \\ 0 & \text{otherwise.} \end{cases}$$

But if $f = 0$ this means that $f(0) = f(1) = \cdots = f(n) = 0$, in other words $\lambda_0 = \lambda_1 = \cdots = \lambda_n = 0$ as required. □

2.5   Coordinates

If V is a finite dimensional vector space over $\mathbb{R}$ or $\mathbb{C}$ then it has a finite basis $B \subseteq V$. Since B spans V, every vector v from V can be written as a linear combination of elements of B. Thus if $B = \{v_1, v_2, \ldots, v_n\}$, each $v \in V$ can be written as

$$\lambda_1 v_1 + \lambda_2 v_2 + \cdots + \lambda_n v_n$$

for some scalars $\lambda_1, \lambda_2, \ldots, \lambda_n$. By Proposition 2.18, these scalars $\lambda_1, \lambda_2, \ldots, \lambda_n$ are unique, so *providing the ordering of B as $v_1, v_2, \ldots, v_n$ is understood* the column vector $(\lambda_1, \lambda_2, \ldots, \lambda_n)^T$ from $\mathbb{R}^n$ or $\mathbb{C}^n$ determines v uniquely; the λ_i are called the *coordinates* of v with respect to the *ordered basis* $v_1, v_2, \ldots, v_n$ of V.

Example 2.30 For the real vector space $\mathbb{R}^3$ we may take ordered basis $\mathbf{v}_1, \mathbf{v}_2, \mathbf{v}_3$ where

$$\mathbf{v}_1 = \begin{pmatrix} 1 \\ 1 \\ 0 \end{pmatrix} \qquad \mathbf{v}_2 = \begin{pmatrix} 1 \\ -1 \\ 0 \end{pmatrix} \qquad \mathbf{v}_3 = \begin{pmatrix} 1 \\ 1 \\ 1 \end{pmatrix}.$$

The coordinates of the vector $\mathbf{v} = (1, 0, 1)^T$ with respect to this ordered basis can be found by solving

$$\begin{pmatrix} 1 \\ 0 \\ 1 \end{pmatrix} = \lambda_1 \begin{pmatrix} 1 \\ 1 \\ 0 \end{pmatrix} + \lambda_2 \begin{pmatrix} 1 \\ -1 \\ 0 \end{pmatrix} + \lambda_3 \begin{pmatrix} 1 \\ 1 \\ 1 \end{pmatrix}$$

as three simultaneous equations in $\lambda_1, \lambda_2, \lambda_3$. This gives $\lambda_1 = \lambda_2 = 1/2$, $\lambda_3 = 1$, so $\mathbf{v}$ has coordinates $(1/2, 1/2, 1)^T$ with respect to the ordered basis $\mathbf{v}_1, \mathbf{v}_2, \mathbf{v}_3$. Similarly, it has coordinates $(1/2, 1, 1/2)^T$ with respect to the ordered basis $\mathbf{v}_1, \mathbf{v}_3, \mathbf{v}_2$—changing the ordering of the basis changes the order of the coordinates. Finally, $\mathbf{v}$ has coordinates $(1, 0, 1)^T$ with respect to the usual basis $\mathbf{e}_1, \mathbf{e}_2, \mathbf{e}_3$ of $\mathbb{R}^3$, just as you would expect.

The convention we shall use in this book is that when the ordering of a basis is important we shall say so, and also omit the curly brackets, as in the phrase 'the ordered basis $v_1, v_2, \ldots, v_n$'. If the ordering is unimportant (so the basis can be just thought of as a set) we use curly brackets, as in 'the basis $\{v_1, v_2, \ldots, v_n\}$'.

Ordered bases and coordinates are used to show that two vector spaces V, W of the same dimension over the same scalar field $\mathbb{R}$ or $\mathbb{C}$ are *isomorphic*, or in other words look the same. Two isomorphic spaces V and W might not have exactly the same vectors, but vectors can be paired off, one from V with one from W, so that the operations of addition and scalar multiplication in W do the same to the paired vectors as the operations of addition and scalar multiplication in V do to the original vectors.

Example 2.31 The real vector space $\mathbb{R}^2$ of column vectors (with the usual addition and scalar multiplication) looks rather similar to the complex numbers, $\mathbb{C}$, regarded as a real vector space. We can represent both diagrammatically as a plane (with x, y coordinates in the case of $\mathbb{R}^2$, the Argand diagram in the case of $\mathbb{C}$). What's more, the column vector $(x, y)^T$ gives precisely the coordinates of the complex number $x + iy \in \mathbb{C}$ with respect to the ordered basis $1, i$ of $\mathbb{C}$. The idea is to pair these two vectors off with each other. This done, there are certain obvious similarities between the vector space operations. Compare

$$\begin{pmatrix} x \\ y \end{pmatrix} + \begin{pmatrix} x' \\ y' \end{pmatrix} = \begin{pmatrix} x + x' \\ y + y' \end{pmatrix}$$

with

$$(x + iy) + (x' + iy') = (x + x') + i(y + y'),$$

and

$$\lambda \begin{pmatrix} x \\ y \end{pmatrix} = \begin{pmatrix} \lambda x \\ \lambda y \end{pmatrix}$$

with

$$\lambda(x + iy) = (\lambda x) + i(\lambda y).$$

Note that we ignore here the fact that in the complex numbers we have a multiplication operation $w \cdot z$ combining $z, w \in \mathbb{C}$ whereas there is no such multiplication of two vectors giving another vector defined on $\mathbb{R}^2$, since we are looking at the two spaces purely as vector spaces.

A 'pairing off' of vectors in V and W is called a *one-to-one correspondence* or a *bijection*, and is really a function $f : V \to W$ such that f is *injective*, i.e.

$$v \neq w \text{ implies } f(v) \neq f(w),$$

and *surjective*, i.e.

$$w \in W \text{ implies } w = f(v) \text{ for some } v \in V.$$

Bijections are used to give the complete definition of isomorphisms of vector spaces.

Definition 2.32 *Two vector spaces V, W, both over $\mathbb{R}$ or both over $\mathbb{C}$, are **isomorphic** if there is a bijection $f : V \to W$ such that*

$$f(u + v) = f(u) + f(v) \tag{13}$$
$$f(\lambda v) = \lambda f(v) \tag{14}$$

for all $u, v \in V$ and all scalars λ.

*The bijection f is said to be an **isomorphism** from V to W. We write $V \cong W$ or $f : V \overset{\sim}{\to} W$.*

It is particularly important to realise that there are two different addition operations here, and two different scalar multiplication operations. So it might be better to write

$$f(u + v) = f(u) \oplus f(v)$$

instead of (13), where $+$ is addition of vectors in V and $\oplus$ is addition of vectors in W. Similarly,

$$f(\lambda \cdot v) = \lambda \odot f(v)$$

would be a more accurate representation of (14), where $\cdot$ is scalar multiplication in V and $\odot$ is scalar multiplication in W. Similarly, we should distinguish between the zero vector 0_V of V and the zero vector 0_W of W as these really are different vectors.

Note that if $f\colon V \rightrightarrows W$, then $f(0_V) = 0_W$ (or as we shall usually write, $f(0) = 0$, ignoring the fact that these two zero vectors are actually different). Indeed, $f(0v) = 0f(v)$ for any vector $v \in V$. But $0v$ is the zero vector of V, $f(v)$ is a vector in W, and any vector in W multiplied by 0 (according to scalar multiplication in W) is the zero vector of W.

Note also that (13) and (14) together imply that

$$f(\lambda_1 a_1 + \lambda_2 a_2 + \cdots + \lambda_k a_k) = \lambda_1 f(a_1) + \lambda_2 f(a_2) + \cdots + \lambda_k f(a_k)$$

for all $a_1, a_2, \ldots, a_k \in V$ and all scalars $\lambda_1, \lambda_2, \ldots, \lambda_k$, so an isomorphism f takes linear combinations of $a_1, a_2, \ldots, a_k \in V$ to linear combinations of $f(a_1), f(a_2), \ldots, f(a_k)$.

The following theorem, when carefully formulated, is actually true for vector spaces of infinite dimension too, but we restrict our discussion here to the finite case to avoid the more difficult issues in set theory that would otherwise be needed.

Theorem 2.33 *Suppose V is a vector space over $\mathbb{R}$ with finite dimension $n \geqslant 0$. Then $V \cong \mathbb{R}^n$ as real vector spaces. Similarly, if V is a vector space over $\mathbb{C}$ with dimension n, then $V \cong \mathbb{C}^n$ as complex vector spaces.*

Proof We have already seen that if $n = 0$ then V is the zero space $\{0\}$, so is obviously isomorphic to the zero space $\mathbb{R}^0$. So assume $n > 0$.

By the definition of 'dimension' there is an ordered basis $v_1, v_2, \ldots, v_n$ of V of size n. The idea is to define $f\colon V \to \mathbb{R}^n$ by taking each $v \in V$ to its coordinate form with respect to the ordered basis $v_1, v_2, \ldots, v_n$. Specifically, we define f by

$$f(\lambda_1 v_1 + \lambda_2 v_2 + \cdots + \lambda_n v_n) = (\lambda_1, \lambda_2, \ldots, \lambda_n)^T$$

noting that this definition is valid since every $v \in V$ has precisely one expression of the form $\lambda_1 v_1 + \lambda_2 v_2 + \cdots + \lambda_n v_n$. We just have to check that f is an isomorphism.

The function f is injective, because if $v = \lambda_1 v_1 + \lambda_2 v_2 + \cdots + \lambda_n v_n$, $w = \mu_1 v_1 + \mu_2 v_2 + \cdots + \mu_n v_n$, and $f(v) = f(w)$, then

$$(\lambda_1, \lambda_2, \ldots, \lambda_n)^T = (\mu_1, \mu_2, \ldots, \mu_n)^T$$

so $\lambda_i = \mu_i$ for all i, so $v = w$. Also, f is surjective, since if $\mathbf{r} = (r_1, r_2, \ldots, r_n)^T \in \mathbb{R}^n$, we have $f(r_1 v_1 + r_2 v_2 + \cdots + r_n v_n) = \mathbf{r}$.

Furthermore, if $v = \lambda_1 v_1 + \lambda_2 v_2 + \cdots + \lambda_n v_n$ and $w = \mu_1 v_1 + \mu_2 v_2 + \cdots + \mu_n v_n$, then

$$v + w = (\lambda_1 + \mu_1)v_1 + (\lambda_2 + \mu_2)v_2 + \cdots + (\lambda_n + \mu_n)v_n$$

so

$$f(v + w) = (\lambda_1 + \mu_1, \lambda_2 + \mu_2, \ldots, \lambda_n + \mu_n)^T$$
$$= (\lambda_1, \lambda_2, \ldots, \lambda_n)^T + (\mu_1, \mu_2, \ldots, \mu_n)^T$$
$$= f(v) + f(w),$$

and if ν is a scalar

$$\nu v = (\nu \lambda_1) v_1 + (\nu \lambda_2) v_2 + \cdots + (\nu \lambda_n) v_n$$

so

$$f(\nu v) = (\nu \lambda_1, \nu \lambda_2, \ldots, \nu \lambda_n)^T$$
$$= \nu (\lambda_1, \lambda_2, \ldots, \lambda_n)^T$$
$$= \nu f(v)$$

as required.

For the case of a vector space over $\mathbb{C}$, the argument is the same, but use scalars from $\mathbb{C}$ instead.   □

It follows that any two real vector spaces V, W of dimension n are isomorphic, as are any two complex vector spaces V, W of dimension n.

2.6   Vector spaces over other fields

The observant reader might have noticed that the two kinds of vector spaces we have been considering—over the reals and over the complexes—have much in common and he or she may wonder whether the notion of a vector space makes sense over any other number system other than $\mathbb{R}$ or $\mathbb{C}$. The answer is yes. All we need is a number system in which we can perform the usual operations of addition, subtraction, multiplication, and division, subject to the usual rules, such as $a + 0 = a$, and $a \cdot a^{-1} = 1$, and so on.

Definition 2.34 *A **field** is a set F containing distinct elements 0 and 1, with two binary operations $+$ and $\cdot$, which satisfy the following axioms.*

$$a + b = b + a \tag{15}$$
$$(a + b) + c = a + (b + c) \tag{16}$$
$$a + 0 = a \tag{17}$$
$$\text{for all } a \text{ there exists } -a \text{ such that } a + (-a) = 0 \tag{18}$$

$$a \cdot b = b \cdot a \tag{19}$$
$$(a \cdot b) \cdot c = a \cdot (b \cdot c) \tag{20}$$
$$a \cdot 1 = a \tag{21}$$
$$\text{for all } a \neq 0 \text{ there exists } a^{-1} \text{ such that } a \cdot a^{-1} = 1 \tag{22}$$

$$a \cdot (b + c) = a \cdot b + a \cdot c. \tag{23}$$

*If a field F is finite, its **order** is the number of elements in F.*

The axioms for fields can be thought of as forming three groups: (A) the rules for addition, (15)–(18); (B) the rules for multiplication, (19)–(22); and (C) the *distributivity law*, (23). Plenty of examples are furnished by arithmetic modulo some prime p.

Example 2.35 If p is any prime number, let $\mathbb{F}_p$ denote the set $\{0, 1, \dots, p-1\}$, and define operations $+$ and $\cdot$ on $\mathbb{F}_p$ as follows. First use ordinary integer arithmetic, and then 'reduce modulo p'; in other words, subtract whatever multiple of p is necessary to bring the answer in the range 0 to $p-1$.

Example 2.36 The field $\mathbb{F}_2$ has two elements, 0 and 1, subject to all the ordinary rules of arithmetic except that $1 + 1 = 0$.

Example 2.37 The field $\mathbb{F}_5$ of order 5 can be constructed as follows. First take ordinary arithmetic on the set $\{0, 1, 2, 3, 4\}$:

+	0	1	2	3	4
0	0	1	2	3	4
1	1	2	3	4	5
2	2	3	4	5	6
3	3	4	5	6	7
4	4	5	6	7	8

·	0	1	2	3	4
0	0	0	0	0	0
1	0	1	2	3	4
2	0	2	4	6	8
3	0	3	6	9	12
4	0	4	8	12	16

which on reduction modulo 5 gives

+	0	1	2	3	4
0	0	1	2	3	4
1	1	2	3	4	0
2	2	3	4	0	1
3	3	4	0	1	2
4	4	0	1	2	3

·	0	1	2	3	4
0	0	0	0	0	0
1	0	1	2	3	4
2	0	2	4	1	3
3	0	3	1	4	2
4	0	4	3	2	1

Example 2.38 The integers modulo 4 do not form a field. One way to see this is to observe that the element 2 does not have a multiplicative inverse, so that you cannot divide by 2. This is because for $x = 0, 1, 2, 3$, we have $2 \cdot x = 0, 2, 0, 2$ respectively, so there is no element x with $2 \cdot x = 1$.

In fact there *is* a field of order 4, but it cannot be obtained in this simple way.

Example 2.39 Let $\mathbb{F}_4$ denote the set $\{0, 1, a, a+1\}$ with addition and multiplication defined by the following tables.

+	0	1	a	$a+1$
0	0	1	a	$a+1$
1	1	0	$a+1$	a
a	a	$a+1$	0	1
$a+1$	$a+1$	a	1	0

·	0	1	a	$a+1$
0	0	0	0	0
1	0	1	a	$a+1$
a	0	a	$a+1$	1
$a+1$	0	$a+1$	1	a

Then it can be verified that this is a field with four elements.

Notice that in the last example, the multiplication table tells us that $a \cdot a = a+1$, which we can think of as a polynomial equation, $a^2 = a+1$ or $a^2 + a + 1 = 0$. In fact all finite fields can be defined in a similar way, starting with a field of prime order (in this case $\mathbb{F}_2$), and adjoining some element a satisfying a suitable polynomial equation. We shall see more of this in Chapter 9.

For the moment, we merely state the following important theorem without proof.

Theorem 2.40 *For each prime p and each positive integer n, there is a unique field of order p^n. Moreover, every finite field is of this form.*

Example 2.41 The field of order 9 can be defined by adjoining a 'square root of -1' to the field $\mathbb{F}_3$ of order 3, in the same way that we obtain $\mathbb{C}$ from $\mathbb{R}$. That is,

$$\mathbb{F}_9 = \{a + bi : a, b \in \mathbb{F}_3, i^2 + 1 = 0\}.$$

Vector spaces over finite fields.   The whole of this chapter can be generalized to vector spaces over an arbitrary field F. Simply replace $\mathbb{R}$ or $\mathbb{C}$ in the definitions and theorems by the field F. All the theorems remain true in this more general context.

Example 2.42 Let $F = \mathbb{F}_2 = \{0, 1\}$, the field of order 2. Then

$$F^8 = \{(x_1, \dots, x_8)^T : x_i \in F\}$$

is a vector space of dimension 8 over F, with a basis

$$\{(1,0,0,0,0,0,0,0)^T, (0,1,0,0,0,0,0,0)^T, \dots, (0,0,0,0,0,0,0,1)^T\}.$$

These vectors are very important in computer science, where they are called 'bytes'. The number of such vectors is 2^8, as there are two possibilities for x_1, two possibilities for x_2, and so on.

More generally, if F is a field of order q, then F^n contains exactly q^n vectors.

The generalization of the main theorem, Theorem 2.33, states that any vector space of dimension n over a field F is isomorphic to F^n. As a corollary we have the following.

Corollary 2.43 *Any vector space of dimension n over a field of order q has exactly q^n vectors.*

In fact, we may use this result to prove the second part of Theorem 2.40; that is, that every finite field has order equal to some prime power.

Lemma 2.44 *Let F be a finite field, and let F_0 be the subset*

$$F_0 = \{0, 1, 1+1, 1+1+1, \dots\}$$

of F. Then F_0 is a subfield of F (i.e. is closed under addition and multiplication, and is a field in its own right), and the order of F_0 is a prime number, p.

Proof F_0 is clearly closed under addition, by the associativity of addition. To show it is closed under multiplication, we use distributivity to show that

$$\underbrace{(1+1+\cdots+1)}_{a} \cdot \underbrace{(1+1+\cdots+1)}_{b} = \underbrace{1+1+\cdots+1}_{ab}. \qquad (24)$$

Now F is finite, so the elements $1, 1+1, 1+1+1, \ldots$ cannot be all distinct, which implies that

$$\underbrace{1+1+\cdots+1}_{r} = \underbrace{1+1+\cdots+1}_{s}$$

for some positive integers $r > s$. Subtracting 1 from both sides of this equation s times gives

$$\underbrace{1+1+\cdots+1}_{r-s} = 0.$$

Let p be the smallest positive integer such that

$$\underbrace{1+1+\cdots+1}_{p} = 0,$$

and note that the argument just given shows that for no $0 \leqslant s < r \leqslant p$ does

$$\underbrace{1+1+\cdots+1}_{r} = \underbrace{1+1+\cdots+1}_{s}$$

for else $0 < r - s < p$ and

$$\underbrace{1+1+\cdots+1}_{r-s} = 0.$$

We now prove that p is prime. If not, there are integers a, b with $1 < a < p$, $1 < b < p$, and $p = ab$. Then by (24) we have

$$0 = \underbrace{1+1+\cdots+1}_{p} = \underbrace{(1+1+\cdots+1)}_{a} \cdot \underbrace{(1+1+\cdots+1)}_{b}$$

which implies that both of

$$\underbrace{(1+1+\cdots+1)}_{a} \quad \text{and} \quad \underbrace{(1+1+\cdots+1)}_{b}$$

are nonzero elements of F without multiplicative inverses. This is a contradiction, so p is prime.

It follows that

$$F_0 = \{0, 1, 1+1, \ldots, \underbrace{1+1+\cdots+1}_{p-1}\}$$

and F_0 has p elements. In fact, it is easy to check that since p is prime, F_0 is isomorphic to the field $\mathbb{F}_p$ of Example 2.35, so F_0 is a subfield of F. $\qquad \square$

The subfield constructed in the previous lemma is clearly the smallest subfield of F, and because of this it is given a special name.

Definition 2.45 *If F is a finite field, the subfield*

$$F_0 = \{0, 1, 1+1, 1+1+1, \dots\}$$

*of F is called the **prime subfield** of F, and the order of F_0 is called the **characteristic** of F.*

Theorem 2.46 *Every finite field F is a vector space over its prime subfield.*

Proof   Observe that the vector space axioms are special cases of the field axioms for F. $\square$

Corollary 2.47 *Every finite field F has order p^n for some prime p (the characteristic of F) and some integer $n \geqslant 1$.*

Proof   Immediate from Theorem 2.46 and corollary 2.43. $\square$

Summary

To sum up this chapter, we started with examples of vector spaces and then gave the axioms for vector spaces which in some sense described 'common features' of the examples we had in mind. We then defined the important concept of *dimension* of a vector space. (The verification that this is well-defined required the exchange lemma.) We concluded by showing that (for finite dimensional spaces at least) the dimension of the vector space characterizes it completely up to isomorphism. In other words any two finite dimensional real vector spaces V_1 and V_2 are isomorphic if and only if they have the same dimension—in which case both are isomorphic to the space of column vectors $\mathbb{R}^n$ for $n = \dim V_1 = \dim V_2$. Similarly for complex spaces and the spaces $\mathbb{C}^n$.

In the (optional) Section 2.6, the level of abstraction was taken one step further; the axioms for a *field* were given and the notion of a vector space V over a field F introduced. A particularly simple example of a field F is the set of numbers $\{0, 1, 2, \dots, p-1\}$ for a prime p, and addition and multiplication being taken modulo p. Similar results for vector spaces over F hold: for example, the typical examples of vector spaces over F are the spaces of column vectors F^n with entries from F, and every finite dimensional vector space over F is isomorphic to F^n for some n.

Exercises

Exercise 2.1 If $f\colon U \to V$ and $g\colon V \to W$ are isomorphisms of vector spaces U, V, W, show that $f^{-1}\colon V \to U$ is an isomorphism from V to U, and that the composition $g \circ f\colon U \to W$ is an isomorphism from U to W.

Exercise 2.2 Show that if $n \neq m$ are natural numbers, then the real vector spaces $\mathbb{R}^n$ and $\mathbb{R}^m$ are *not* isomorphic. [Hint: if $f\colon \mathbb{R}^n \to \mathbb{R}^m$ is an isomorphism and $e_1, \dots, e_n$ is the usual basis of $\mathbb{R}^n$, show that $\mathbb{R}^m$ is spanned by

$\{f(\mathbf{e}_1), \dots, f(\mathbf{e}_n)\}$ (since f is surjective) and furthermore $\{f(\mathbf{e}_1), \dots, f(\mathbf{e}_n)\}$ is linearly independent. Deduce $n = m$.]

Exercise 2.3 Find bases for the following subspaces of $\mathbb{R}^4$:

(a) $\{(x, y, z, t)^T : x + y + 2t = 0, y - 3z = 0\}$;

(b) $\{(x, y, z, t)^T : x - 2y + 3z - 4t = 0\}$;

(c) $\{(x, y, z, t)^T : 2x - y = 3y - z = 4z - t = 0\}$.

Exercise 2.4 For each of the following subsets A of $\mathbb{R}^4$, find a basis for span A, and express span A in the form $\{(x, y, z, t)^T : \dots\}$:

(a) $A = \{(1, 0, 2, -1)^T, (0, 1, -1, 2)^T\}$;

(b) $A = \{(1, -1, 0, 0)^T, (0, 1, -1, 0)^T, (0, 0, 1, -1)^T, (-1, 0, 0, 1)^T\}$;

(c) $A = \{(x, y, 0, 0)^T : x > y > 0\}$.

Exercise 2.5 What is the dimension $(0, 1, 2, 3, \dots,$ or infinite) of the following vector spaces? (Give reasons.)

(a) $\mathbb{C}^5$, as a complex vector space.

(b) $\mathbb{C}^5$, as a real vector space.

(c) The set of polynomials $p(X)$ of degree at most 7 with coefficients from $\mathbb{R}$, as a real vector space.

(d) The set of polynomials $p(X)$ of arbitrary degree with coefficients from $\mathbb{R}$, as a real vector space.

Exercise 2.6 For each integer $n \geqslant 0$, let $f_n \colon \mathbb{R} \to \mathbb{R}$ be the function defined by $f_n(x) = x^n$. Show that the set $B = \{f_n : n \geqslant 0\}$ is a basis for the vector space $\mathbb{R}[x]$ of real polynomial functions. Show that the map ϕ defined on $\mathbb{R}[x]$ by

$$\phi : \sum \lambda_i f_i \mapsto \sum \lambda_i f_{i+1}$$

is injective but not surjective. Deduce that $\mathbb{R}[x]$ is isomorphic (as a vector space) to a proper subspace of itself.

Exercise 2.7 Write out the addition and multiplication tables for $\mathbb{F}_3$.

Exercise 2.8 (a) Write out the addition and multiplication tables for $\mathbb{Z}_6$ (the integers modulo 6), and show that $\mathbb{Z}_6$ is not a field.

(b) More generally, show that if a and b are any two integers bigger than 1, then $\mathbb{Z}_{ab}$ is not a field.

Exercise 2.9 Let p be a prime number, and $\mathbb{Z}_p$ be the set of integers modulo p. For any nonzero element $a \in \mathbb{Z}_p$, we define the map 'multiplication by a'

$$m_a : \mathbb{Z}_p \to \mathbb{Z}_p$$

by

$$m_a : b \mapsto ab \pmod{p}$$

(a) Use uniqueness of prime factorization to show that m_a is an injection, and hence a bijection.

(b) Let b be the element such that $m_a(b) = 1 \pmod{p}$. Show that b is an inverse to a, and deduce that $\mathbb{Z}_p$ is a field.

Exercise 2.10 Let V be the set of all functions $f \colon \mathbb{N} \to \mathbb{R}$, and let $\mathbf{0}$ be the function defined by $\mathbf{0}(n) = 0$ (see Example 2.29).

(a) Prove that each of the axioms holds, showing that V is a real vector space.

(b) Let $B = \{e_i : i \in \mathbb{N}\}$ where e_i is as in Example 2.29. Show that B does not span V. [Hint: consider f such that $f(n) = 1$ for all n.]

(c) Show that $W = \operatorname{span} B$ and the real vector space $\mathbb{R}[X]$ of polynomials with coefficients from $\mathbb{R}$ are isomorphic to each other.

(d) Show that V and W are *not* isomorphic to each other, i.e. there is no isomorphism $f \colon V \cong W$. (This is tricky.)

[Hint for (d): given such an f, use induction on n to define a function $g \colon \mathbb{N} \to \mathbb{R}$ so that for each n there is $k \geqslant n$ such that

$$
\begin{pmatrix} g(0) \\ \vdots \\ g(k) \end{pmatrix} \notin \operatorname{span} \left\{ \begin{pmatrix} f^{-1}(e_0)(0) \\ \vdots \\ f^{-1}(e_0)(k) \end{pmatrix}, \dots, \begin{pmatrix} f^{-1}(e_{n-1})(0) \\ \vdots \\ f^{-1}(e_{n-1})(k) \end{pmatrix} \right\}
$$

in $\mathbb{R}^{k+1}$.]

Part II

Bilinear and sesquilinear forms

3

Inner product spaces

In this chapter we consider ways of 'multiplying' two vectors in a vector space V to give a scalar. Such a product is called an inner product, or scalar product, and the theory is based initially on a familiar example (sometimes called the *dot product*) on three-dimensional vectors in $\mathbb{R}^3$. To begin with, we consider real vector spaces, but we will consider complex vector spaces later on.

3.1 The standard inner product

The scalar product, inner product, or dot product, of two vectors in $\mathbb{R}^2$ or $\mathbb{R}^3$ is rather well known. We start this chapter with its definition and description of its main properties. Throughout this section, $\|\mathbf{v}\|$ will denote the *length* of the vector $\mathbf{v}$ in $\mathbb{R}^2$ or $\mathbb{R}^3$.

The standard inner product in $\mathbb{R}^2$. Let $\mathbf{r}, \mathbf{s} \in \mathbb{R}^2$ be nonzero vectors, and let θ be the angle from vector $\mathbf{r}$ to $\mathbf{s}$. The angle $\mathbf{r}$ makes with the x-axis will be denoted α. (See Figure 3.1.) We suppose also that $\mathbf{r} = (r_1, r_2)^T$ and $\mathbf{s} = (s_1, s_2)^T$ in coordinate form, so

$$\|\mathbf{r}\| = \sqrt{r_1^2 + r_2^2} \text{ and } \|\mathbf{s}\| = \sqrt{s_1^2 + s_2^2}.$$

Now, the matrix for rotation about the origin by an angle α in the *clockwise* (or *negative*) direction is the basis-changing matrix

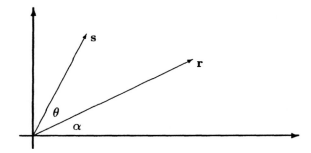

Fig. 3.1 Two vectors and the angle between them

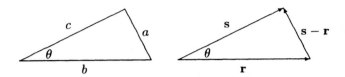

Fig. 3.2 The cosine rule

$$\begin{pmatrix} \cos\alpha & \sin\alpha \\ -\sin\alpha & \cos\alpha \end{pmatrix} = \frac{1}{\|\mathbf{r}\|} \begin{pmatrix} r_1 & r_2 \\ -r_2 & r_1 \end{pmatrix}.$$

In other words, multiplication of a position vector $(x,y)^T$ on the left by this matrix gives the corresponding position after rotation by the angle α as indicated.

Clearly, this matrix moves $\mathbf{s}$ to the vector $\|\mathbf{s}\|(\cos\theta, \sin\theta)^T$, since θ is the angle between $\mathbf{r}$ and $\mathbf{s}$, so

$$\|\mathbf{s}\| \begin{pmatrix} \cos\theta \\ \sin\theta \end{pmatrix} = \frac{1}{\|\mathbf{r}\|} \begin{pmatrix} r_1 & r_2 \\ -r_2 & r_1 \end{pmatrix} \begin{pmatrix} s_1 \\ s_2 \end{pmatrix}$$

$$= \frac{1}{\|\mathbf{r}\|} \begin{pmatrix} r_1 s_1 + r_2 s_2 \\ r_1 s_2 - r_2 s_1 \end{pmatrix}$$

from which we deduce that

$$r_1 s_1 + r_2 s_2 = \|\mathbf{r}\|\|\mathbf{s}\| \cos\theta \tag{1}$$

and

$$r_1 s_2 - r_2 s_1 = \|\mathbf{r}\|\|\mathbf{s}\| \sin\theta. \tag{2}$$

This chapter and the next are concerned with expressions like these. In particular, $\|\mathbf{r}\|\|\mathbf{s}\| \cos\theta$ is called the *standard inner product* of $\mathbf{r}$ and $\mathbf{s}$, and will be denoted $\mathbf{r} \cdot \mathbf{s}$ or $\langle \mathbf{r}|\mathbf{s} \rangle$. Note that as $\cos 0 = 1$ we have

$$\langle \mathbf{r}|\mathbf{r} \rangle = \|\mathbf{r}\|^2.$$

The cosine rule.   Before we look at inner products in three or more dimensions, we will deduce the familiar identity $a^2 = b^2 + c^2 - 2bc\cos\theta$ for a triangle with sides of length a, b, c and angle θ opposite the side a.

Let $\mathbf{r}, \mathbf{s}$ be the vectors in $\mathbb{R}^2$ with angle between them θ and lengths $\|\mathbf{r}\| = b$ and $\|\mathbf{s}\| = c$, and $\|\mathbf{s} - \mathbf{r}\| = a$, as in Figure 3.2; suppose that $\mathbf{r} = (r_1, r_2)^T$, $\mathbf{s} = (s_1, s_2)^T$. Then

$$\|\mathbf{s} - \mathbf{r}\| = (s_1 - r_1)^2 + (s_2 - r_2)^2$$
$$= s_1^2 - 2s_1 r_1 + r_1^2 + s_2^2 - 2s_2 r_2 + r_2^2$$
$$= s_1^2 + s_2^2 + r_1^2 + r_2^2 - 2(s_1 r_1 + s_2 r_2)$$
$$= \|\mathbf{s}\|^2 + \|\mathbf{r}\|^2 - 2\|\mathbf{r}\|\|\mathbf{s}\|\cos\theta.$$

So

$$a^2 = b^2 + c^2 - 2bc\cos\theta.$$

The standard inner product in $\mathbb{R}^3$.   As in $\mathbb{R}^2$, the standard inner product $\mathbf{r}\cdot\mathbf{s}$ or $\langle\mathbf{r}|\mathbf{s}\rangle$ is defined to be $\|\mathbf{r}\|\|\mathbf{s}\|\cos\theta$ where θ is the angle between the two vectors. Suppose $\mathbf{r} = (r_1, r_2, r_3)^T$ and $\mathbf{s} = (s_1, s_2, s_3)^T$; then by the cosine rule (referring to Figure 3.2 if necessary) we have

$$2\|\mathbf{r}\|\|\mathbf{s}\|\cos\theta = \|\mathbf{r}\|^2 + \|\mathbf{s}\|^2 - \|\mathbf{s} - \mathbf{r}\|^2$$
$$= r_1^2 + r_2^2 + r_3^2 + s_1^2 + s_2^2 + s_3^2$$
$$- ((s_1 - r_1)^2 + (s_2 - r_2)^2 + (s_3 - r_3)^2)$$
$$= 2(r_1 s_1 + r_2 s_2 + r_3 s_3),$$

by cancelling terms. Therefore we have an expression very similar to that obtained in two dimensions for our inner product, i.e.

$$\langle\mathbf{r}|\mathbf{s}\rangle = r_1 s_1 + r_2 s_2 + r_3 s_3. \tag{3}$$

3.2   Inner products

Although it is more difficult to interpret the idea of angle in four or more dimensions, the expression in (3) suggests we *define* $\langle\mathbf{v}|\mathbf{w}\rangle$ in $\mathbb{R}^n$ by

$$\langle\mathbf{v}|\mathbf{w}\rangle = \sum_{i=1}^{n} v_i w_i$$

for two vectors $\mathbf{v} = (v_1, v_2, \ldots, v_n)^T$ and $\mathbf{w} = (w_1, w_2, \ldots, w_n)^T$ in $\mathbb{R}^n$. This is called the *standard inner product* on $\mathbb{R}^n$, but there are many other possible inner products, as we shall see.

What are the essential properties of an inner product? We obviously have

$$\langle\mathbf{w}|\mathbf{v}\rangle = \sum_{i=1}^{n} w_i v_i = \sum_{i=1}^{n} v_i w_i = \langle\mathbf{v}|\mathbf{w}\rangle,$$

$$\langle\mathbf{v}|\lambda\mathbf{w}\rangle = \sum_{i=1}^{n} v_i \lambda w_i = \lambda\sum_{i=1}^{n} v_i w_i = \lambda\langle\mathbf{v}|\mathbf{w}\rangle,$$

and

$$\langle \mathbf{u}|\mathbf{v} + \mathbf{w}\rangle = \sum_{i=1}^{n} u_i(v_i + w_i) = \sum_{i=1}^{n} u_iv_i + \sum_{i=1}^{n} u_iw_i = \langle \mathbf{u}|\mathbf{v}\rangle + \langle \mathbf{u}|\mathbf{w}\rangle.$$

A slightly less obvious property which is nevertheless important is that

$$\langle \mathbf{v}|\mathbf{v}\rangle = \sum_{i=1}^{n} v_i^2 \geqslant 0,$$

and also that if $\langle \mathbf{v}|\mathbf{v}\rangle = 0$ then $\sum_{i=1}^{n} v_i^2 = 0$ and hence $v_i = 0$ for all i, which implies $\mathbf{v} = \mathbf{0}$. It is these properties which we (somewhat arbitrarily) choose as the defining properties of an inner product.

Definition 3.1 *If V is a vector space over $\mathbb{R}$, then an **inner product** on V is a map (written $\langle \;|\; \rangle$) from $V \times V$ to $\mathbb{R}$ (taking a pair of vectors (v, w) to a real number $\langle v|w\rangle$) with the following properties.*

(a) *(**Symmetry**) $\langle v|w\rangle = \langle w|v\rangle$ for all vectors v, w in V.*

(b) *(**Linearity**) $\langle u|\lambda v + \mu w\rangle = \lambda\langle u|v\rangle + \mu\langle u|w\rangle$, for all vectors u, v, w in V and all scalars λ, μ.*

(c) *(**Positive definiteness**)*
      i. *$\langle v|v\rangle \geqslant 0$, and*
      ii. *if $\langle v|v\rangle = 0$ then $v = 0$ (equivalently, if $v \neq 0$, then $\langle v|v\rangle \neq 0$)*
   for all vectors $v \in V$.

Notice that because it is symmetric, the linearity in the second variable implies linearity in the first variable. That is,

$$\langle \lambda u + \mu v|w\rangle = \lambda\langle u|w\rangle + \mu\langle v|w\rangle.$$

Thus an inner product is linear in both variables, so we call it *bilinear*.

Definition 3.2 *A finite dimensional vector space over $\mathbb{R}$ with an inner product defined on it is called a **Euclidean space**.*

We can define all sorts of different inner products on a vector space, not just the standard ones given above.

Example 3.3 In $\mathbb{R}^2$ we could define

$$\langle (a, b)^T|(c, d)^T\rangle = ac + bc + ad + 3bd.$$

This satisfies the above three properties, so is an inner product. The first two are easy to verify. To check the third one, observe first that $\langle (a, b)^T|(a, b)^T\rangle = a^2 + 2ab + 3b^2 = (a + b)^2 + 2b^2 \geqslant 0$. Moreover, if $\langle (a, b)^T|(a, b)^T\rangle = 0$, then $(a + b)^2 + 2b^2 = 0$, so $a + b = 0$ and $b = 0$, and therefore $(a, b)^T = (0, 0)^T$.

Example 3.4 Another example is given by the vector space $\mathscr{C}[a, b]$ of all continuous functions[1] from the closed interval $[a, b]$ to $\mathbb{R}$. Here, addition and scalar

---

[1]We shall not give a completely rigorous account of 'continuous functions' and integration. Instead, the student is directed to any standard textbook in analysis. For now, the basic properties needed here can be take on trust. See also Appendix A.

multiplication are the usual pointwise addition and scalar multiplication of functions (similar to that in Example 2.29), and the zero element is the function which takes the value 0 everywhere. If f and g are two functions in $\mathscr{C}[a, b]$, we can define their inner product to be

$$\langle f | g \rangle = \int_a^b f(x)g(x)\, dx.$$

To prove that this is an inner product according to Definition 3.1, we first need the facts that

$$\int_a^b f(x)g(x)\, dx = \int_a^b g(x)f(x)\, dx,$$

$$\int_a^b f(x)(\alpha g(x) + \beta h(x))\, dx = \alpha \int_a^b f(x)g(x)\, dx + \beta \int_a^b f(x)h(x)\, dx,$$

and

$$\int_a^b (f(x))^2\, dx \geqslant 0,$$

which immediately give properties (a), (b), and (c i).

To prove (c ii), we need the (somewhat more difficult) result that if $g: [a, b] \to \mathbb{R}$ is a continuous function which is nonnegative and not identically 0, then $\int_a^b g(x)\, dx \neq 0$. This result is proved in Appendix A, as Lemma A.2. Now suppose that f is not identically 0, so that $(f(x))^2$ defines a continuous function which is everywhere nonnegative and not identically zero. Then

$$\langle f | f \rangle = \int_a^b (f(x))^2\, dx \neq 0,$$

giving (c ii), so we have now proved that this is an inner product.

Example 3.5 The vector space $\mathscr{C}[a, b]$ in Example 3.4 is a very large infinite dimensional space, but we can construct a similar finite dimensional example, by taking the space $\mathbb{R}_n[x]$ of all polynomials in x of degree less than n, and defining an inner product by

$$\langle f | g \rangle = \int_0^1 f(x)g(x)\, dx.$$

The ordinary scalar product on $\mathbb{R}^2$ and $\mathbb{R}^3$ is related to concepts of *length* and *distance* in a way that can easily be generalized to arbitrary Euclidean spaces. If we take $\mathbf{v} = (x, y, z)^T \in \mathbb{R}^3$ then for the standard inner product, $\mathbf{v} \cdot \mathbf{v} = x^2 + y^2 + z^2$, which is the square of the length of the vector $\mathbf{v}$. The distance between two vectors $\mathbf{v}$ and $\mathbf{w}$ is then naturally defined as the length of $\mathbf{v} - \mathbf{w}$. In general we define:

Definition 3.6 *The* **norm** *(or length) of a vector v is written $\|v\|$ and defined by $\|v\| = \sqrt{\langle v|v \rangle}$, the positive square root of the inner product of v with itself. The* **distance** *between two vectors v and w is written $d(v, w)$ and defined by $d(v, w) = \|v - w\|$.*

As a consequence, we note that $\|-v\| = \|v\|$; the 'length' of the vector $-v$ equals the length of v. More generally, if you multiply a vector by a scalar, then its length is multiplied by the absolute value of the same scalar.

Proposition 3.7 *For all vectors v in a Euclidean space V, and for all $\lambda \in \mathbb{R}$, we have $\|\lambda v\| = |\lambda| \cdot \|v\|$.*

Proof $\|\lambda v\| = \sqrt{\langle \lambda v | \lambda v \rangle} = \sqrt{(\lambda^2 \langle v|v \rangle)} = |\lambda| \cdot \sqrt{\langle v|v \rangle} = |\lambda| \cdot \|v\|.$ $\qquad\square$

If you *add* two vectors together, the relationship between the lengths is not so simple, but we still get the familiar *triangle inequality*, i.e. $\|v + w\| \leqslant \|v\| + \|w\|$. To prove this in general requires the following very important result.

Proposition 3.8 (The Cauchy–Schwarz inequality) *For all vectors v, w in a Euclidean space V,*

$$|\langle v|w \rangle| \leqslant \|v\| \cdot \|w\|.$$

Proof For every real value of λ we have

$$\begin{aligned}
0 \leqslant \|v + \lambda w\|^2 &= \langle v + \lambda w | v + \lambda w \rangle \\
&= \langle v|v \rangle + \lambda \langle v|w \rangle + \lambda \langle w|v \rangle + \lambda^2 \langle w|w \rangle \\
&= \lambda^2 \|w\|^2 + 2\lambda \langle v|w \rangle + \|v\|^2.
\end{aligned}$$

Now regard the right-hand side as a quadratic polynomial in the variable λ. This polynomial is always nonnegative, so it has at most one real root. Therefore the discriminant ('$b^2 - 4ac$' for the polynomial $ax^2 + bx + c$) is nonpositive. In symbols,

$$4\langle v|w \rangle^2 - 4\|v\|^2 \|w\|^2 \leqslant 0;$$

hence

$$\langle v|w \rangle^2 \leqslant \|v\|^2 \|w\|^2.$$

Now take the positive square root of both sides to obtain the result. $\qquad\square$

A slightly different proof is obtained by saying that the first inequality in the above proof, namely

$$0 \leqslant \lambda^2 \|w\|^2 + 2\lambda \langle v|w \rangle + \|v\|^2,$$

holds for all values of λ, in particular if $w \neq 0$ then the inequality above holds for

$$\lambda = -\frac{\langle v|w \rangle}{\|w\|^2}.$$

Substituting in and simplifying yields the Cauchy–Schwarz inequality. Of course, if $w = 0$ then $\langle v|w \rangle = 0$ so both sides of the inequality are zero and so the inequality is true here too. (Compare this also with the proof of the complex version given in Proposition 3.21 below.)

The Cauchy–Schwarz inequality has many forms, as it can be applied to all sorts of spaces. For example, applying it to the standard inner product on $\mathbb{R}^n$ we obtain the following.

Corollary 3.9 *If $x_1, \ldots, x_n, y_1, \ldots, y_n$ are any real numbers, then*

$$\left(\sum_{i=1}^{n} x_i y_i \right)^2 \leqslant \left(\sum_{i=1}^{n} x_i^{\,2} \right) \left(\sum_{i=1}^{n} y_i^{\,2} \right).$$

Similarly, we can apply it to Example 3.4 to obtain:

Corollary 3.10 *If f and g are continuous real-valued functions on the closed interval $[a, b]$, then*

$$\left(\int_a^b f(x)g(x)\, dx \right)^2 \leqslant \int_a^b (f(x))^2\, dx \int_a^b (g(x))^2\, dx.$$

Proposition 3.11 (The triangle inequality) *For any vectors v, w in a Euclidean space V,*

$$\|v + w\| \leqslant \|v\| + \|w\|.$$

Proof Expanding directly,

$$
\begin{aligned}
\|v + w\|^2 &= \langle v + w | v + w \rangle \\
&= \langle v | v \rangle + 2\langle v | w \rangle + \langle w | w \rangle \\
&\leqslant \|v\|^2 + 2|\langle v | w \rangle| + \|w\|^2 \\
&\leqslant \|v\|^2 + 2\|v\|\|w\| + \|w\|^2 \qquad \text{by Proposition 3.8} \\
&= (\|v\| + \|w\|)^2.
\end{aligned}
$$

Now take the positive square root of each side. $\square$

We have just seen how to generalize the concept of distance from ordinary three-dimensional Euclidean space to arbitrary real inner product spaces, in such a way that the basic theorems like the triangle inequality still hold in this more general context. Another concept that can be easily generalized is that of *angle*. Recall that two vectors in $\mathbb{R}^3$ are perpendicular (at right angles, or *orthogonal*) if their inner product is zero. More generally, if $\mathbf{v}$ and $\mathbf{w}$ are two nonzero vectors in $\mathbb{R}^3$ then the angle θ between them is given by $\mathbf{v} \cdot \mathbf{w} = \|\mathbf{v}\|\|\mathbf{w}\|\cos\theta$. These properties can be used as *definitions* in arbitrary spaces with inner products defined on them.

Definition 3.12 *If V is a Euclidean space, and v and w are elements of V, then v and w are said to be **orthogonal** if $\langle v | w \rangle = 0$. If both v and w are nonzero, the **angle** between v and w is defined to be θ where $0 \leqslant \theta \leqslant \pi$ and*

$$\cos\theta = \frac{\langle v | w \rangle}{\|v\| \cdot \|w\|}.$$

Note that the Cauchy–Schwarz inequality (Proposition 3.8) implies that

$$-1 \leqslant \frac{\langle v|w \rangle}{\|v\| \cdot \|w\|} \leqslant 1$$

and so this definition of θ makes sense.

Example 3.13 In $\mathbb{R}^3$ with the standard inner product, these definitions coincide with the ordinary geometrical definitions.

Example 3.14 In $\mathscr{C}[-\pi, \pi]$ with the inner product

$$\langle f|g \rangle = \int_{-\pi}^{\pi} f(x)g(x)\,dx,$$

define functions f_k and g_k by $f_k(x) = \cos(kx)$ for integers $k \geqslant 0$ and $g_k(x) = \sin(kx)$ for integers $k \geqslant 1$. Then you can check that any two of these functions are orthogonal to each other. For example,

$$\langle f_k|g_m \rangle = \int_{-\pi}^{\pi} \cos(kx)\sin(mx)\,dx = 0$$

since the integrand is an odd function of x, i.e. a function $f(x)$ such that $f(-x) = -f(x)$ for all $x \in \mathbb{R}$. Also, provided $k \neq m$,

$$
\begin{aligned}
\langle f_k|f_m \rangle &= \int_{-\pi}^{\pi} \cos(kx)\cos(mx)\,dx \\
&= \tfrac{1}{2}\int_{-\pi}^{\pi} (\cos(kx + mx) + \cos(kx - mx))\,dx \\
&= \tfrac{1}{2}\left[\frac{\sin(k\pi + m\pi)}{k+m} - \frac{\sin-(k\pi + m\pi)}{k+m}\right.\\
&\qquad \left. + \frac{\sin(k\pi - m\pi)}{k-m} - \frac{\sin-(k\pi - m\pi)}{k-m}\right] \\
&= 0.
\end{aligned}
$$

Similarly, if $k \neq m$,

$$
\begin{aligned}
\langle g_k|g_m \rangle &= \int_{-\pi}^{\pi} \sin(kx)\sin(mx)\,dx \\
&= \tfrac{1}{2}\int_{-\pi}^{\pi} (\cos(kx - mx) - \cos(kx + mx))\,dx \\
&= 0.
\end{aligned}
$$

(If $k = m$ then $\cos(kx - mx) = 1$ and both integrals come to π.)

This is an important example which you will most likely meet elsewhere as well. It is the foundation of the theory of *Fourier series*.

3.3   Inner products over $\mathbb{C}$

So far in this chapter we have been working with *real* vector spaces. However, the whole theory goes through for *complex* vector spaces with very little change. In what follows we write $\bar{z}$ for the complex conjugate of z, and $|z|$ for the absolute value of z, so that $|z|^2 = \bar{z}z$. We shall also use $\mathrm{Re}(z)$ for the real part of z, and $\mathrm{Im}(z)$ for the imaginary part of z, so $z + \bar{z} = 2\,\mathrm{Re}(z)$ and $z - \bar{z} = 2i\,\mathrm{Im}(z)$.

Example 3.15 Let $V = \mathbb{C}$, regarded as a one-dimensional complex vector space. It would be nice if our definition of an inner product gave rise to $\|z - w\|$ being the distance from z to w in the Argand diagram, i.e. $|z - w|$, which would mean that $\|z\|^2 = |z|^2 = \bar{z}z$. This suggests we should define the inner product by $\langle z|w \rangle = \bar{z}w$. This has slightly different properties from the real inner products we have seen so far. For example,

$$\langle z|w \rangle = \bar{z}\,w = \overline{\bar{w}\,z} = \overline{\langle w|z \rangle}.$$

We still have

$$\langle z|\lambda w \rangle = \bar{z}\,\lambda w = \lambda \bar{z}\,w = \lambda \langle z|w \rangle,$$

but now

$$\langle \lambda z|w \rangle = \overline{\lambda z}\,w = \bar{\lambda}\bar{z}\,w = \bar{\lambda}\langle z|w \rangle.$$

Based on this example, we give the following formal definition of a complex inner product.

Definition 3.16 *If V is a vector space over $\mathbb{C}$, then a map $\langle\,|\,\rangle$ from $V \times V$ to $\mathbb{C}$ (taking (v, w) to $\langle v|w \rangle$) is an **inner product** if the following are true for all $u, v, w \in V$ and $\lambda, \mu \in \mathbb{C}$.*

(a) *(**Conjugate-symmetry**) $\langle v|w \rangle = \overline{\langle w|v \rangle}$.*

(b) *(**Linearity**) $\langle u|\lambda v + \mu w \rangle = \lambda \langle u|v \rangle + \mu \langle u|w \rangle$.*

(c) *(**Positive definiteness**)*
   i. *$\langle v|v \rangle \in \mathbb{R}$ with $\langle v|v \rangle \geqslant 0$, and*
   ii. *if $\langle v|v \rangle = 0$ then $v = 0$*
      (equivalently, if $v \neq 0$ then $\langle v|v \rangle \neq 0$).

Notice that (a) and (b) imply that

$$\begin{aligned}
\langle \lambda u + \mu v|w \rangle &= \overline{\langle w|\lambda u + \mu v \rangle} \\
&= \overline{\lambda \langle w|u \rangle + \mu \langle w|v \rangle} \\
&= \bar{\lambda}\,\overline{\langle w|u \rangle} + \bar{\mu}\,\overline{\langle w|v \rangle} \\
&= \bar{\lambda}\,\langle u|w \rangle + \bar{\mu}\,\langle v|w \rangle.
\end{aligned}$$

Note in particular $\langle ru + sv|w \rangle = r \langle u|w \rangle + s \langle v|w \rangle$ for any *real* r, s; in particular, putting $r = s = 1$ we have $\langle u + v|w \rangle = \langle u|w \rangle + \langle v|w \rangle$. Although this type of inner product is not bilinear it is nearly so. It is sometimes called *sesquilinear*, i.e. 'one-and-a-half-linear', because of this and part (b) of Definition 3.16, and it is also called *conjugate-symmetric* in view of part (a) of Definition 3.16.

Definition 3.17 *A finite dimensional vector space over* $\mathbb{C}$ *with an inner product defined on it is called a **unitary space**.*

We will also need to be able to refer to vector spaces with an inner product without specifying whether these spaces are over $\mathbb{R}$ or over $\mathbb{C}$, and without any assumption that they have finite dimension. Such spaces will be referred to in the rest of this book as *inner product spaces*.

Example 3.18 Let $V = \mathbb{C}^n$ and let $\mathbf{v}$ and $\mathbf{w}$ be any two vectors in V, say $\mathbf{v} = (v_1, \ldots, v_n)^T$ and $\mathbf{w} = (w_1, \ldots, w_n)^T$. Then define

$$\langle \mathbf{v}|\mathbf{w}\rangle = \sum_{i=1}^{n} \overline{v_i} w_i.$$

Then $\langle\,|\,\rangle$ is an inner product (as you should check), called the *standard inner product on* $\mathbb{C}^n$.

Most of the results for *real* inner product spaces also work for *complex* inner product spaces, with a sprinkling of complex conjugates and absolute values inserted.

Definition 3.19 *If V is a unitary space and $v \in V$, we define the **norm** of v to be $\|v\| = \sqrt{\langle v|v\rangle}$, and the **distance** between two vectors v and w to be $d(v,w) = \|v - w\|$.*

(The concept of *angle* is not so easy to interpret in the complex case, so we will not try to do so.)

Proposition 3.20 *For any vector v in a complex inner product space and any $\lambda \in \mathbb{C}$, we have $\|\lambda v\| = |\lambda| \cdot \|v\|$.*

Proof $\|\lambda v\| = \sqrt{\langle \lambda v|\lambda v\rangle} = \sqrt{(\overline{\lambda}\lambda\langle v|v\rangle)} = |\lambda| \cdot \sqrt{\langle v|v\rangle} = |\lambda| \cdot \|v\|$.   □

Proposition 3.21 (The Cauchy–Schwarz inequality) *For any vectors v, w in a complex inner product space V,*

$$|\langle v|w\rangle| \leqslant \|v\| \cdot \|w\|.$$

Proof For variety, we give here a slightly different proof from the one we gave for the real case. Note carefully the places in the proof where we have to insert an absolute value, complex conjugate, or real part. For every *complex* value of λ we have

$$\begin{aligned}
0 \leqslant \|v + \lambda w\|^2 &= \langle v + \lambda w|v + \lambda w\rangle \\
&= \langle v|v + \lambda w\rangle + \overline{\lambda}\langle w|v + \lambda w\rangle \\
&= \langle v|v\rangle + \lambda\langle v|w\rangle + \overline{\lambda}\langle w|v\rangle + \overline{\lambda}\lambda\langle w|w\rangle \\
&= \langle v|v\rangle + \lambda\langle v|w\rangle + \overline{\lambda}\,\overline{\langle v|w\rangle} + \overline{\lambda}\lambda\langle w|w\rangle \\
&= \|v\|^2 + 2\operatorname{Re}(\lambda\langle v|w\rangle) + |\lambda|^2\|w\|^2.
\end{aligned}$$

Now this is true for all λ, in particular if $w \neq 0$ it is true for

$$\lambda = -\frac{\langle w|v \rangle}{\|w\|^2}.$$

Since $\langle v|w \rangle \langle w|v \rangle = \langle v|w \rangle \overline{\langle v|w \rangle} = |\langle v|w \rangle|^2$, we have

$$0 \leqslant \|v\|^2 + 2\,\mathrm{Re}\left(\frac{-|\langle v|w \rangle|^2}{\|w\|^2}\right) + \frac{|\langle v|w \rangle|^2}{\|w\|^4}\|w\|^2 = -\frac{|\langle v|w \rangle|^2}{\|w\|^2} + \|v\|^2.$$

Therefore

$$|\langle v|w \rangle|^2 \leqslant \|w\|^2 \|v\|^2,$$

from which the result for $w \neq 0$ follows by taking the positive square root of both sides. For $w = 0$ the inequality is trivial as $\langle v|w \rangle = \|w\| = 0$.   □

Proposition 3.22 (The triangle inequality) *For any v, w in a complex inner product space V,*

$$\|v + w\| \leqslant \|v\| + \|w\|.$$

Proof   As before,

$$\begin{aligned}
\|v + w\|^2 &= \langle v + w|v + w \rangle \\
&= \langle v|v \rangle + \langle v|w \rangle + \langle w|v \rangle + \langle w|w \rangle \\
&= \langle v|v \rangle + 2\,\mathrm{Re}\langle v|w \rangle + \langle w|w \rangle \\
&\leqslant \|v\|^2 + 2|\langle v|w \rangle| + \|w\|^2 \\
&\leqslant \|v\|^2 + 2\|v\| \cdot \|w\| + \|w\|^2 \qquad \text{by Proposition 3.21} \\
&= (\|v\| + \|w\|)^2.
\end{aligned}$$

Now take the positive square root of each side.   □

Example 3.23 For $a < b$ from $\mathbb{R}$, the space $\mathscr{C}_{\mathbb{C}}[a, b]$ is the complex vector space of all continuous functions (see Appendix A) $f : [a, b] \to \mathbb{C}$ with pointwise addition and scalar multiplication, analogous to the real case (Example 3.4). As in the real case, $\mathscr{C}_{\mathbb{C}}[a, b]$ can be given an inner product,

$$\langle f|g \rangle = \int_a^b \overline{f(x)} g(x)\, dx.$$

The axioms of linearity in the second argument and conjugate-symmetry are straightforward to verify; positive definiteness is proved just as in the real case, using $\overline{f(x)}f(x) = |f(x)|^2 \in \mathbb{R}$, and the fact that $|f(x)|$ defines a continuous real-valued function on $[a, b]$.

The Cauchy–Schwarz inequality for this space says that whenever $f, g \in \mathscr{C}_{\mathbb{C}}[a, b]$ then $|\langle f|g \rangle| \leqslant \|f\|\,\|g\|$, or

$$\left| \int_a^b \overline{f(x)} g(x)\, dx \right| \leqslant \left(\int_a^b |f(x)|^2\, dx \right)^{1/2} \left(\int_a^b |g(x)|^2\, dx \right)^{1/2}.$$

Example 3.24 Let V be the set of all continuous functions $f \colon \mathbb{R} \to \mathbb{C}$ such that the integral

$$\int_{-\infty}^{\infty} |f(x)|^2 \, dx = \lim_{\substack{a \to \infty \\ b \to \infty}} \int_{-a}^{b} |f(x)|^2 \, dx$$

exists in $\mathbb{R}$. We show that V becomes a vector space over $\mathbb{C}$ when addition and scalar multiplication are defined pointwise, as in the previous example.

The most difficult part is in showing that V is closed under addition, i.e. that if $f, g \in V$ then $f + g \in V$, in particular that the integral

$$\int_{-\infty}^{\infty} |f(x) + g(x)|^2 \, dx$$

is finite. For this, let $a, b > 0$ and note that

$$
\begin{aligned}
|f(x) + g(x)|^2 &= (\overline{f(x)} + \overline{g(x)})(f(x) + g(x)) \\
&= \overline{f(x)}f(x) + \overline{f(x)}g(x) + \overline{g(x)}f(x) + \overline{g(x)}g(x) \\
&= |f(x)|^2 + 2\operatorname{Re}(\overline{f(x)}g(x)) + |g(x)|^2.
\end{aligned}
$$

So $\int_{-a}^{b} |f(x) + g(x)|^2 \, dx$ equals

$$
\int_{-a}^{b} |f(x)|^2 \, dx + 2 \int_{-a}^{b} \operatorname{Re}(\overline{f(x)}g(x)) \, dx + \int_{-a}^{b} |g(x)|^2 \, dx
$$
$$
\leqslant \langle f|f \rangle_{[-a,b]} + 2|\langle f|g \rangle_{[-a,b]}| + \langle g|g \rangle_{[-a,b]}
$$

where $\langle\,|\,\rangle_{[-a,b]}$ denotes the inner product in $\mathscr{C}_{\mathbb{C}}[-a,b]$ of the previous example. Now put $r = \int_{-\infty}^{\infty} |f(x)|^2 \, dx$, and $s = \int_{-\infty}^{\infty} |g(x)|^2 \, dx$; note also that

$$|\langle f|g \rangle_{[-a,b]}| \leqslant |\langle f|f \rangle_{[-a,b]}|^{1/2} |\langle g|g \rangle_{[-a,b]}|^{1/2}$$

by Cauchy–Schwarz in $\mathscr{C}_{\mathbb{C}}[-a,b]$. Hence $\int_{-a}^{b} |f(x) + g(x)|^2 \, dx$ is at most

$$\langle f|f \rangle_{[-a,b]} + 2|\langle f|f \rangle_{[-a,b]}|^{1/2} |\langle g|g \rangle_{[-a,b]}|^{1/2} + \langle g|g \rangle_{[-a,b]} \leqslant r + 2\sqrt{rs} + s.$$

This gives an upper bound on $I(a,b) = \int_{-a}^{b} |f(x) + g(x)|^2 \, dx$ which is independent of a, b. Since $I(a,b)$ is nondecreasing as $a, b \to \infty$, this shows that $\lim_{a,b \to \infty} I(a,b) = \int_{-\infty}^{\infty} |f(x) + g(x)|^2 \, dx$ exists, as required.

To prove closure of V under scalar multiplication, just note that for $\lambda \in \mathbb{C}$,

$$\int_{-a}^{b} |\lambda f(x)|^2 \, dx = |\lambda|^2 \int_{-a}^{b} |f(x)|^2 \, dx \leqslant |\lambda|^2 \int_{-\infty}^{\infty} |f(x)|^2 \, dx.$$

The vector space axioms are then verified in the usual way and all are now straightforward.

In fact,

$$\langle f|g \rangle = \int_{-\infty}^{\infty} \overline{f(x)} g(x) \, dx$$

defines an inner product on V. Once again, the only difficult part here is to show that if $f, g \in V$ then the integral $\int_{-\infty}^{\infty} \overline{f(x)} g(x) \, dx$ exists in $\mathbb{C}$. This can be done by considering the real and imaginary parts of this integral separately, and using the Cauchy–Schwarz inequality in $\mathscr{C}_{\mathbb{C}}[-a, b]$, as before. The axioms for an inner product can be verified as for $\mathscr{C}_{\mathbb{C}}[a, b]$.

Exercises

Exercise 3.1 For each of the following pairs of vectors in $\mathbb{R}^3$, calculate the angle between them.

(a) $(1/\sqrt{2}, 1/\sqrt{2}, 0)^T$, $(0, 1, 0)^T$
(b) $(1, 1, 0)^T$, $(1, -1, 0)^T$
(c) $(1, 0, 1)^T$, $(-1, 1, 0)^T$.

Exercise 3.2 In each of the following cases compute the projection of $\mathbf{s}$ in the direction of $\mathbf{r}$. Then write down a nonzero vector $\mathbf{t}$ in the space spanned by $\mathbf{s}$ and $\mathbf{r}$ orthogonal to $\mathbf{r}$. Verify your answer by computing $\langle \mathbf{t}|\mathbf{r} \rangle$.

(a) $\mathbf{s} = (5, 2, 1)^T$, $\mathbf{r} = (-1, 2, 3)^T$
(b) $\mathbf{s} = (1, 1, 1)^T$, $\mathbf{r} = (-1, 0, 1)^T$.

Exercise 3.3 From equation (2), the value $r_1 s_2 - r_2 s_1 = \|\mathbf{r}\|\|\mathbf{s}\| \sin \theta$ is the two-dimensional determinant

$$\begin{vmatrix} r_1 & s_1 \\ r_2 & s_2 \end{vmatrix}.$$

Give a geometrical interpretation of this determinant in terms of the area of the parallelogram drawn out by $\mathbf{r}$ and $\mathbf{s}$. Explain the significance of its sign in terms of the direction travelled as you go from $\mathbf{r}$ to $\mathbf{s}$.

Exercise 3.4 The triangle inequality is often used in one of the following alternative forms:

$$\|v - w\| \leqslant \|v\| + \|w\|$$
$$\|v + w\| \geqslant \|v\| - \|w\|$$
$$\|v - w\| \geqslant \|v\| - \|w\|.$$

Show that these forms follow from Proposition 3.11 by simple substitutions. (Remember that $\|-w\| = \|w\|$.)

Exercise 3.5 By expanding $\langle v + w|v + w \rangle$ and $\langle v - w|v - w \rangle$, prove that, for any vectors v, w in a real inner product space,

(a) $\langle v|w\rangle = \frac{1}{2}(\|v+w\|^2 - \|v\|^2 - \|w\|^2)$,

(b) $\|v+w\|^2 + \|v-w\|^2 = 2(\|v\|^2 + \|w\|^2)$.

Exercise 3.6 Prove that in any complex inner product space

(a) $\langle v|w\rangle = \frac{1}{2}(\|v+w\|^2 + i\|v-iw\|^2 - (1+i)(\|v\|^2 + \|w\|^2))$,

(b) $\langle v|w\rangle = \frac{1}{4}(\|v+w\|^2 - \|v-w\|^2 + i\|v-iw\|^2 - i\|v+iw\|^2)$.

Exercise 3.7 In each of (a), (b), (c) below, determine whether $\langle\,|\,\rangle$ is an inner product on V. Justify your answers. In every case which is an inner product, write down the appropriate form of the Cauchy–Schwarz inequality.

(a) $V = \mathbb{R}^2$, $\langle \mathbf{v}|\mathbf{w}\rangle = (v_1 - w_1)^2 + (v_2 - w_2)^2$ where $\mathbf{v} = \begin{pmatrix} v_1 \\ v_2 \end{pmatrix}$ and $\mathbf{w} = \begin{pmatrix} w_1 \\ w_2 \end{pmatrix}$.

(b) $V = \mathbb{R}^3$, $\langle \mathbf{x}|\mathbf{y}\rangle = 2x_1y_1 + 3x_2y_2 + 2x_3y_3 + x_1y_3 + x_3y_1 - 2x_2y_3 - 2x_3y_2$, where $\mathbf{x} = (x_1, x_2, x_3)^T$ and $\mathbf{y} = (y_1, y_2, y_3)^T$.

(c) $V = \mathbb{R}^3$, $\langle \mathbf{x}|\mathbf{y}\rangle = x_1y_1 + 2x_2y_2 - 2x_1y_3 - 2x_3y_1 - x_2y_3 - x_3y_2$, where $\mathbf{x} = (x_1, x_2, x_3)^T$ and $\mathbf{y} = (y_1, y_2, y_3)^T$.

Exercise 3.8 Do the same as the last question for the following.

(a) V is the set of all $n \times n$ matrices with entries from $\mathbb{R}$, and $\langle \mathbf{A}|\mathbf{B}\rangle = \text{tr}(\mathbf{AB})$. (Recall: if $\mathbf{C}$ is the matrix (c_{ij}), then $\text{tr}(\mathbf{C})$ denotes the *trace* of $\mathbf{C}$, defined by $\text{tr}(\mathbf{C}) = \sum_{i=1}^{n} c_{ii}$.)

(b) V is the set of all $n \times n$ matrices with entries from $\mathbb{R}$, and $\langle \mathbf{A}|\mathbf{B}\rangle = \text{tr}(\mathbf{A}^T\mathbf{B})$.

(For one of them, try to write down an expression for $\langle \mathbf{A}|\mathbf{A}\rangle$ where $\mathbf{A} = (a_{ij})$ in terms of the a_{ij}. It should remind you of $\mathbb{R}^{n^2}$. For the other, try to show that $\langle \mathbf{A}|\mathbf{B}\rangle$ is not positive definite.)

Exercise 3.9 Complete the proof that

$$\langle f|g\rangle = \int_{-\infty}^{\infty} \overline{f(x)} g(x)\, dx$$

is an inner product on the complex vector space $\mathscr{C}_{\mathbb{C}}[-a, a]$ of Example 3.24.

Exercise 3.10 Let V be the real vector space $\mathbb{R}^3$ with the standard inner product. Describe the set of vectors in V which are orthogonal to $(-1, 0, 2)^T$. In particular, show that this is a subspace of V, find a basis for it, and hence deduce its dimension.

4

Bilinear and sesquilinear forms

In the last chapter we introduced the idea of an inner product on a vector space. Our main example was the standard inner product on $\mathbb{R}^3$ (and, more generally, on $\mathbb{R}^n$). This inner product has a particularly elegant geometric meaning, but we saw other important examples (involving vector spaces of continuous functions and integration, for example) where the geometric interpretation isn't so clear.

It turns out that there are many other interesting forms defined like inner products which are not positive definite—some of which, like Minkowski space (Example 4.1), have significant physical interpretations. This chapter starts off the study of these forms, and in particular shows how they may be represented by matrices.

4.1 Bilinear forms

Example 4.1 (Minkowski space) This example is very important in the theory of special relativity. Take $V = \mathbb{R}^4$, with three 'space' coordinates x, y, z and one 'time' coordinate t, and define a function $F \colon V \times V \to \mathbb{R}$ by

$$F\big((x_1, y_1, z_1, t_1)^T, (x_2, y_2, z_2, t_2)^T\big) = x_1 x_2 + y_1 y_2 + z_1 z_2 - c^2 t_1 t_2,$$

where c is the speed of light. Then F is symmetric,

$$\begin{aligned} F\big((x_1, y_1, z_1, t_1)^T, (x_2, y_2, z_2, t_2)^T\big) &= x_1 x_2 + y_1 y_2 + z_1 z_2 - c^2 t_1 t_2 \\ &= F\big((x_2, y_2, z_2, t_2)^T, (x_1, y_1, z_1, t_1)^T\big), \end{aligned}$$

and linear in both arguments, e.g.

$$\begin{aligned} F\big(\lambda(x_1, & y_1, z_1, t_1)^T + \mu(x_2, y_2, z_2, t_2)^T, (x_3, y_3, z_3, t_3)^T\big) \\ &= (\lambda x_1 + \mu x_2)x_3 + (\lambda y_1 + \mu y_2)y_3 + (\lambda z_1 + \mu z_2)z_3 - c^2(\lambda t_1 + \mu t_2)t_3 \\ &= \lambda(x_1 x_3 + y_1 y_3 + z_1 z_3 - c^2 t_1 t_3) + \mu(x_2 x_3 + y_2 y_3 + z_2 z_3 - c^2 t_2 t_3) \\ &= \lambda F\big((x_1, y_1, z_1, t_1)^T, (x_3, y_3, z_3, t_3)^T\big) \\ &\quad + \mu F\big((x_2, y_2, z_2, t_2)^T, (x_3, y_3, z_3, t_3)^T\big), \end{aligned}$$

but if we try to define a 'norm' by $\|v\|^2 = F(v, v)$ and a 'distance' by $d(v, w) = \|v - w\|$, then we run into problems since F is not positive definite. In fact $F(v - w, v - w)$ can be positive, giving a so-called *space-like* distance

$$d(v, w) = \sqrt{F(v - w, v - w)},$$

negative, giving a *time-like* distance

$$d(v, w) = \frac{1}{c}\sqrt{-F(v - w, v - w)},$$

or even zero, as is for example $d\big((0, 0, 0, 0)^T, (c, 0, 0, 1)^T\big)$.

In this chapter we will generalize the idea of inner product by dropping the condition of being positive definite, and sometimes also the symmetry condition.

Definition 4.2 *A **bilinear form** on a real vector space V is a map F from $V \times V$ to $\mathbb{R}$ which for all $u, v, w \in V$ and $\alpha, \beta \in \mathbb{R}$ satisfies*

(a) $F(\alpha u + \beta v, w) = \alpha F(u, w) + \beta F(v, w)$, *and*

(b) $F(u, \alpha v + \beta w) = \alpha F(u, v) + \beta F(u, w)$.

Definition 4.3 *A bilinear form on V is **symmetric** if also*

(c) $F(u, v) = F(v, u)$ *for all $u, v \in V$.*

Example 4.4 (The Lorentz plane) Take $V = \mathbb{R}^2$ and define the map F from $V \times V$ to $\mathbb{R}$ by $F\big((a, b)^T, (c, d)^T\big) = ac - bd$. Then F is a symmetric bilinear form on V (as you should check), but is not an inner product since $F\big((0, 1)^T, (0, 1)^T\big) = -1$, and therefore property (c i) in Definition 3.1 of an inner product fails.

We also have $F\big((1, 1)^T, (1, 1)^T\big) = 0$, so there is a nonzero vector whose 'norm' is zero, and property (c ii) fails also.

The next example is really a whole family of examples. It is particularly important since it will turn out that all bilinear forms on finite dimensional real vector spaces can be regarded as essentially just one of these examples, in the same way that every finite dimensional real vector space is isomorphic to $\mathbb{R}^n$ for some n.

Example 4.5 Let V be the real vector space $\mathbb{R}^n$, and suppose $\mathbf{A}$ is an $n \times n$ matrix with entries from $\mathbb{R}$. Define an operation F on $V \times V$ by

$$F(\mathbf{x}, \mathbf{y}) = \mathbf{x}^T \mathbf{A} \mathbf{y}.$$

Since $\mathbf{x}^T$ is a $1 \times n$ matrix, $\mathbf{A}$ is $n \times n$, and $\mathbf{y}$ is $n \times 1$, this is well-defined matrix multiplication and also $F(\mathbf{x}, \mathbf{y})$ is a 1×1 matrix, which we can consider the same as the real number which is its only entry. Thus $F \colon V \times V \to \mathbb{R}$ and it turns out that F is a bilinear form on V. Furthermore, if $\mathbf{A}$ is symmetric, i.e. if $\mathbf{A}^T = \mathbf{A}$, then F is symmetric.

Proof To prove all these claims, we must check the two axioms for a bilinear form. We have

$$\begin{aligned}
F(\alpha \mathbf{x} + \beta \mathbf{y}, \mathbf{z}) &= (\alpha \mathbf{x} + \beta \mathbf{y})^T \mathbf{A} \mathbf{z} \\
&= (\alpha \mathbf{x}^T + \beta \mathbf{y}^T) \mathbf{A} \mathbf{z} \\
&= \alpha \mathbf{x}^T \mathbf{A} \mathbf{z} + \beta \mathbf{y}^T \mathbf{A} \mathbf{z} \\
&= \alpha F(\mathbf{x}, \mathbf{z}) + \beta F(\mathbf{y}, \mathbf{z}),
\end{aligned}$$

and similarly,

$$F(\mathbf{x}, \alpha\mathbf{y} + \beta\mathbf{z}) = \mathbf{x}^T\mathbf{A}(\alpha\mathbf{y} + \beta\mathbf{z})$$
$$= \mathbf{x}^T\mathbf{A}(\alpha\mathbf{y}) + \mathbf{x}^T\mathbf{A}(\beta\mathbf{z})$$
$$= \alpha\mathbf{x}^T\mathbf{A}\mathbf{y} + \beta\mathbf{x}^T\mathbf{A}\mathbf{z}$$
$$= \alpha F(\mathbf{x}, \mathbf{y}) + \beta F(\mathbf{x}, \mathbf{z}).$$

Now suppose $\mathbf{A}$ is symmetric. Then since the transpose of a 1×1 matrix is itself, $F(\mathbf{x}, \mathbf{y}) = F(\mathbf{x}, \mathbf{y})^T$. Using the fact that the transpose of a product of matrices is equal to the product of the transposes taken in the opposite order (Proposition 1.3), we have

$$F(\mathbf{x}, \mathbf{y}) = F(\mathbf{x}, \mathbf{y})^T$$
$$= (\mathbf{x}^T\mathbf{A}\mathbf{y})^T$$
$$= (\mathbf{y}^T\mathbf{A}^T(\mathbf{x}^T)^T)$$
$$= (\mathbf{y}^T\mathbf{A}\mathbf{x})$$
$$= F(\mathbf{y}, \mathbf{x}),$$

since $\mathbf{A}^T = \mathbf{A}$ and $(\mathbf{x}^T)^T = \mathbf{x}$. $\qquad\qquad\qquad\qquad\qquad\qquad\qquad$ □

Example 4.6 If $V = \mathbb{R}^n$ and $\mathbf{A}$ is the $n \times n$ identity matrix

$$\begin{pmatrix} 1 & 0 & \cdots & 0 \\ 0 & 1 & \cdots & 0 \\ \vdots & \vdots & \ddots & \vdots \\ 0 & 0 & \cdots & 1 \end{pmatrix}$$

then

$$F\big((x_1, \ldots, x_n)^T, (y_1, \ldots, y_n)^T\big) = (x_1, \ldots, x_n)\mathbf{A}(y_1, \ldots, y_n)^T$$
$$= (x_1, \ldots, x_n)(y_1, \ldots, y_n)^T$$
$$= x_1 y_1 + \cdots + x_n y_n,$$

so F is the standard inner product on V.

4.2 Representation by matrices

Suppose we have a real vector space V with an ordered basis $e_1, \ldots, e_n$, and a bilinear form F defined on V. Given two vectors v and w we can write them as linear combinations of the basis vectors, say $v = \sum_{i=1}^n v_i e_i$ and $w = \sum_{i=1}^n w_i e_i$. Then, using bilinearity,

$$F(v, w) = F\left(\sum_{i=1}^n v_i e_i, \sum_{j=1}^n w_j e_j\right) = \sum_{i=1}^n v_i F\left(e_i, \sum_{j=1}^n w_j e_j\right)$$
$$= \sum_{i=1}^n \sum_{j=1}^n v_i w_j F(e_i, e_j).$$

Thus if we know the values the form F takes on the basis elements, i.e. $F(e_i, e_j)$ for all i, j, then we can work out the form on any pair of vectors.

For convenience, we may put the values $F(e_i, e_j)$ of the basis vectors into a matrix,

$$\begin{pmatrix} F(e_1, e_1) & \dots & F(e_1, e_n) \\ \vdots & \ddots & \vdots \\ F(e_n, e_1) & \dots & F(e_n, e_n) \end{pmatrix}.$$

Definition 4.7 *The matrix* $\mathbf{A} = (a_{ij})$ *defined by* $a_{ij} = F(e_i, e_j)$ *is called the **matrix of the bilinear form** F **with respect to the ordered basis** $e_1, \dots, e_n$ **of** V.*

Notice that if F is a symmetric bilinear form, then $F(e_i, e_j) = F(e_j, e_i)$ for all i, j, so the matrix $\mathbf{A}$ is symmetric in the sense that $a_{ij} = a_{ji}$ for all i, j, or in other words $\mathbf{A} = \mathbf{A}^T$.

Now, in V, any vector v is a (unique) linear combination

$$v = v_1 e_1 + v_2 e_2 + \dots + v_n e_n$$

of basis vectors. Thus with respect to the basis $e_1, e_2, \dots, e_n$ the vector v is determined by its coordinate form, the column vector $(v_1, v_2, \dots, v_n)^T$. This representation can be used to give an elegant formula for the form F, just as in Example 4.5.

Proposition 4.8 *Suppose V is a real vector space with ordered basis $e_1, \dots, e_n$, and F is a bilinear form defined on V, with matrix $\mathbf{A}$ with respect to this basis. Then for any vectors $v, w \in V$ and their corresponding coordinate forms $\mathbf{v} = (v_1, \dots, v_n)^T$, $\mathbf{w} = (w_1, \dots, w_n)^T$ with respect to the same basis $e_1, e_2, \dots, e_n$ we have*

$$F(v, w) = \mathbf{v}^T \mathbf{A} \mathbf{w}.$$

Proof   We have

$$\mathbf{v}^T \mathbf{A} \mathbf{w} = \begin{pmatrix} v_1 & \dots & v_n \end{pmatrix} \begin{pmatrix} F(e_1, e_1) & \dots & F(e_1, e_n) \\ \vdots & \ddots & \vdots \\ F(e_n, e_1) & \dots & F(e_n, e_n) \end{pmatrix} \begin{pmatrix} w_1 \\ \vdots \\ w_n \end{pmatrix}$$

$$= \sum_{i=1}^{n} \sum_{j=1}^{n} v_i F(e_i, e_j) w_j$$

$$= F\left(\sum_{i=1}^{n} v_i e_i, \sum_{j=1}^{n} w_j e_j \right)$$

by bilinearity.   □

The conclusion is that every bilinear form on a finite dimensional real vector space is 'the same as' (more precisely, isomorphic to) one of the examples described in Example 4.5. Thus matrices can be applied to calculations in examples of inner products on vector spaces other than $\mathbb{R}^n$, as in the next example.

Example 4.9 Let V be the three-dimensional vector space $\mathbb{R}_3[x]$ of all polynomials of degree less than 3, with the inner product

$$\langle f|g\rangle = \int_0^1 f(x)g(x)\,dx$$

defined in Example 3.5. Then $1, x, x^2$ is an ordered basis of V, and the corresponding matrix of inner products is

$$\mathbf{A} = \begin{pmatrix} 1 & 1/2 & 1/3 \\ 1/2 & 1/3 & 1/4 \\ 1/3 & 1/4 & 1/5 \end{pmatrix}.$$

This matrix can be used to find $\langle x - 1|2x^2 + x\rangle$, for example. We write $x - 1 = (-1)\cdot 1 + 1\cdot x$ and $2x^2 + x = 1\cdot x + 2\cdot x^2$ in coordinate form with respect to this basis as $(-1, 1, 0)^T$ and $(0, 1, 2)^T$, and work out $(-1, 1, 0)\mathbf{A}(0, 1, 2)^T = -1/3$. If you like, you can check this by direct integration, as follows.

$$\int_0^1 (x-1)(2x^2+x)\,dx = \int_0^1 (2x^3 - x^2 - x)\,dx = \tfrac{1}{2} - \tfrac{1}{3} - \tfrac{1}{2} = -\tfrac{1}{3}.$$

Being able to represent bilinear forms as matrices is not the end of the story, however, as the next two examples show.

Example 4.10 If we take the standard inner product on $\mathbb{R}^2$, then with respect to the standard basis $(1, 0)^T, (0, 1)^T$ the matrix of this inner product is the identity matrix $\begin{pmatrix} 1 & 0 \\ 0 & 1 \end{pmatrix}$. If we choose a different basis, however, then we get a different matrix representing the inner product.

For example, we calculate the matrix of the standard inner product with respect to the ordered basis $(1, 2)^T, (-1, 0)^T$. Here, $\langle (1, 2)^T|(1, 2)^T\rangle = 1^2 + 2^2 = 5$, $\langle (1, 2)^T|(-1, 0)^T\rangle = -1 + 0 = -1$, $\langle (-1, 0)^T|(1, 2)^T\rangle = -1$ by symmetry of $\langle | \rangle$, and $\langle (-1, 0)^T|(-1, 0)^T\rangle = 1$. So the representing matrix is $\begin{pmatrix} 5 & -1 \\ -1 & 1 \end{pmatrix}$.

Example 4.11 Take $V = \mathbb{R}^2$ and $F(\mathbf{x}, \mathbf{y}) = x_1 y_1 + x_1 y_2 + x_2 y_1$ where $\mathbf{x} = (x_1, x_2)^T$ and $\mathbf{y} = (y_1, y_2)^T$. With respect to the ordered basis $\mathbf{e}_1, \mathbf{e}_2$ given by

$$\mathbf{e}_1 = \begin{pmatrix} 0 \\ 2 \end{pmatrix}, \quad \mathbf{e}_2 = \begin{pmatrix} 1 \\ -1 \end{pmatrix},$$

we get the matrix $\mathbf{B} = \begin{pmatrix} 0 & 2 \\ 2 & -1 \end{pmatrix}$, while with respect to the basis $\mathbf{e}_1', \mathbf{e}_2'$, where

$$e_1' = \begin{pmatrix} 1 \\ 1 \end{pmatrix}, \ e_2' = \begin{pmatrix} -1 \\ 0 \end{pmatrix},$$

we get $\mathbf{B}' = \begin{pmatrix} 3 & -2 \\ -2 & 1 \end{pmatrix}.$

Notice that the form F and hence both the matrices $\mathbf{B}$ and $\mathbf{B}'$ are symmetric.

4.3 The base-change formula

Since we can choose all sorts of different bases, it is important to know what happens to the matrix of an inner product if we change the basis.

Let V be a real vector space with basis $e_1, \dots, e_n$ and bilinear form F. Suppose we choose a new ordered basis $f_1, \dots, f_n$, and write our new basis vectors in terms of the old ones, as

$$f_i = \sum_{k=1}^n p_{ki} e_k,$$

say. Write $\mathbf{P}$ for the matrix (p_{ij}), the *base-change matrix*.

Example 4.12 Let $V = \mathbb{R}^3$ and take the standard basis $e_1 = (1, 0, 0)^T$, $e_2 = (0, 1, 0)^T$, $e_3 = (0, 0, 1)^T$. If we now choose for a new basis $\mathbf{f}_1 = (1, 0, 1)^T$, $\mathbf{f}_2 = (2, -1, 0)^T$, $\mathbf{f}_3 = (0, 1, -1)^T$ we have

$$\mathbf{f}_1 = 1e_1 + 0e_2 + 1e_3$$
$$\mathbf{f}_2 = 2e_1 + (-1)e_2 + 0e_3$$
$$\mathbf{f}_3 = 0e_1 + 1e_2 + (-1)e_3$$

so the base-change matrix from basis e_1, e_2, e_3 to $\mathbf{f}_1, \mathbf{f}_2, \mathbf{f}_3$ is

$$\begin{pmatrix} 1 & 2 & 0 \\ 0 & -1 & 1 \\ 1 & 0 & -1 \end{pmatrix}.$$

In this example, the base-change matrix is the matrix formed from columns equal to the coordinate forms of the vectors $\mathbf{f}_1, \mathbf{f}_2, \mathbf{f}_3$ with respect to the old basis e_1, e_2, e_3.

In general,

The base-change matrix is the matrix $\mathbf{P}$ formed from columns giving the coordinate forms of the new ordered basis $f_1, f_2, \dots, f_n$ with respect to the old ordered basis $e_1, e_2, \dots, e_n$.

To understand the importance of this matrix in calculations, consider a vector $v \in V$ written in coordinate form as $\sum_{k=1}^n v_i f_i$ with respect to the new basis $\{f_1, f_2, \dots, f_n\}$. We have

$$\begin{pmatrix} v_1 \\ v_2 \\ \vdots \\ v_n \end{pmatrix} = v_1 \begin{pmatrix} 1 \\ 0 \\ \vdots \\ 0 \end{pmatrix} + v_2 \begin{pmatrix} 0 \\ 1 \\ \vdots \\ 0 \end{pmatrix} + \cdots + v_n \begin{pmatrix} 0 \\ 0 \\ \vdots \\ 1 \end{pmatrix}$$

so

$$\mathbf{P}\begin{pmatrix} v_1 \\ v_2 \\ \vdots \\ v_n \end{pmatrix} = v_1 \mathbf{P}\begin{pmatrix} 1 \\ 0 \\ \vdots \\ 0 \end{pmatrix} + v_2 \mathbf{P}\begin{pmatrix} 0 \\ 1 \\ \vdots \\ 0 \end{pmatrix} + \cdots + v_n \mathbf{P}\begin{pmatrix} 0 \\ 0 \\ \vdots \\ 1 \end{pmatrix}.$$

Now $\mathbf{P}(0,\ldots,0,1,0,\ldots,0)^T$ (where the 1 is in the ith position) is the column vector formed from the ith column of $\mathbf{P}$, i.e. the coordinate form of f_i with respect to the old basis $e_1, e_2, \ldots, e_n$. Therefore $\mathbf{P}(v_1, v_2, \ldots, v_n)^T$ is the coordinate form of v with respect to $e_1, e_2, \ldots, e_n$. In other words,

Multiplication of a column vector by the base-change matrix $\mathbf{P}$ *converts coordinate forms from the new ordered basis* $f_1, f_2, \ldots, f_n$ *to the old ordered basis* $e_1, e_2, \ldots, e_n$.

Also, it turns out that the inverse $\mathbf{P}^{-1}$ of the base-change matrix always exists, and

Multiplication of a column vector by the base-change matrix $\mathbf{P}^{-1}$ *converts coordinate forms from the old ordered basis* $e_1, e_2, \ldots, e_n$ *to the new ordered basis* $f_1, f_2, \ldots, f_n$.

Example 4.13 In the complex vector space $\mathbb{C}^2$, take ordered bases $\mathbf{a}_1, \mathbf{a}_2$ and $\mathbf{b}_1, \mathbf{b}_2$ where

$$\mathbf{a}_1 = \begin{pmatrix} 1 \\ i \end{pmatrix}, \quad \mathbf{a}_2 = \begin{pmatrix} 1 \\ 1+i \end{pmatrix}, \qquad \mathbf{b}_1 = \begin{pmatrix} i \\ -i \end{pmatrix}, \quad \mathbf{b}_2 = \begin{pmatrix} 1 \\ 1 \end{pmatrix}.$$

Then the base-change matrices from the usual basis $(1,0)^T, (0,1)^T$ to $\mathbf{a}_1, \mathbf{a}_2$ is

$$\mathbf{P} = \begin{pmatrix} 1 & 1 \\ i & 1+i \end{pmatrix},$$

and from the usual basis $(1,0)^T, (0,1)^T$ to $\mathbf{b}_1, \mathbf{b}_2$ is

$$\mathbf{Q} = \begin{pmatrix} i & 1 \\ -i & 1 \end{pmatrix}.$$

This can be used to find the base-change matrix from $\mathbf{a}_1, \mathbf{a}_2$ to $\mathbf{b}_1, \mathbf{b}_2$ as follows. If $(v_1, v_2)^T$ is the coordinate form of a vector $\mathbf{v}$ with respect to $\mathbf{a}_1, \mathbf{a}_2$, then $\mathbf{P}(v_1, v_2)^T$ is the coordinate form of $\mathbf{v}$ with respect to the usual basis. Then $\mathbf{Q}^{-1}\mathbf{P}(v_1, v_2)^T$ is the coordinate form of $\mathbf{v}$ with respect to $\mathbf{b}_1, \mathbf{b}_2$, so the base-change matrix from $\mathbf{a}_1, \mathbf{a}_2$ to $\mathbf{b}_1, \mathbf{b}_2$ is

$$\mathbf{Q}^{-1}\mathbf{P} = \begin{pmatrix} i & 1 \\ -i & 1 \end{pmatrix}^{-1} \begin{pmatrix} 1 & 1 \\ i & 1+i \end{pmatrix} = \frac{1}{2i}\begin{pmatrix} 1 & -1 \\ i & i \end{pmatrix}^{-1} \begin{pmatrix} 1 & 1 \\ i & 1+i \end{pmatrix}$$

$$= \frac{1}{2i}\begin{pmatrix} 1-i & -i \\ i-1 & 2i-1 \end{pmatrix} = \frac{1}{2}\begin{pmatrix} -1-i & -1 \\ 1+i & 2+i \end{pmatrix}.$$

The base-change matrix also enables us to convert between the matrices for bilinear forms. Given a bilinear form F,

$$
\begin{aligned}
F(f_i, f_j) &= F\left(\sum_{k=1}^{n} p_{ki} e_k, \sum_{l=1}^{n} p_{lj} e_l \right) \\
&= \sum_{k=1}^{n} \sum_{l=1}^{n} p_{ki} p_{lj} F(e_k, e_l) \\
&= \sum_{k=1}^{n} p_{ki} \sum_{l=1}^{n} F(e_k, e_l) p_{lj}.
\end{aligned}
$$

Now, p_{ki} is the (i, k)th entry of $\mathbf{P}^T$, so this expression is equal to the (i, j)th entry of $\mathbf{P}^T \mathbf{A} \mathbf{P}$. We have proved the following.

Proposition 4.14 (The base-change formula) *Given two ordered bases of a Euclidean space V, $e_1, \ldots, e_n$ and $f_1, \ldots, f_n$, related by the base-change matrix $\mathbf{P}$ from basis $e_1, \ldots, e_n$ to $f_1, \ldots, f_n$, suppose $\mathbf{A}$ and $\mathbf{B}$ are the matrices of the inner product with respect to $e_1, \ldots, e_n$ and $f_1, \ldots, f_n$. Then $\mathbf{B} = \mathbf{P}^T \mathbf{A} \mathbf{P}$.*

Example 4.15 Take the standard inner product on $\mathbb{R}^2$, with respect to the two bases $\mathbf{e}_1, \mathbf{e}_2$ where

$$
\mathbf{e}_1 = \begin{pmatrix} 2 \\ 1 \end{pmatrix}, \ \mathbf{e}_2 = \begin{pmatrix} 1 \\ -1 \end{pmatrix},
$$

and $\mathbf{f}_1, \mathbf{f}_2$ where

$$
\mathbf{f}_1 = \begin{pmatrix} 1 \\ -2 \end{pmatrix}, \ \mathbf{f}_2 = \begin{pmatrix} 0 \\ 1 \end{pmatrix}.
$$

Then with the notation of Proposition 4.14 we have

$$
\mathbf{A} = \begin{pmatrix} 5 & 1 \\ 1 & 2 \end{pmatrix} \qquad \mathbf{B} = \begin{pmatrix} 5 & -2 \\ -2 & 1 \end{pmatrix}.
$$

The base-change matrix $\mathbf{P}$ can be calculated as

$$
\mathbf{P} = \begin{pmatrix} 2 & 1 \\ 1 & -1 \end{pmatrix}^{-1} \begin{pmatrix} 1 & 0 \\ -2 & 1 \end{pmatrix} = \tfrac{1}{3} \begin{pmatrix} 1 & 1 \\ 1 & -2 \end{pmatrix} \begin{pmatrix} 1 & 0 \\ -2 & 1 \end{pmatrix} = \tfrac{1}{3} \begin{pmatrix} -1 & 1 \\ 5 & -2 \end{pmatrix}.
$$

(Alternatively, we can calculate $\mathbf{P}$ from first principles, by writing $\mathbf{P} = \begin{pmatrix} a & b \\ c & d \end{pmatrix}$, so $f_1 = a e_1 + c e_2$, giving

$$
\begin{pmatrix} 1 \\ -2 \end{pmatrix} = a \begin{pmatrix} 2 \\ 1 \end{pmatrix} + c \begin{pmatrix} 1 \\ -1 \end{pmatrix},
$$

and hence $2a + c = 1$ and $a - c = -2$, which can be solved to give $a = -\tfrac{1}{3}$ and $c = \tfrac{5}{3}$. Similarly we find $b = \tfrac{1}{3}$ and $d = -\tfrac{2}{3}$.)

We can check that

$$\mathbf{P}^T \mathbf{AP} = \tfrac{1}{9} \begin{pmatrix} -1 & 5 \\ 1 & -2 \end{pmatrix} \begin{pmatrix} 5 & 1 \\ 1 & 2 \end{pmatrix} \begin{pmatrix} -1 & 1 \\ 5 & -2 \end{pmatrix}$$

$$= \tfrac{1}{9} \begin{pmatrix} 0 & 9 \\ 3 & -3 \end{pmatrix} \begin{pmatrix} -1 & 1 \\ 5 & -2 \end{pmatrix} = \tfrac{1}{9} \begin{pmatrix} 45 & -18 \\ -18 & 9 \end{pmatrix} = \begin{pmatrix} 5 & -2 \\ -2 & 1 \end{pmatrix} = \mathbf{B}.$$

4.4   Sesquilinear forms over $\mathbb{C}$

So far in this chapter, everything has been for vector spaces over the field of real numbers $\mathbb{R}$. If we replaced $\mathbb{R}$ throughout by any other field F, e.g. the field of complex numbers, everything remains valid, and we would have the beginnings of the theory of bilinear forms over the field F. In particular, a bilinear form on a finite dimensional vector space over F is represented by a matrix $\mathbf{A}$ with entries from F and is isomorphic to one of the canonical examples $F(\mathbf{v}, \mathbf{w}) = \mathbf{v}^T \mathbf{A} \mathbf{w}$ on F^n. But we are interested also in inner products over $\mathbb{C}$ also, and these forms are *not* bilinear, but are instead *sesquilinear*, so the previous sections do not apply. This last section concerns matrix representations of such forms over $\mathbb{C}$, and generalizes the notion of bilinear form in a different way, using the complex-conjugate operation in $\mathbb{C}$.

Definition 4.16 *A* **sesquilinear form** *on a complex vector space V is a map F from $V \times V$ to $\mathbb{C}$ which for all $u, v, w \in V$ and $\alpha, \beta \in \mathbb{C}$ satisfies*

(a) $F(\alpha u + \beta v, w) = \overline{\alpha} F(u, w) + \overline{\beta} F(v, w)$, *and*

(b) $F(u, \alpha v + \beta w) = \alpha F(u, v) + \beta F(u, w)$.

Definition 4.17 *A sesquilinear form $F \colon V \times V \to \mathbb{C}$ is* **conjugate-symmetric** *if also*

(c) $F(u, v) = \overline{F(v, u)}$ *for all $u, v \in V$.*

As examples, note that any complex inner product (as in the last chapter) is a conjugate-symmetric sesquilinear form.

Everything we have done here for bilinear forms on vector spaces over $\mathbb{R}$ applies equally well to sesquilinear forms on vector spaces over $\mathbb{C}$, provided complex-conjugate signs are added in the appropriate place. First, we have a family of 'canonical examples' exactly analogous to Example 4.5.

Example 4.18 *Let V be the complex vector space $\mathbb{C}^n$, and suppose $\mathbf{A}$ is an $n \times n$ matrix with entries from $\mathbb{C}$. Define F on $V \times V$ by*

$$F(\mathbf{x}, \mathbf{y}) = \overline{\mathbf{x}}^T \mathbf{A} \mathbf{y},$$

where $\overline{\mathbf{y}}$ is the complex conjugate of the column vector $\mathbf{y}$, formed by taking the complex conjugate of each entry. As before, $F(\mathbf{x}, \mathbf{y})$ is a 1×1 matrix, which we consider as the same as the complex number which is its only entry. Thus $F \colon V \times V \to \mathbb{C}$ and F is a sesquilinear form on V. Also, if $\mathbf{A}$ is conjugate-symmetric, i.e. if $\overline{\mathbf{A}}^T = \mathbf{A}$, then F is conjugate-symmetric.

Proof This is just the same as before with a few complex conjugates added.

$$
\begin{aligned}
F(\alpha\mathbf{x} + \beta\mathbf{y}, \mathbf{z}) &= (\overline{\alpha}\,\overline{\mathbf{x}} + \overline{\beta}\,\overline{\mathbf{y}})^T \mathbf{A}\mathbf{z} \\
&= (\overline{\alpha}\,\overline{\mathbf{x}}^T + \overline{\beta}\,\overline{\mathbf{y}}^T)\mathbf{A}\mathbf{z} \\
&= \overline{\alpha}\,\overline{\mathbf{x}}^T\mathbf{A}\mathbf{z} + \overline{\beta}\,\overline{\mathbf{y}}^T\mathbf{A}\mathbf{z} \\
&= \overline{\alpha}F(\mathbf{x}, \mathbf{z}) + \overline{\beta}F(\mathbf{y}, \mathbf{z}),
\end{aligned}
$$

and

$$
\begin{aligned}
F(\mathbf{x}, \alpha\mathbf{y} + \beta\mathbf{z}) &= \overline{\mathbf{x}}^T\mathbf{A}(\alpha\mathbf{y} + \beta\mathbf{z}) \\
&= \overline{\mathbf{x}}^T\mathbf{A}(\alpha\mathbf{y}) + \overline{\mathbf{x}}^T\mathbf{A}(\beta\mathbf{z}) \\
&= \alpha\overline{\mathbf{x}}^T\mathbf{A}\mathbf{y} + \beta\overline{\mathbf{x}}^T\mathbf{A}\mathbf{z} \\
&= \alpha F(\mathbf{x}, \mathbf{y}) + \beta F(\mathbf{x}, \mathbf{z}).
\end{aligned}
$$

Now suppose $\mathbf{A}$ is conjugate-symmetric. Then $F(\mathbf{x}, \mathbf{y}) = F(\mathbf{x}, \mathbf{y})^T$ and we have

$$
\begin{aligned}
F(\mathbf{x}, \mathbf{y}) &= (\overline{\mathbf{x}}^T\mathbf{A}\mathbf{y})^T \\
&= \mathbf{y}^T\mathbf{A}^T\overline{\mathbf{x}} \\
&= \overline{\mathbf{y}^T\,\overline{\mathbf{A}}^T\mathbf{x}} \\
&= \overline{\mathbf{y}^T\mathbf{A}\mathbf{x}} \\
&= \overline{F(\mathbf{y}, \mathbf{x})},
\end{aligned}
$$

similar to the real case.   □

Definition 4.19 *Given an ordered basis $e_1, \dots, e_n$ of a complex inner product space V, and a sesquilinear form F on V, the **matrix of the form** F with respect to this ordered basis is the matrix $\mathbf{B}$ whose (i, j)th entry is $b_{ij} = F(e_i, e_j)$.*

If F is conjugate-symmetric, $b_{ji} = F(e_j, e_i) = \overline{F(e_i, e_j)} = \overline{b_{ij}}$, so $\mathbf{B}$ is conjugate-symmetric.

Notice that a diagonal entry b_{ii} of a conjugate-symmetric matrix $\mathbf{B}$ is always real, since $b_{ii} = \overline{b_{ii}}$.

Proposition 4.20 *With the same notation, if*

$$
v = \sum_{i=1}^{n} v_i e_i \text{ and } w = \sum_{j=1}^{n} w_j e_j,
$$

then

$$
\begin{aligned}
F(v, w) &= \sum_{i=1}^{n}\sum_{j=1}^{n} \overline{v_i} F(e_i, e_j) w_j \\
&= \begin{pmatrix} \overline{v_1} & \cdots & \overline{v_n} \end{pmatrix} \begin{pmatrix} F(e_1, e_1) & \cdots & F(e_1, e_n) \\ \vdots & \ddots & \vdots \\ F(e_n, e_1) & \cdots & F(e_n, e_n) \end{pmatrix} \begin{pmatrix} w_1 \\ \vdots \\ w_n \end{pmatrix},
\end{aligned}
$$

which we can interpret in matrix terms as $\overline{\mathbf{v}}^T \mathbf{B} \mathbf{w}$.

Proposition 4.21 (The base-change formula) *With the same notation, if we change to the new basis* $f_1, \ldots, f_n$, *given in terms of the old basis* $e_1, \ldots, e_n$ *by* $f_i = \sum_{k=1}^{n} p_{ki} e_k$, *then*

$$F(f_i, f_j) = F\left(\sum_{k=1}^{n} p_{ki} e_k, \sum_{l=1}^{n} p_{lj} e_l\right) = \sum_{k=1}^{n} \sum_{l=1}^{n} \overline{p_{ki}} F(e_k, e_l) p_{lj},$$

which is the (i, j)*th entry of the matrix* $\overline{\mathbf{P}}^T \mathbf{B} \mathbf{P}$.

Exercises

Exercise 4.1 Calculate the base-change matrix **P** for change of basis from e_1, e_2 to e'_1, e'_2 in Example 4.11, and verify the formula $\mathbf{B}' = \mathbf{P}^T \mathbf{B} \mathbf{P}$ in this case.

Exercise 4.2 Which of the functions $F\big((x_1, x_2, x_3)^T, (y_1, y_2, y_3)^T\big)$ here are inner products on $\mathbb{R}^3$?

(a) $x_1^2 + y_1^2 + x_2 y_2 + x_3 y_3$.
[Hint: what could the matrix representation be?]
(b) $3x_1 y_1 + 5x_2 y_2 + 4x_3 y_3 + 2x_1 y_2 + 2x_2 y_1 + 3x_2 y_3 + 3y_2 x_3 - x_1 y_3 - x_3 y_1$.
[Consider $2(x_1 + x_2)^2 + 3(x_2 + x_3)^2 + (x_1 - x_3)^2$.]
(c) $3x_1 y_1 + 5x_2 y_2 + 4x_3 y_3 + 2x_1 y_2 + 2x_2 y_1 - 3x_2 y_3 - 3y_2 x_3 - x_1 y_3 - x_3 y_1$.
[Consider $2(x_1 + x_2)^2 + 3(x_2 - x_3)^2 + (x_1 - x_3)^2$.]

Exercise 4.3 An inner product,

$$\langle (x_1, x_2)^T | (y_1, y_2)^T \rangle = 4x_1 y_1 + 2x_2 y_1 + 2x_1 y_2 + 2x_2 y_2,$$

is defined on $\mathbb{R}^2$.

(a) Calculate the matrix **A** representing $\langle \,|\, \rangle$ with respect to the standard basis $e_1 = (1, 0)^T$, $e_2 = (0, 1)^T$.
(b) Calculate the matrix representing $\langle \,|\, \rangle$ with respect to the basis $f_1 = (1/2, 0)^T$, $f_2 = (-1/2, 1)^T$.
(c) Calculate matrices representing the bilinear form

$$F\big((x_1, x_2)^T, (y_1, y_2)^T\big) = 4x_1 y_1 + 2x_2 y_1 + 2x_1 y_2 + x_2 y_2$$

with respect to e_1, e_2 and f_1, f_2. Is F an inner product?

Exercise 4.4 Let

$$\mathbf{A} = \begin{pmatrix} 1 & 1 & 1 \\ 1 & 3 & 2 \\ 1 & 2 & 4 \end{pmatrix}.$$

Perform row operations on **A** of the form $\rho_i := \rho_i + \lambda \rho_j$, $1 \leqslant j < i \leqslant 3$ and $\lambda \in \mathbb{R}$, to write **A** in the form

$$\begin{pmatrix} a & b & c \\ 0 & d & e \\ 0 & 0 & f \end{pmatrix}.$$

Compute a matrix **P** with

$$\mathbf{PA} = \begin{pmatrix} a & b & c \\ 0 & d & e \\ 0 & 0 & f \end{pmatrix}.$$

Why is **P** invertible? Now calculate $\mathbf{PAP}^T$. What does this matrix represent?

Exercise 4.5 A sesquilinear form on $\mathbb{C}^3$ is defined by

$$F((x,y,z)^T,(x',y',z')^T)) = (\overline{x},\overline{y},\overline{z}) \begin{pmatrix} 0 & -1-2i & 0 \\ -1+2i & 0 & i \\ 0 & -i & 1 \end{pmatrix} \begin{pmatrix} x' \\ y' \\ z' \end{pmatrix}.$$

Compute the matrix of F with respect to:
(a) the basis $(i,1,0)^T$, $(1,1+i,1)^T$, $(0,0,i)^T$;
(b) the basis $(-1-3i,2+i,-1+2i)^T$, $(2+i,1-2i,2+i)^T$, $(0,0,-i)^T$.
Is F an inner product?

Exercise 4.6 Let F be a sesquilinear form on the complex vector space $\mathbb{C}^3$ and e_1, e_2, e_3 is the usual ordered basis of $\mathbb{C}^3$. Suppose that $F(\mathbf{f}_1, \mathbf{f}_1) = 1$, $F(\mathbf{f}_2, \mathbf{f}_2) = 2$, $F(\mathbf{f}_3, \mathbf{f}_3) = 2$, and $F(\mathbf{f}_i, \mathbf{f}_j) = 0$ for $i \neq j$, where

$$\mathbf{f}_1 = (1,i,0)^T, \quad \mathbf{f}_2 = (0,i,1)^T, \quad \mathbf{f}_1 = (0,0,1+i)^T.$$

Calculate the matrix of F with respect to the basis e_1, e_2, e_3. Is F an inner product?

Exercise 4.7 Write down the matrices $\mathbf{A}, \mathbf{B}, \mathbf{C}$ (respectively) of the standard inner product on $\mathbb{R}^3$ with respect to:
(a) the standard basis;
(b) the ordered basis $(2,1,1)^T, (1,2,0)^T, (0,1,-1)^T$;
(c) the ordered basis $(-1,1,1)^T, (3,0,0)^T, (1,2,-1)^T$.
Also write down base-change matrices $\mathbf{P}$ and $\mathbf{Q}$ (respectively) for:
(d) base change from (a) to (b);
(e) base change from (b) to (c);
Verify that $\mathbf{B} = \mathbf{P}^T\mathbf{AP}$ and $\mathbf{C} = \mathbf{Q}^T\mathbf{BQ}$ both hold.

Exercise 4.8 An *alternating form* F is a bilinear form on a vector space V satisfying $F(v,v) = 0$ for all $v \in V$.
(a) Show by expanding $F(v+w,v+w)$ that if F is an alternating form on a vector space V then F is *skew-symmetric*, i.e. $F(v,w) = -F(w,v)$ for all v and w in V.
(b) Clearly the zero function is an alternating form; give another example. [Hint: find a suitable matrix $\mathbf{A}$.]

5

Orthogonal bases

One important use of the standard inner product of $\mathbb{R}^3$ is that it allows us to distinguish between 'nice' bases such as the usual one where the basis vectors are orthogonal to each other and each has length 1, and others such as $(1,0,0)^T, (1,1,0)^T, (1,1,1)^T$ where this is not true.

The main objective of this chapter is to show how to find such nice (or, as we shall call them, *orthonormal*) bases for a given finite dimensional vector space with an arbitrary inner product, and to study the properties of these orthonormal bases.

This chapter concerns inner products on vector spaces, and all our vector spaces are over $\mathbb{R}$ or $\mathbb{C}$ throughout.

5.1 Orthonormal bases

In a vector space V over $\mathbb{R}$, any two bases look the 'same'. Indeed, by Theorem 2.33 there is an isomorphism of real vector spaces taking one basis onto the other. In inner product spaces this is no longer true; certain bases turn out to be easier to calculate with than others, and isomorphisms of vector spaces do not necessarily preserve inner products.

Example 5.1 Let $V = \mathbb{R}^3$ with the standard inner product

$$\langle (x_1, x_2, x_3)^T | (y_1, y_2, y_3)^T \rangle = x_1 y_1 + x_2 y_2 + x_3 y_3$$

and let $\mathbf{e}_1 = (1,0,0)^T$, $\mathbf{e}_2 = (0,1,0)^T$, $\mathbf{e}_3 = (0,0,1)^T$ be the vectors of the standard basis. These basis vectors have important properties not true for all bases. For example,

$$\langle \mathbf{e}_1 | \mathbf{e}_2 \rangle = 0,$$
$$\langle \mathbf{e}_1 | \mathbf{e}_3 \rangle = 0,$$

and

$$\langle \mathbf{e}_2 | \mathbf{e}_3 \rangle = 0.$$

So $\langle \mathbf{e}_i | \mathbf{e}_j \rangle = 0$ whenever $i \neq j$. This property is referred to by saying that $\{ \mathbf{e}_1, \mathbf{e}_2, \mathbf{e}_3 \}$ is an *orthogonal set*. Not all bases are orthogonal; for example,

$\mathbf{f}_1 = (1,0,0)^T$, $\mathbf{f}_2 = (1,1,0)^T$, $\mathbf{f}_3 = (1,1,1)^T$ form a basis which is not orthogonal, since $\langle \mathbf{f}_1 | \mathbf{f}_2 \rangle = \langle \mathbf{f}_1 | \mathbf{f}_3 \rangle = 1$, and $\langle \mathbf{f}_2 | \mathbf{f}_3 \rangle = 2$.

Moreover,

$$\langle \mathbf{e}_1 | \mathbf{e}_1 \rangle = \langle \mathbf{e}_2 | \mathbf{e}_2 \rangle = \langle \mathbf{e}_3 | \mathbf{e}_3 \rangle = 1,$$

i.e. $\langle \mathbf{e}_i | \mathbf{e}_i \rangle = 1$ for all i. Orthogonal bases with this extra property are said to be *orthonormal*. Again, not all bases have this property. For example, $\mathbf{g}_1 = (1,1,1)^T$, $\mathbf{g}_2 = (1,-1,0)^T$, and $\mathbf{g}_3 = (1/2,1/2,-1)^T$ define an orthogonal basis which is not orthonormal, as $\langle \mathbf{g}_1 | \mathbf{g}_1 \rangle = 3$, $\langle \mathbf{g}_2 | \mathbf{g}_2 \rangle = 2$, and $\langle \mathbf{g}_3 | \mathbf{g}_3 \rangle = 3/2$.

One familiar fact about the standard basis $\mathbf{e}_1, \mathbf{e}_2, \mathbf{e}_3$ is that any vector $\mathbf{v} = (v_1, v_2, v_3)^T \in \mathbb{R}^3$ is equal to the sum of its projections along the axes. That is,

$$\begin{pmatrix} v_1 \\ v_2 \\ v_3 \end{pmatrix} = \begin{pmatrix} v_1 \\ 0 \\ 0 \end{pmatrix} + \begin{pmatrix} 0 \\ v_2 \\ 0 \end{pmatrix} + \begin{pmatrix} 0 \\ 0 \\ v_3 \end{pmatrix} = \langle \mathbf{e}_1 | \mathbf{v} \rangle \mathbf{e}_1 + \langle \mathbf{e}_2 | \mathbf{v} \rangle \mathbf{e}_2 + \langle \mathbf{e}_3 | \mathbf{v} \rangle \mathbf{e}_3,$$

and the scalar coefficients $\langle \mathbf{e}_i | \mathbf{v} \rangle$ are easy to compute. Compare this with the much harder task of calculating the (unique!) scalars λ_1, λ_2, and λ_3 such that

$$\begin{pmatrix} 23 \\ -34 \\ 9 \end{pmatrix} = \lambda_1 \begin{pmatrix} 2 \\ 3 \\ 8 \end{pmatrix} + \lambda_2 \begin{pmatrix} -2 \\ 1 \\ 1 \end{pmatrix} + \lambda_3 \begin{pmatrix} 1 \\ 1 \\ 7 \end{pmatrix},$$

i.e. of computing

$$\begin{pmatrix} \lambda_1 \\ \lambda_2 \\ \lambda_3 \end{pmatrix} = \begin{pmatrix} 2 & -2 & 1 \\ 3 & 1 & 1 \\ 8 & 1 & 7 \end{pmatrix}^{-1} \begin{pmatrix} 23 \\ -34 \\ 9 \end{pmatrix}.$$

Our object here is to show how such orthonormal bases can be found in an arbitrary inner product space, and to describe some of their properties and applications.

Definition 5.2 *Two vectors v and w in an inner product space V (over $\mathbb{R}$ or $\mathbb{C}$) are said to be **orthogonal** if $\langle v | w \rangle = 0$.*

*The set of vectors $\{v_1, v_2, \ldots, v_i, \ldots\}$ is said to be **orthogonal**, and the individual vectors $v_1, v_2, \ldots, v_i, \ldots$ in the set are said to be **mutually orthogonal**, if each pair of distinct vectors v_i, v_j ($i \neq j$) in the set is an orthogonal pair.*

A consequence of this definition is that if one of v or w is zero then v, w are orthogonal: if v is the zero vector then $0v = 0 = v$ so $\langle v | w \rangle = \langle 0v | w \rangle = 0 \langle v | w \rangle = 0$ and similarly for w. This special case is somewhat uninteresting, and we shall usually specifically exclude it in the statements of our theorems.

One very important fact about orthogonal vectors is that provided they are all nonzero then they are linearly independent. This is true for any number of orthogonal vectors.

Proposition 5.3 *If V is an inner product space over $\mathbb{R}$ or $\mathbb{C}$, $v_1, \ldots, v_n \in V$, $v_i \neq 0$ for all i, and the v_i are mutually orthogonal, then $\{v_1, \ldots, v_n\}$ is a linearly independent set.*

Proof Suppose $v_1, \ldots, v_n$ satisfy $\sum_{i=1}^{n} \lambda_i v_i = 0$; we need to show that each λ_i is zero. We can take the inner product with each vector v_j in turn to get

$$
\begin{aligned}
0 &= \langle v_j | 0 \rangle \\
&= \left\langle v_j \Big| \sum_{i=1}^{n} \lambda_i v_i \right\rangle \\
&= \sum_{i=1}^{n} \lambda_i \langle v_j | v_i \rangle \\
&= \lambda_j \langle v_j | v_j \rangle,
\end{aligned}
$$

since $\langle v_i | v_j \rangle = 0$ whenever $i \neq j$. But $v_j \neq 0$, so $\langle v_j | v_j \rangle \neq 0$ since the inner product is positive definite, and so $\lambda_j = 0$ for each j. Therefore $\{v_1, \ldots, v_n\}$ is linearly independent. $\qquad\square$

Given a set $\{v_1, \ldots, v_n\}$ of mutually orthogonal vectors, we can *normalize* them so that they all have norm 1 (in other words, unit length), by defining

$$
w_i = \frac{v_i}{\|v_i\|} = \frac{v_i}{\sqrt{\langle v_i | v_i \rangle}}. \tag{1}
$$

The resulting vectors $w_1, \ldots, w_n$ are then orthonormal, according to the next definition.

Definition 5.4 *A set $\{w_1, \ldots, w_n\}$ of vectors in an inner product space V is said to be **orthonormal** if $\langle w_i | w_j \rangle = 0$ whenever $i \neq j$ and also $\langle w_i | w_i \rangle = 1$ for each i. If an orthonormal set is also a basis of V, it is called an **orthonormal basis** of V.*

5.2 The Gram–Schmidt process

Another way of thinking of orthonormal bases is to consider those bases of a vector space V for which the matrix representing the inner product is as simple as possible.

If we have an *orthogonal* basis $\{v_1, v_2, \ldots, v_n\}$ for an inner product space V, then the matrix of the inner product with respect to this basis is diagonal, since the entries off the diagonal are $\langle v_i | v_j \rangle$ for some $i \neq j$. If we have an *orthonormal* basis, then the matrix for the inner product is the identity matrix since the diagonal entries are $\langle v_i | v_i \rangle = 1$.

In fact, for any finite dimensional inner product space, we can always find an orthonormal basis. It is usually easiest to describe (and carry out) the method in two steps: first we find an orthogonal basis, and then we normalize the basis vectors to get an orthonormal basis as in (1).

The key step in the method is the following lemma, which is illustrated in Figure 5.1 for the case $V = \mathbb{R}^2$ with the standard inner product. The idea of

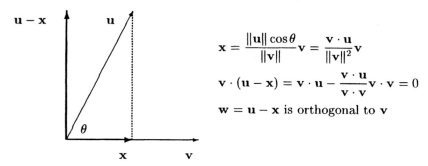

$$\mathbf{x} = \frac{\|\mathbf{u}\|\cos\theta}{\|\mathbf{v}\|}\mathbf{v} = \frac{\mathbf{v}\cdot\mathbf{u}}{\|\mathbf{v}\|^2}\mathbf{v}$$

$$\mathbf{v}\cdot(\mathbf{u}-\mathbf{x}) = \mathbf{v}\cdot\mathbf{u} - \frac{\mathbf{v}\cdot\mathbf{u}}{\mathbf{v}\cdot\mathbf{v}}\mathbf{v}\cdot\mathbf{v} = 0$$

$$\mathbf{w} = \mathbf{u}-\mathbf{x} \text{ is orthogonal to } \mathbf{v}$$

Fig. 5.1   Orthogonal projections of a vector

this lemma is just as for $\mathbb{R}^2$: given two vectors u, v, form w by taking away the 'projection' of u onto the direction of v.

Lemma 5.5 *If u, v are any two vectors in an inner product space V with $v \neq 0$, then the vector*

$$w = u - \frac{\langle v|u\rangle}{\langle v|v\rangle}v$$

is orthogonal to v.

Proof   First, $\langle v|v\rangle \neq 0$ since $v \neq 0$, so $u - (\langle v|u\rangle/\langle v|v\rangle)v$ is defined, and

$$\langle v|w\rangle = \left\langle v \,\middle|\, u - \frac{\langle v|u\rangle}{\langle v|v\rangle}v \right\rangle$$

$$= \langle v|u\rangle - \frac{\langle v|u\rangle}{\langle v|v\rangle}\langle v|v\rangle$$

$$= 0$$

as required.   □

More generally, we can make the vector u orthogonal to as many vectors as we like, provided these vectors are orthogonal to each other.

Lemma 5.6 *If V is an inner product space, $u, v_1, \ldots, v_k \in V$, and $v_1, \ldots, v_k$ are mutually orthogonal nonzero vectors, then the vector*

$$w = u - \sum_{i=1}^{k} \frac{\langle v_i|u\rangle}{\langle v_i|v_i\rangle}v_i$$

is orthogonal to $v_1, \ldots, v_k$.

Proof   Note that we must have the condition $v_i \neq 0$ in order to be able to divide by $\langle v_i|v_i\rangle$ in the formula for w.

For each j we have

$$
\begin{aligned}
\langle v_j | w \rangle &= \left\langle v_j \,\middle|\, u - \sum_{i=1}^{k} \frac{\langle v_i | u \rangle}{\langle v_i | v_i \rangle} v_i \right\rangle \\
&= \langle v_j | u \rangle - \sum_{i=1}^{k} \frac{\langle v_j | u \rangle}{\langle v_i | v_i \rangle} \langle v_j | v_i \rangle \\
&= \langle v_j | u \rangle - \frac{\langle v_j | u \rangle}{\langle v_j | v_j \rangle} \langle v_j | v_j \rangle \\
&= 0,
\end{aligned}
$$

as claimed. $\qquad\qquad\qquad\qquad\qquad\qquad\qquad\qquad\qquad\qquad\qquad\qquad$ □

To see how the process of producing an orthogonal basis works, let us first consider a small example. We take the standard inner product on $\mathbb{R}^3$, and pretend we do not already know an orthonormal basis.

Example 5.7 Let V be the vector space $\mathbb{R}^3$ with the standard inner product, and take the ordered basis $\mathbf{v}_1, \mathbf{v}_2, \mathbf{v}_3$ where $\mathbf{v}_1 = (0, 1, 1)^T$, $\mathbf{v}_2 = (0, 2, 0)^T$, and $\mathbf{v}_3 = (-1, 1, 0)^T$. Take our first vector of the new basis to be $\mathbf{w}_1 = \mathbf{v}_1 = (0, 1, 1)^T$. The second vector $\mathbf{w}_2$ has to be orthogonal to $\mathbf{v}_1$, so using Lemma 5.5 we can take

$$
\mathbf{w}_2 = \mathbf{v}_2 - \frac{\langle \mathbf{w}_1 | \mathbf{v}_2 \rangle}{\langle \mathbf{w}_1 | \mathbf{w}_1 \rangle} \mathbf{w}_1 = \begin{pmatrix} 0 \\ 2 \\ 0 \end{pmatrix} - \frac{2}{2} \begin{pmatrix} 0 \\ 1 \\ 1 \end{pmatrix} = \begin{pmatrix} 0 \\ 1 \\ -1 \end{pmatrix}.
$$

The third vector $\mathbf{w}_3$ has to be orthogonal to both $\mathbf{w}_1$ and $\mathbf{w}_2$. Using Lemma 5.6 we take

$$
\begin{aligned}
\mathbf{w}_3 &= \mathbf{v}_3 - \frac{\langle \mathbf{w}_2 | \mathbf{v}_3 \rangle}{\langle \mathbf{w}_2 | \mathbf{w}_2 \rangle} \mathbf{w}_2 - \frac{\langle \mathbf{w}_1 | \mathbf{v}_3 \rangle}{\langle \mathbf{w}_1 | \mathbf{w}_1 \rangle} \mathbf{w}_1 \\
&= \begin{pmatrix} -1 \\ 1 \\ 0 \end{pmatrix} - \frac{1}{2} \begin{pmatrix} 0 \\ 1 \\ 1 \end{pmatrix} - \frac{1}{2} \begin{pmatrix} 0 \\ 1 \\ -1 \end{pmatrix} = \begin{pmatrix} -1 \\ 0 \\ 0 \end{pmatrix}.
\end{aligned}
$$

Thus $\{\mathbf{w}_1, \mathbf{w}_2, \mathbf{w}_3\}$ is an orthogonal basis of V (as you should check).

We can normalize the vectors to get an orthonormal basis

$$
\left\{ \begin{pmatrix} 0 \\ 1/\sqrt{2} \\ 1/\sqrt{2} \end{pmatrix}, \begin{pmatrix} 0 \\ 1/\sqrt{2} \\ -1/\sqrt{2} \end{pmatrix}, \begin{pmatrix} -1 \\ 0 \\ 0 \end{pmatrix} \right\}.
$$

In fact what we have just done is a general method, called *Gram–Schmidt orthonormalization*. It can be stated either as an algorithm, or as a theorem. What's more, it works for inner product spaces over either the real or the complex numbers, with exactly the same proof.

Theorem 5.8 (The Gram–Schmidt process) *If* $\{v_1, \dots, v_n\}$ *is a basis of a finite dimensional inner product space* V, *then* $\{w_1, \dots, w_n\}$ *obtained by*

$$w_1 = v_1$$

$$w_2 = v_2 - \frac{\langle w_1 | v_2 \rangle}{\langle w_1 | w_1 \rangle} w_1$$

$$\vdots$$

$$w_k = v_k - \sum_{i=1}^{k-1} \frac{\langle w_i | v_k \rangle}{\langle w_i | w_i \rangle} w_i$$

$$\vdots$$

is an orthogonal basis of V.

Proof   We prove this theorem by induction, using the following induction hypothesis $\mathscr{P}(k)$:

$\{w_1, \dots, w_k\}$ is an orthogonal set of nonzero vectors in the space spanned by $\{v_1, \dots, v_k\}$.

For $k = 1$, $w_1 = v_1 \neq 0$, so $\mathscr{P}(1)$ is clear. Now suppose $n > k \geqslant 1$ and $\mathscr{P}(k)$ holds. Then by induction the vectors $w_1, \dots, w_k$ are all nonzero and mutually orthogonal, so, by Lemma 5.6, w_{k+1} is orthogonal to all the vectors $w_1, \dots, w_k$.

Moreover, w_{k+1} is nonzero, for otherwise

$$w_{k+1} = 0 = v_{k+1} - \sum_{i=1}^{k} \frac{\langle w_i | v_{k+1} \rangle}{\langle w_i | w_i \rangle} w_i$$

gives

$$v_{k+1} = \sum_{i=1}^{k} \frac{\langle w_i | v_{k+1} \rangle}{\langle w_i | w_i \rangle} w_i = \sum_{i=1}^{k} \lambda_i w_i$$

where $\lambda_i = \langle w_i | v_{k+1} \rangle / \langle w_i | w_i \rangle$, so v_{k+1} is in the subspace spanned by $w_1, \dots, w_k$. But this is impossible since, by the induction hypothesis $\mathscr{P}(k)$, $w_1, \dots, w_k$ are in the space spanned by $v_1, \dots, v_k$, so v_{k+1} is in this space also; hence there are scalars μ_j with

$$v_{k+1} = \mu_1 v_1 + \mu_2 v_2 + \dots + \mu_k v_k,$$

so $\{v_1, \dots, v_k, v_{k+1}\}$ is linearly dependent, a contradiction. Hence $w_k \neq 0$.

Finally, by its definition, w_{k+1} is in the space spanned by $w_1, \dots, w_k$ and v_{k+1}, and by the induction hypothesis each w_i ($i \leqslant k$) is in the space spanned by $v_1, \dots, v_k$; hence w_{k+1} is in the space spanned by $v_1, \dots, v_{k+1}$, and we have proved $\mathscr{P}(k+1)$. □

Corollary 5.9 *Any finite dimensional inner product space V has an orthonormal basis. In fact, given the orthogonal basis $\{w_1, \ldots, w_n\}$ from the previous theorem, the set $\{y_1, \ldots, y_n\}$ defined by*

$$y_i = \frac{w_i}{\|w_i\|} = \frac{w_i}{\sqrt{\langle w_i | w_i \rangle}}$$

is an orthonormal basis.

Example 5.10 (Legendre polynomials) In the vector space $\mathbb{R}_3[x]$ of all polynomials of degree less than 3, with the inner product

$$\langle f | g \rangle = \int_{-1}^{1} f(x) g(x)\, dx,$$

take the 'natural' basis $\{1, x, x^2\}$ and apply the Gram–Schmidt process (without normalization).

First $\langle 1 | 1 \rangle = \int_{-1}^{1} 1\, dx = 2$, and $\langle x | 1 \rangle = \int_{-1}^{1} x\, dx = 0$, so 1 and x are already orthogonal. Now $\langle x^2 | 1 \rangle = \int_{-1}^{1} x^2\, dx = \frac{2}{3}$ and $\langle x^2 | x \rangle = 0$, so we replace x^2 by $x^2 - \frac{2/3}{2} . 1 = x^2 - 1/3$. Indeed, this process can be continued indefinitely, to produce polynomials

$$1, \; x, \; x^2 - \tfrac{1}{3}, \; x^3 - \tfrac{3}{5}x, \; x^4 - \tfrac{6}{7}x^2 + \tfrac{3}{35}, \; \cdots$$

We can normalize these polynomials so that they have norm 1. First, $\langle 1 | 1 \rangle = 2$ so we replace 1 by $\sqrt{2}/2$. Next, $\langle x | x \rangle = 2/3$ so we replace x by $\frac{\sqrt{6}}{2}x$. Then

$$\left\langle x^2 - \tfrac{1}{3} \middle| x^2 - \tfrac{1}{3} \right\rangle = \int_{-1}^{1} \left(x^4 - \tfrac{2}{3}x^2 + \tfrac{1}{9} \right) dx = \tfrac{2}{5} - \tfrac{4}{9} + \tfrac{2}{9} = \tfrac{8}{45}$$

so we replace $x^2 - \frac{1}{3}$ by $\frac{3\sqrt{10}}{4}(x^2 - \frac{1}{3})$, and so on.

These polynomials are scalar multiples of the *Legendre polynomials*. Since the square roots get rather cumbersome, the Legendre polynomials $P_n(x)$ are usually defined instead by taking the appropriate scalar multiple so that $P_n(1) = 1$. Thus $P_0(x) = 1$, $P_1(x) = x$, $P_2(x) = \frac{3}{2}x^2 - \frac{1}{2}$, $P_3(x) = \frac{5}{2}x^3 - \frac{3}{2}x$, $P_4(x) = \frac{35}{8}x^4 - \frac{15}{4}x^2 + \frac{3}{8}$, and so on.

There is another way of looking at what we have proved in Corollary 5.9. Suppose V is a finite dimensional real vector space with inner product $\langle | \rangle$. Suppose V has dimension n; then we know from Theorem 2.33 that V and $\mathbb{R}^n$ are isomorphic as real vector spaces. But our vector space V has an inner product, and Theorem 2.33 says nothing about these.

Instead, we make the following definition which imposes as a further condition on an isomorphism that it should preserve a bilinear (or sesquilinear) form too.

Definition 5.11 *Two real vector spaces V, W with forms $F \colon V \times V \to \mathbb{R}$ and $G \colon W \times W \to \mathbb{R}$ respectively are* **isomorphic** *if there is a bijection $f \colon V \to W$ such that*

$$f(u + v) = f(u) + f(v) \tag{2}$$
$$f(\lambda v) = \lambda f(v) \tag{3}$$
$$F(u, v) = G(f(u), f(v)) \tag{4}$$

for all $u, v \in V$ and all scalars $\lambda \in \mathbb{R}$.

Similarly, two complex vector spaces V, W with forms $F \colon V \times V \to \mathbb{C}$ and $G \colon W \times W \to \mathbb{C}$ respectively are **isomorphic** *if there is a bijection $f \colon V \to W$ such that*

$$f(u + v) = f(u) + f(v) \tag{5}$$
$$f(\lambda v) = \lambda f(v) \tag{6}$$
$$F(u, v) = G(f(u), f(v)) \tag{7}$$

for all $u, v \in V$ and all scalars $\lambda \in \mathbb{C}$.

Corollary 5.12 *Let V be a Euclidean space of dimension n. Then V is isomorphic to $\mathbb{R}^n$ with the standard inner product as an inner product space. Similarly, each unitary space V of dimension n is isomorphic to $\mathbb{C}^n$ with the standard inner product.*

Proof We do the real case, the complex case being identical. Denote the inner product in V by F, and the standard inner product in $\mathbb{R}^n$ by $\langle \, | \, \rangle$.

By Corollary 5.9 there is an orthonormal basis $v_1, v_2, \ldots, v_n$ of V. By the proof of Theorem 2.33

$$f(\lambda_1 v_1 + \lambda_2 v_2 + \cdots + \lambda_n v_n) = (\lambda_1, \lambda_2, \ldots, \lambda_n)^T$$

defines an isomorphism of real vector spaces from V to $\mathbb{R}^n$. We must show (4) holds too. Let $v = \lambda_1 v_1 + \lambda_2 v_2 + \cdots + \lambda_n v_n$ and $w = \mu_1 v_1 + \mu_2 v_2 + \cdots + \mu_n v_n$ be two typical elements of V, and let $\mathbf{a} = (\lambda_1, \lambda_2, \ldots, \lambda_n)^T$ and $\mathbf{b} = (\mu_1, \mu_2, \ldots, \mu_n)^T$ be the coordinate form of v, w with respect to the orthonormal basis $v_1, v_2, \ldots, v_n$ of V. Note that $f(v) = \mathbf{a}$ and $f(w) = \mathbf{b}$. Since the matrix representing the inner product F on V is the identity matrix $\mathbf{I}$ (because $v_1, v_2, \ldots, v_n$ is orthonormal), by Proposition 4.8 we have

$$\begin{aligned} F(v, w) &= \mathbf{a}^T \mathbf{I} \mathbf{b} \\ &= \mathbf{a}^T \mathbf{b} \\ &= \lambda_1 \mu_1 + \lambda_2 \mu_2 + \cdots + \lambda_n \mu_n \\ &= \langle f(v) | f(w) \rangle \end{aligned}$$

as required. $\square$

5.3   Properties of orthonormal bases

There are a number of useful results about orthonormal bases, all using the same basic idea we used in proving Proposition 5.3.

Most of the results here concern finite dimensional spaces and are rather familiar in the case of $\mathbb{R}^n$, and probably also $\mathbb{C}^n$: these could be proved by using Corollary 5.12 to show the space V in question is isomorphic to $\mathbb{R}^n$ or $\mathbb{C}^n$, with the isomorphism taking an orthonormal basis to the standard basis of $\mathbb{R}^n$ or $\mathbb{C}^n$, and then proving the result for the standard basis in $\mathbb{R}^n$ or $\mathbb{C}^n$. However, in the cases of these particular results, it is just as easy to prove them directly, and this is what we shall do here, starting with the real case.

Inner product spaces over $\mathbb{R}$.   We start by considering spaces over the real numbers.

Proposition 5.13 (Fourier expansion) *Suppose that $\{e_1, \dots, e_n\}$ is an orthonormal basis of a Euclidean space V. Then for any $v \in V$ we have*

$$v = \sum_{i=1}^{n} \langle e_i | v \rangle e_i.$$

Proof   We can write v in terms of the basis, say $v = \sum_{i=1}^{n} \lambda_i e_i$. Then take the inner product of both sides of this equation with e_j, to obtain

$$\langle e_j | v \rangle = \sum_{i=1}^{n} \lambda_i \langle e_j | e_i \rangle = \lambda_j.$$

Substituting back $\lambda_i = \langle e_i | v \rangle$ into $v = \sum_{i=1}^{n} \lambda_i e_i$ we obtain the equation required. □

Example 5.14 You are probably familiar with this result in $\mathbb{R}^3$. If $\{\mathbf{i}, \mathbf{j}, \mathbf{k}\}$ is the standard (orthonormal) basis, $\mathbf{u} \cdot \mathbf{v}$ is the standard inner product, and if $\mathbf{v} \in \mathbb{R}^3$, then

$$\mathbf{v} = (\mathbf{i} \cdot \mathbf{v})\mathbf{i} + (\mathbf{j} \cdot \mathbf{v})\mathbf{j} + (\mathbf{k} \cdot \mathbf{v})\mathbf{k}.$$

That is, $\mathbf{v}$ is the sum of its projections onto the three coordinate axes.

Example 5.15 We consider the orthonormal basis $\{e_1, e_2, e_3, e_4\}$ of $\mathbb{R}^4$ with the standard inner product, where

$$e_1 = (1/2, 1/2, 1/2, 1/2)^T,$$
$$e_2 = (1/2, 1/2, -1/2, -1/2)^T,$$
$$e_3 = (1/2, -1/2, 1/2, -1/2)^T, \text{ and}$$
$$e_4 = (1/2, -1/2, -1/2, 1/2)^T,$$

and express $v = (1, 2, 3, 4)^T$ in terms of this basis. We calculate $\langle e_1 | v \rangle = 5$, $\langle e_2 | v \rangle = -2$, $\langle e_3 | v \rangle = -1$, and $\langle e_4 | v \rangle = 0$, and therefore $v = 5e_1 - 2e_2 - e_3$.

The coefficients $\langle e_i | v \rangle$ in Proposition 5.13 are the coordinates of the vector v with respect to the ordered basis $e_1, \ldots, e_n$. They are also sometimes called *Fourier coefficients* because of their use in Example 3.14. That example, however, is an infinite dimensional space, for which Proposition 5.13 does not hold in general. With infinitely many dimensions, there may be convergence problems for infinite series which you might learn about elsewhere.

Example 5.16 Recall from Example 3.14 that the functions $\sin(kx)$ for different positive integer values of k are mutually orthogonal, with respect to the inner product $\langle f | g \rangle = \int_{-\pi}^{\pi} f(x) g(x) \, dx$, and have norm $\sqrt{\pi}$. Thus if $f : V \to V$ is defined by

$$f(x) = \sum_{k=1}^{n} \lambda_k \sin(kx)$$

then we can recover the coefficients λ_k from the function $f(x)$ as follows. First normalize the functions to give an orthonormal set of functions $f_k(x)$, where $f_k(x) = (1/\sqrt{\pi}) \sin(kx)$, so that $f(x) = \sum_{k=1}^{n} (\lambda_k \sqrt{\pi}) f_k(x)$. Then by Proposition 5.13 the coefficients $\lambda_k \sqrt{\pi}$ are given by

$$\lambda_k \sqrt{\pi} = \langle f_k | f \rangle = \int_{-\pi}^{\pi} \frac{1}{\sqrt{\pi}} \sin(kx) f(x) \, dx$$

so

$$\lambda_k = \frac{1}{\pi} \int_{-\pi}^{\pi} f(x) \sin(kx) \, dx.$$

Proposition 5.17 (Pythagoras's theorem) *Suppose $e_1, \ldots, e_n$ is an orthonormal basis of a Euclidean space V. Then, for all $v \in V$,*

$$\|v\|^2 = \sum_{i=1}^{n} \langle e_i | v \rangle^2.$$

Proof From Proposition 5.13, $v = \sum_{i=1}^{n} \langle e_i | v \rangle e_i$. Now take the inner product of both sides with v. $\qquad\qquad\square$

In words, the square of the length of a vector is equal to the sum of the squares of its coordinates. This should make it clear why it is called Pythagoras's theorem.

Example 5.18 In Example 5.10 we saw that the polynomials $\frac{\sqrt{2}}{2}$, $\frac{\sqrt{6}}{2}x$, and $\frac{3\sqrt{10}}{4}(x^2 - \frac{1}{3})$ form an orthonormal basis of the space $\mathbb{R}_3[x]$ of polynomials of degree less than 3 with inner product

$$\langle f | g \rangle = \int_{-1}^{1} f(x) g(x) \, dx.$$

Applying this to the polynomial x^2, we have

$$x^2 = \frac{2\sqrt{10}}{15}\left(\frac{3\sqrt{10}}{4}\left(x^2 - \frac{1}{3}\right)\right) + \frac{\sqrt{2}}{3}\frac{\sqrt{2}}{2}.$$

So x^2 has 'coordinates' $\sqrt{2}/3$, 0, and $2\sqrt{10}/15$. By Pythagoras's theorem

$$\begin{aligned}
\|x^2\|^2 &= \left(\frac{\sqrt{2}}{3}\right)^2 + \left(\frac{2\sqrt{10}}{15}\right)^2 \\
&= \frac{2}{9} + \frac{8}{45} = \frac{2}{5}
\end{aligned}$$

which you can check directly by computing $\int_{-1}^{1} x^4\,dx$.

Pythagoras's theorem is the special case $v = w$ of the following.

Corollary 5.19 (Parseval's identity) *If $\{e_1, \ldots, e_n\}$ is an orthonormal basis of the Euclidean space V, and $v, w \in V$, then*

$$\langle v|w\rangle = \sum_{i=1}^{n}\langle v|e_i\rangle\langle e_i|w\rangle.$$

Proof From Proposition 5.13, $w = \sum_{i=1}^{n}\langle e_i|w\rangle e_i$, and taking the inner product of both sides of this with v gives the required identity. □

In $\mathbb{R}^3$ with the standard inner product and standard basis, $\langle \mathbf{v}|e_i\rangle$ is the ith coordinate v_i of $\mathbf{v}$. In this case,

$$\langle \mathbf{v}|\mathbf{w}\rangle = \sum_{i=1}^{3} v_i w_i.$$

In other words, Parseval's identity is another way of stating the isomorphism in Corollary 5.12.

So far, the results concern finite dimensional spaces. The next result is a version of Pythagoras's theorem for infinite dimensional spaces. But since we cannot in general form an infinite sum, $\sum_{i=1}^{\infty}\langle v|e_i\rangle^2$, the infinite dimensional version of Pythagoras's theorem becomes an inequality.

Proposition 5.20 (Bessel's inequality) *If $\{e_1, \ldots, e_k\}$ is an orthonormal set (not necessarily a basis) of vectors in a real inner product space V, and $v \in V$, then*

$$\sum_{i=1}^{k}\langle e_i|v\rangle^2 \leqslant \|v\|^2.$$

Proof Let $w = v - \sum_{i=1}^{n} \langle e_i | v \rangle e_i$, and compute $\|w\|$ as follows.

$$0 \leqslant \|w\|^2 = \langle w | w \rangle = \Big\langle v - \sum_{i=1}^{n} \langle e_i | v \rangle e_i \Big| v - \sum_{j=1}^{n} \langle e_j | v \rangle e_j \Big\rangle$$

$$= \langle v | v \rangle - 2 \sum_{i=1}^{n} \langle e_i | v \rangle^2 + \sum_{i=1}^{n} \sum_{j=1}^{n} \langle e_i | v \rangle \langle e_j | v \rangle \langle e_i | e_j \rangle$$

$$= \langle v | v \rangle - \sum_{i=1}^{n} \langle e_i | v \rangle^2,$$

since $\langle e_i | e_j \rangle = 0$ except when $i = j$, and $\langle e_i | e_i \rangle = 1$, as required. $\square$

Example 5.21 We can apply Bessel's inequality to Example 5.16. If g is any continuous function on $[-\pi, \pi]$, we obtain

$$\sum_{k=1}^{n} \left(\int_{-\pi}^{\pi} f_k(x) g(x) \, dx \right)^2 \leqslant \int_{-\pi}^{\pi} (g(x))^2 \, dx$$

so

$$\sum_{k=1}^{n} \left(\int_{-\pi}^{\pi} g(x) \sin(kx) \, dx \right)^2 \leqslant \pi \int_{-\pi}^{\pi} (g(x))^2 \, dx.$$

In the special case when $g(x) = x$, we get

$$\sum_{k=1}^{n} \left(\int_{-\pi}^{\pi} x \sin(kx) \, dx \right)^2 \leqslant \pi \int_{-\pi}^{\pi} x^2 \, dx = 2\pi^4/3.$$

Now integrating $\int_{-\pi}^{\pi} x \sin(kx) \, dx$ by parts gives $(-1)^{k+1} 2\pi/k$, as you can check, so we have proved

$$\sum_{k=1}^{n} \frac{4\pi^2}{k^2} \leqslant \frac{2}{3} \pi^4$$

or

$$\sum_{k=1}^{n} \frac{1}{k^2} \leqslant \frac{1}{6} \pi^2.$$

In fact this is one of the special cases when the convergence problems can be overcome (although we will not prove it in this book), and it turns out that $\sum_{k=1}^{\infty} (1/k^2) = \pi^2/6$.

Inner product spaces over $\mathbb{C}$.   All of the above results have analogues for spaces over $\mathbb{C}$. In all cases the proofs are the same, perhaps with some complex conjugates added.

Proposition 5.22 *If $\{e_1, \ldots, e_n\}$ is an orthonormal basis of a complex inner product space V, and $v, w \in V$, then*

(a) *(**Fourier expansion**) $v = \sum\limits_{i=1}^{n} \langle e_i | v \rangle e_i$.*

(b) *(**Pythagoras's theorem**) $\|v\|^2 = \sum\limits_{i=1}^{n} |\langle e_i | v \rangle|^2$.*

(c) *(**Parseval's identity**)*

$$\langle v | w \rangle = \sum_{i=1}^{n} \langle v | e_i \rangle \langle e_i | w \rangle = \sum_{i=1}^{n} \overline{\langle e_i | v \rangle} \langle e_i | w \rangle.$$

Proof (a) is exactly as before.

For (b), note $\|v\|^2 = \langle v|v \rangle = \sum\limits_{i=1}^{n} \langle v|e_i \rangle \langle e_i|v \rangle = \sum\limits_{i=1}^{n} \langle v|e_i \rangle \overline{\langle v|e_i \rangle} = \sum\limits_{i=1}^{n} |\langle v|e_i \rangle|^2$.

For (c), consider $w = \sum\limits_{i=1}^{n} \langle e_i|w \rangle e_i$ and take the inner product with v of both sides of this equation. $\qquad\square$

Proposition 5.23 (Bessel's inequality) *If $\{e_1, \ldots, e_k\}$ is an orthonormal set (not necessarily a basis!) in a complex inner product space, then*

$$\sum_{i=1}^{k} |\langle e_i | v \rangle|^2 \leqslant \|v\|^2.$$

Proof Let $w = v - \sum\limits_{i=1}^{n} \langle e_i|v \rangle e_i$, and compute $\|w\|$ as follows.

$$\begin{aligned}
\|w\|^2 = \langle w|w \rangle &= \Big\langle v - \sum_{i=1}^{n} \langle e_i|v \rangle e_i \Big| v - \sum_{j=1}^{n} \langle e_j|v \rangle e_j \Big\rangle \\
&= \langle v|v \rangle - \sum_{i=1}^{n} \overline{\langle e_i|v \rangle} \langle e_i|v \rangle - \sum_{j=1}^{n} \langle e_j|v \rangle \langle v|e_j \rangle \\
&\quad + \sum_{i=1}^{n} \sum_{j=1}^{n} \overline{\langle e_i|v \rangle} \langle e_j|v \rangle \langle e_i|e_j \rangle \\
&= \langle v|v \rangle - \sum_{i=1}^{n} |\langle e_i|v \rangle|^2,
\end{aligned}$$

since $\langle e_i|e_j \rangle = 0$ except when $i = j$. The result now follows as $\|w\|^2 \geqslant 0$. $\qquad\square$

5.4 Orthogonal complements

This section describes the structure of vector spaces with an inner product in greater detail, and in particular introduces the notion of orthogonal subspaces. The main result uses the Gram–Schmidt method, and many of the ideas here are used implicitly in Chapter 7, especially in the proof of Theorem 7.8; also,

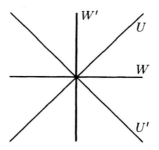

Fig. 5.2 Orthogonal complements in $\mathbb{R}^2$

in Chapter 13, Theorem 13.16 gives an alternative short proof using orthogonal complements of results that will be proved by alternative methods. Nothing, however, in this section is actually required later, so the section can be safely skipped on a first reading if necessary.

Example 5.24 Let $V = \mathbb{R}^2$ and $U = \{(x,x) : x \in \mathbb{R}\}$. Then the subspace $W = \{(x,0) : x \in \mathbb{R}\}$ has the property that every vector in V is the sum of a vector from U and a vector from W:

$$(x,y)^T = (y,y)^T + (x - y, 0)^T.$$

The same is true for the subspace $W' = \{(0,x) : x \in \mathbb{R}\}$:

$$(x,y)^T = (x,x)^T + (0, y - x)^T.$$

We will say that W and W' are *complements* to U in the space V. In fact there are infinitely many complements, and in terms of the vector space structure on V there is nothing to choose between them. But in the presence of the standard inner product one complement stands out, namely the *perpendicular* line $U' = \{(x, -x) : x \in \mathbb{R}\}$. (See Figure 5.2.)

We start by giving definitions covering the ideas of 'complement' and 'orthogonal complement'.

Definition 5.25 *If U and W are subspaces of a vector space V, then the **sum** of U and W is defined as*

$$U + W = \{u + w : u \in U, w \in W\}.$$

Proposition 5.26 *With this definition, $U + W$ is a subspace of V.*

Proof If v_1 and v_2 are elements of $U + W$, then there are elements u_1, u_2 of U and w_1, w_2 of W such that $v_1 = u_1 + w_1$ and $v_2 = u_2 + w_2$. Then for any scalar α we have

$$\begin{aligned} v_1 + \alpha v_2 &= (u_1 + w_1) + \alpha(u_2 + w_2) \\ &= (u_1 + \alpha u_2) + (w_1 + \alpha w_2) \in U + W, \end{aligned}$$

since $u_1 + \alpha u_2 \in U$ and $w_1 + \alpha w_2 \in W$. $\qquad\square$

Example 5.27 Let $V = \mathbb{R}^2$, and consider the two subspaces U and W defined by $U = \{(x,0) : x \in \mathbb{R}\}$ and $W = \{(0,y) : y \in \mathbb{R}\}$. Then

$$\begin{aligned}
U + W &= \{u + w : u \in U, w \in W\} \\
&= \{(x,0) + (0,y) : x \in \mathbb{R}, y \in \mathbb{R}\} \\
&= \{(x,y) : x \in \mathbb{R}, y \in \mathbb{R}\} \\
&= V.
\end{aligned}$$

On the other hand $U \cup W$ consists *only* of the points which are on one of the two coordinate axes.

Definition 5.28 *If V is a vector space and U is a subspace of V, then W is called* **a complement** *to U in V if*

(a) *W is a subspace of V,*
(b) *$V = U + W$, and*
(c) *$U \cap W = \{0\}$.*

If these three conditions are satisfied, we write $V = U \oplus W$, and say that V is the **direct sum** *of U and W.*

The point of Example 5.24 was to say that although there may be many vector space complements to a subspace U, in an *inner product space* there is a 'special' one, the *orthogonal complement*. This is defined next.

Definition 5.29 *If V is an inner product space and U is a subspace of V, we define*

$$U^\perp = \{v \in V : \langle u|v \rangle = 0 \text{ for all } u \in U\}.$$

This is called **the orthogonal complement** *to U in V, or 'U perp' for short.*

In view of the name, we ought to prove at once that $U^\perp$ is actually a complement to U, in the sense of Definition 5.28. In fact, this is not true in general, but it *is* true if U is finite dimensional. Before we prove this, we prove an easy lemma which will be very useful in practice when calculating orthogonal complements of subspaces.

Lemma 5.30 *If V is an inner product space, U is a subspace of V, and U has a basis $\{u_1, \ldots, u_k\}$, then*

$$U^\perp = \{v \in V : \langle u_i|v \rangle = 0 \text{ for all } i \text{ with } 1 \leqslant i \leqslant k\}.$$

Proof Let $W = \{v \in V : \langle u_i|v \rangle = 0 \text{ for all } i \text{ with } 1 \leqslant i \leqslant k\}$. If $v \in U^\perp$ then $\langle u|v \rangle = 0$ for all $u \in U$, so in particular $\langle u_i|v \rangle = 0$ for all i, so $v \in W$. Conversely, if $w \in W$, then $\langle u_i|w \rangle = 0$ for all i, and if $u \in U$, then u can be written as $u = \sum_{i=1}^k \lambda_i u_i$ for some scalars λ_i. Thus

$$\langle u|w \rangle = \left\langle \sum_{i=1}^k \lambda_i u_i \middle| w \right\rangle = \sum_{i=1}^k \lambda_i \langle u_i|w \rangle = 0,$$

so $w \in U^\perp$. Hence $U^\perp = W$, as required. $\qquad\qquad\square$

This result means that in order to check whether a vector w is in $U^\perp$ it is sufficient to check that w is orthogonal to all vectors in some basis for U. We can now use this to show that $U^\perp$ is indeed a complement to U in the sense of Definition 5.28.

Proposition 5.31 *If V is an inner product space, and U is a finite dimensional subspace of V, then*

(a) $U^\perp$ *is a subspace of V,*
(b) $U \cap U^\perp = \{0\}$, and
(c) $U + U^\perp = V$.
(Thus $V = U \oplus U^\perp$.)

Proof For (a), note that if $v, w \in U^\perp$, then $\langle u|v \rangle = \langle u|w \rangle = 0$ for all $u \in U$. Therefore

$$\langle u|v + \alpha w \rangle = \langle u|v \rangle + \alpha \langle u|w \rangle = 0$$

for all $u \in U$ and all scalars α, so $v + \alpha w \in U^\perp$. So $U^\perp$ is a subspace of V.

For (b), observe that if $u \in U \cap U^\perp$ then in particular $\langle u|u \rangle = 0$, so $u = 0$. Thus $U \cap U^\perp = \{0\}$.

We show for (c) that any vector can be resolved into two components, v_S in U and v_P orthogonal to U. This is very similar to Lemma 5.6 (see also Figure 5.1). We first use Gram–Schmidt orthonormalization (Theorem 5.8) to find an orthonormal basis for U, say $\{u_1, \ldots, u_k\}$. Then for any $v \in V$, define

$$v_S = \sum_{j=1}^{k} \langle v|u_j \rangle u_j$$

and $v_P = v - v_S$. By definition $v_S \in U$, and for all i

$$\langle v_P|u_i \rangle = \langle v|u_i \rangle - \langle v_S|u_i \rangle$$
$$= \langle v|u_i \rangle - \sum_{j=1}^{k} \langle v|u_j \rangle \langle u_j|u_i \rangle$$
$$= \langle v|u_i \rangle - \langle v|u_i \rangle = 0.$$

Therefore $v_P \in U^\perp$. Thus we have proved that any vector $v \in V$ can be written as $v = v_S + v_P \in U + U^\perp$, so every vector in V is in $U + U^\perp$, and therefore $U + U^\perp = V$, as required. $\qquad\square$

Proposition 5.32 *If $V = U \oplus W$, then $\dim(V) = \dim(U) + \dim(W)$.*

Proof Choose a basis $\{u_1, \ldots, u_m\}$ of U and a basis $\{w_1, \ldots, w_n\}$ of W. By the hypothesis $V = U + W$, any vector v in V is of the form $v = u + w$ for some $u \in U$ and $w \in W$, so can be written as

$$v = \sum_{i=1}^{m} \lambda_i u_i + \sum_{j=1}^{n} \mu_j w_j.$$

Thus $\{u_1, \ldots, u_m, w_1, \ldots, w_n\}$ spans V. On the other hand, if

$$\sum_{i=1}^{m} \lambda_i u_i + \sum_{j=1}^{n} \mu_j w_j = 0$$

then

$$\sum_{i=1}^{m} \lambda_i u_i = -\sum_{j=1}^{n} \mu_j w_j \in U \cap W = \{0\},$$

whence $\lambda_i = 0$ for all i since $\{u_1, \ldots, u_m\}$ is a basis of U and $\mu_j = 0$ for all j since $\{w_1, \ldots, w_n\}$ is a basis of W. Thus $\{u_1, \ldots, u_m, w_1, \ldots, w_n\}$ is a linearly independent set, so is a basis for V, and

$$\dim(V) = m + n = \dim(U) + \dim(W),$$

as required. $\qquad\square$

Corollary 5.33 *If V is a finite dimensional inner product space, and U is a subspace of V, then*

(a) $\dim(U) + \dim(U^{\perp}) = \dim(V)$, *and*
(b) $(U^{\perp})^{\perp} = U$.

Proof The first part is immediate from Propositions 5.31 and 5.32. Now if $x \in U$ then $\langle x|v \rangle = 0$ for all $v \in U^{\perp}$. Therefore $x \in (U^{\perp})^{\perp}$, and hence $U \subseteq (U^{\perp})^{\perp}$.

But from the first part applied to $U^{\perp}$, we have $\dim(U^{\perp}) + \dim((U^{\perp})^{\perp}) = \dim(V)$, so $\dim(U) = \dim((U^{\perp})^{\perp})$, and therefore $U = (U^{\perp})^{\perp}$. $\qquad\square$

Exercises

Exercise 5.1 Use the Gram–Schmidt process with the following inner products, starting with the ordered bases given, to obtain a new basis which is orthogonal.

(a) $V = \mathbb{R}^3$ with the standard inner product, and with the basis $(0, 1, -1)^T$, $(1, 1, 0)^T$, $(2, -2, 1)^T$.

(b) $V = \mathbb{R}^2$, with inner product defined by

$$\langle \mathbf{x}|\mathbf{y} \rangle = \begin{pmatrix} x_1 & x_2 \end{pmatrix} \begin{pmatrix} 5 & -1 \\ -1 & 1 \end{pmatrix} \begin{pmatrix} y_1 \\ y_2 \end{pmatrix},$$

and basis $(1, 0)^T$, $(0, 1)^T$.

(c) $V = \mathbb{R}_3[x]$, with inner product $\langle f|g \rangle = \int_0^1 f(x)g(x)\,dx$, and basis f_0, f_1, f_2 defined by $f_0(x) = 1$, $f_1(x) = x$, and $f_2(x) = x^2$.

Exercise 5.2 If $\{e_1, \ldots, e_n\}$ is any orthogonal basis of a real inner product space V, and $v \in V$, prove that

$$v = \sum_{i=1}^{n} \frac{\langle e_i|v \rangle}{\langle e_i|e_i \rangle} e_i.$$

Use this to write x^3 as a linear combination of the Legendre polynomials $P_0(x) = 1$, $P_1(x) = x$, $P_2(x) = \frac{3}{2}x^2 - \frac{1}{2}$, and $P_3(x) = \frac{5}{2}x^3 - \frac{3}{2}x$.

Exercise 5.3 A bilinear form on $\mathbb{R}^3$ is given by

$$F(\mathbf{x}, \mathbf{y}) = \begin{pmatrix} x_1 & x_2 & x_3 \end{pmatrix} \begin{pmatrix} 2 & 0 & -1 \\ 0 & 4 & -1 \\ -1 & -1 & 3 \end{pmatrix} \begin{pmatrix} y_1 \\ y_2 \\ y_3 \end{pmatrix}.$$

Apply the Gram–Schmidt process to F, to find a basis with respect to which the matrix of F is diagonal. (You could start with the standard basis. Don't forget to check that the set of vectors that Gram–Schmidt gives you *is* a basis.) Is F an inner product? Explain your answer.

Exercise 5.4 Let $E = \{e_1, e_2, \dots\}$ be an infinite orthogonal set of nonzero vectors in a real inner product space V. By normalizing E or otherwise, prove that for all $v \in V$

$$\|v\|^2 \geqslant \sum_{i=1}^{n} \frac{\langle v|e_i \rangle^2}{\|e_i\|^2}$$

for all natural numbers n.

Exercise 5.5 By considering $g(x) = x^2$, Bessel's inequality, and the method of Example 5.21, find an upper bound for $\sum_{n=1}^{N}(1/n^4)$ which is independent of N.

Exercise 5.6 Let V be the real vector space $\mathbb{R}[x]$ of polynomials in x with real coefficients, and let $\langle | \rangle$ be an inner product on V. Suppose that the sets $\{P_0, P_1, \dots, P_n, \dots\} \subseteq V$ and $\{Q_0, Q_1, \dots, Q_n, \dots\} \subseteq V$ are both orthogonal, with P_n and Q_n both having degree n for each $n \in \mathbb{N}$. Prove that $\langle P_n|R \rangle = \langle Q_n|R \rangle = 0$ for all polynomials R with degree $k < n$. [Hint: consider Fourier expansions of R.] Hence show that there are scalars λ_n with $P_n = \lambda_n Q_n$.

Exercise 5.7 (Legendre polynomials) Let V be the real inner product space $\mathbb{R}[x]$ of polynomials over $\mathbb{R}$ with inner product

$$\langle P(x)|Q(x) \rangle = \int_{-1}^{1} P(x)Q(x)\,dx.$$

The Legendre polynomials are defined to be the unique sequence of polynomials $P_n(x) \in V$ with P_n having degree n, $\langle P_i|P_j \rangle = 0$ for all $i \neq j$, and $P_n(1) = 1$ for all n. (See Exercise 5.6 to see why this specifies them uniquely.)

(a) Prove that for all $n \in \mathbb{N}$ there is $\alpha_n \in \mathbb{R}$ such that

$$P_{n+1}(x) = \alpha_n x P_n(x) + (1 - \alpha_n)P_{n-1}(x).$$

[Consider the Fourier expansion of $xP_n(x)$.]

(b) Use integration by parts to show that

$$\langle xP_n(x)|P_n'(x) \rangle = 1 - \tfrac{1}{2}\|P_n\|^2$$

and

$$\langle P_{n-1}(x)|P_n'(x) \rangle = 2,$$

where $'$ denotes differentiation with respect to x.

(c) Using part (a), show that

$$(\alpha_n - 1)\langle P_{n-1}(x)|P'_n(x)\rangle = \alpha_n\langle x P_n(x)|P'_n(x)\rangle.$$

Hence deduce from (b) that

$$P_{n+1}(x) = \frac{4}{\|P_n\|^2 + 2} x P_n(x) + \frac{\|P_n\|^2 - 2}{\|P_n\|^2 + 2} P_{n-1}(x)$$

for $n \geqslant 1$.

(d) Using part (a) (twice!) and $\langle x P_n(x)|P_{n+1}(x)\rangle = \langle P_n(x)|x P_{n+1}(x)\rangle$, show that

$$\|P_{n+1}\|^2 = \frac{\alpha_n(\alpha_{n+1} - 1)}{\alpha_{n+1}}\|P_n\|^2.$$

Hence, by induction on n, show that

$$\|P_n\|^2 = \frac{2}{2n + 1}$$

and

$$P_{n+1}(x) = \frac{2n + 1}{n + 1} x P_n(x) - \frac{n}{n + 1} P_{n-1}.$$

(e) Show that

$$(1 - x^2) P''_n(x) - 2x P'_n(x) + n(n + 1) P_n = 0$$

by taking inner products with P_i for $i \leqslant n$.

Exercise 5.8 (Hermite polynomials) Let V be the real vector space $\mathbb{R}[x]$ with $\langle P(x)|Q(x)\rangle = \int_{-\infty}^{\infty} e^{-x^2/2} P(x)Q(x)\,dx$. Let $I_n = \langle x^n|1\rangle$.

(a) Prove that $I_0 = \sqrt{2\pi}$. [Hint: write $I_0^2 = \iint e^{-(x^2+y^2)/2}\,dx\,dy$ and change to polar coordinates to evaluate this multiple integral.]

(b) Show that $I_{2n+1} = 0$ and $I_{2n} = ((2n)!/(2^n n!))I_0$. (Use induction on n.) Hence show that $\langle\,|\,\rangle$ is an inner product on V.

(c) The Hermite polynomials are defined to be the unique sequence of polynomials $H_n(x) \in V$ with H_n having degree n, $\{H_n : n \in \mathbb{N}\}$ is orthogonal for this inner product, and the leading coefficient of $H_n(X)$ is 1. Use Gram–Schmidt or otherwise to show that

$H_0(x) = 1$	$H_3(x) = x^3 - 3x$
$H_1(x) = x$	$H_4(x) = x^4 - 6x^2 + 3$
$H_2(x) = x^2 - 1$	$H_5(x) = x^5 - 10x^3 + 15x.$

(d) Show that $H_{n+1}(x) = x H_n(x) - \beta_n H_{n-1}(x)$, where $\beta_n = \|H_n\|^2/\|H_{n-1}\|^2$.

(e) Show that

$$\langle xH_n(x)|H'_n(x)\rangle = \tfrac{1}{2}(-\|H_n\|^2 + \|H_{n+1}\|^2 + \beta_n^2\|H_{n-1}\|^2)$$

and

$$\langle H_{n-1}(x)|H'_n(x)\rangle = \|H_n\|^2,$$

by using integration by parts.

(f) By taking the inner product of the equation in (d) with $H'_n(x)$, show that $\beta_{n+1} = 1 + \beta_n$.

(g) Hence derive the following properties of the polynomials H_n:
   i. $\|H_n\|^2 = n!\sqrt{2\pi}$;
   ii. $H_{n+1}(x) = xH_n(x) - nH_{n-1}(x)$;
   iii. $H''_n(x) - xH'_n(x) + nH_n(x) = 0$.

Exercise 5.9 (Laguerre polynomials) Let V be the real vector space $\mathbb{R}[x]$ with $\langle P(x)|Q(x)\rangle = \int_0^\infty e^{-x}P(x)Q(x)\,dx$.

(a) Prove that $\int_0^\infty x^n e^{-x}\,dx = n!$ for all n.

(b) Hence show that $\langle\,|\,\rangle$ is an inner product on V.

(c) The Laguerre polynomials are the unique sequence of polynomials $L_n(x) \in V$ with L_n having degree n, $\{L_n : n \in \mathbb{N}\}$ orthogonal for this inner product, and the leading coefficient of $L_n(x)$ is 1. Use Gram–Schmidt to show that

$$
\begin{aligned}
L_0(x) &= 1 & L_3(x) &= x^3 - 9x^2 + 18x - 6 \\
L_1(x) &= x - 1 & L_4(x) &= x^4 - 16x^3 + 72x^2 - 96x + 24 \\
L_2(x) &= x^2 - 4x + 2 & L_5(x) &= x^5 - 25x^4 + 200x^3 - 600x^2 + 600x - 120.
\end{aligned}
$$

(d) Derive the following properties of the polynomials L_n:
   i. $\|L_n\|^2 = (n!)^2$;
   ii. $L_{n+1}(x) + (2n + 1 - x)L_n(x) + n^2 L_{n-1}(x) = 0$; and
   iii. $xL''_n(x) + (1 - x)L'_n(x) + nL_n(x) = 0$.

Exercise 5.10 Prove the following generalization of Proposition 5.32: if U and W are subspaces of a vector space V, then

$$\dim(U) + \dim(W) = \dim(U + W) + \dim(U \cap W).$$

Exercise 5.11 If U and W are subspaces of a finite dimensional inner product space V, show that

(a) if $U \subseteq W$, then $W^\perp \subseteq U^\perp$,

(b) $(U + W)^\perp = U^\perp \cap W^\perp$,

(c) $U^\perp + W^\perp \subseteq (U \cap W)^\perp$, and if also V is finite dimensional then $U^\perp + W^\perp = (U \cap W)^\perp$.

Exercise 5.12 In which of the following is $V = U \oplus W$?

(a) $V = \mathbb{C}^3$, $U = \{(a,b,c)^T : a + b = 0\}$, $W = \{(a,b,c)^T : a = b = c\}$;
(b) $V = \mathbb{R}^4$, $U = \text{span}((1,0,1,0)^T, (0,1,0,1)^T)$,
   $W = \text{span}((1,0,2,0)^T, (0,1,0,2)^T)$;
(c) $V = \mathbb{R}_4[x]$, $U = \{$polynomials in V of even degree$\}$,
   $W = \{$polynomials in V of odd degree$\}$;
(d) $V = \mathbb{R}_n[x]$, $U = \{$polynomials in V of degree less than $k\}$,
   $W = \{$polynomials in V divisible by $x^k\}$.

Exercise 5.13 For Exercise 5.12, parts (a) and (b), find $U^\perp$, where the inner product is the standard inner product.

Exercise 5.14 (Hard!) Let $V = \ell^2$ be the set of all real sequences (a_n) such that $\sum_{n=1}^\infty a_n^2$ converges.

(a) Prove that V is a vector space under componentwise addition and scalar multiplication.
(b) Define a suitable inner product on V, and prove that it *is* an inner product.
(c) Deduce that if (a_n) and (b_n) are in V then

$$\left(\sum_{n=1}^\infty (a_n + b_n)^2\right)^{\frac{1}{2}} \leqslant \left(\sum_{n=1}^\infty (a_n)^2\right)^{\frac{1}{2}} + \left(\sum_{n=1}^\infty (b_n)^2\right)^{\frac{1}{2}}.$$

(d) Let U be the subspace of all *finite* sequences (i.e. those which are 0 from some point on). Show that $U^\perp = \{0\}$, and deduce that $(U^\perp)^\perp = V$ and $(U^\perp)^\perp \neq U$.

6

When is a form definite?

In the last chapter we used the Gram–Schmidt process to obtain orthonormal bases for vector spaces with inner products. In this chapter we will present several variations of the idea behind Gram–Schmidt, this time applied to a form which need not be positive definite.

Sections 6.1 and 6.2 both address the same question: given a real symmetric (or complex conjugate-symmetric) matrix $\mathbf{A}$, how can we determine if the form $\mathbf{x}^T\mathbf{A}\mathbf{x}$ is positive definite? And, if it isn't, how can we find a nonzero vector $\mathbf{x}$ such that $\mathbf{x}^T\mathbf{A}\mathbf{x} \leqslant 0$? In this chapter, we consider bilinear forms on vector spaces over $\mathbb{R}$ and sesquilinear forms on vector spaces over $\mathbb{C}$ only, since the notion of 'positive definite' as we have defined it only really applies to these forms. Section 6.2 requires knowledge of determinants, and can be safely omitted or postponed until later if the reader is not familiar with these at this stage.

6.1 The Gram–Schmidt process revisited

In this section we study symmetric bilinear forms (or in the complex case, conjugate-symmetric sesquilinear forms) over a finite dimensional vector space V, and try to answer the question: when is it an inner product?

The idea behind the proof of these results is as follows: if our bilinear form $F(,)$ is an inner product we should be able to find a basis for which F has a diagonal representation, by the orthogonalization process. So we will try to mimic this process and see what we get.

Looking back at Theorem 5.8 we see that we have to replace $\langle w_i | w_i \rangle$ by $F(w_i, w_i)$, and use the formula

$$w_k = v_k - \sum_{i=1}^{k-1} \frac{F(w_i, v_k)}{F(w_i, w_i)} w_i. \tag{1}$$

But in the more general context of symmetric bilinear forms, $F(w_i, w_i)$ could be 0, and we might not be able to divide by it.

Example 6.1 Let $\mathbf{A}$ be the matrix

$$\begin{pmatrix} 2 & 1 & -1 \\ 1 & 2 & 0 \\ -1 & 0 & 2/3 \end{pmatrix}$$

and let $F(\mathbf{x}, \mathbf{y}) = \mathbf{x}^T A \mathbf{y}$ be the corresponding form on $\mathbb{R}^3$. Take the usual basis $\mathbf{v}_1 = (1, 0, 0)^T$, $\mathbf{v}_2 = (0, 1, 0)^T$, $\mathbf{v}_3 = (0, 0, 1)^T$ of $\mathbb{R}^3$ and apply (1). This gives $\mathbf{w}_1 = \mathbf{v}_1 = (1, 0, 0)^T$, $F(\mathbf{w}_1, \mathbf{w}_1) = 2$,

$$\mathbf{w}_2 = \mathbf{v}_2 - \frac{F(\mathbf{w}_1, \mathbf{v}_2)}{F(\mathbf{w}_1, \mathbf{w}_1)} \mathbf{w}_1 = \begin{pmatrix} 0 \\ 1 \\ 0 \end{pmatrix} - \frac{1}{2} \begin{pmatrix} 1 \\ 0 \\ 0 \end{pmatrix} = \begin{pmatrix} -1/2 \\ 1 \\ 0 \end{pmatrix},$$

$F(\mathbf{w}_2, \mathbf{w}_2) = F(\mathbf{v}_2, \mathbf{v}_2) - 2F(\mathbf{v}_2, (1/2)\mathbf{v}_1) + F(\mathbf{v}_1, \mathbf{v}_1) = 2 - 1 + (1/2) = 3/2$, and

$$\mathbf{w}_3 = \mathbf{v}_3 - \frac{F(\mathbf{w}_1, \mathbf{v}_3)}{F(\mathbf{w}_1, \mathbf{w}_1)} \mathbf{w}_1 - \frac{F(\mathbf{w}_2, \mathbf{v}_3)}{F(\mathbf{w}_2, \mathbf{w}_2)} \mathbf{w}_2$$

$$= \begin{pmatrix} 0 \\ 0 \\ 1 \end{pmatrix} - \frac{-1}{2} \begin{pmatrix} 1 \\ 0 \\ 0 \end{pmatrix} - \frac{1/2}{3/2} \begin{pmatrix} -1/2 \\ 1 \\ 0 \end{pmatrix} = \begin{pmatrix} 2/3 \\ -1/3 \\ 1 \end{pmatrix}.$$

Now, however,

$$F(\mathbf{w}_3, \mathbf{w}_3) = \begin{pmatrix} 2/3 & -1/3 & 1 \end{pmatrix} \begin{pmatrix} 2 & 1 & -1 \\ 1 & 2 & 0 \\ -1 & 0 & 2/3 \end{pmatrix} \begin{pmatrix} 2/3 \\ -1/3 \\ 1 \end{pmatrix} = 0$$

so we have found a nonzero vector, $\mathbf{w}_3 = (2/3, -1/3, 1)^T$ with $F(\mathbf{w}_3, \mathbf{w}_3) = 0$, so F is *not* positive definite.

Used as in this example, the Gram–Schmidt process gives a method of determining whether a form is positive definite or not; it does not always give a basis though. The formula (1) applies only as long as $F(w_k, w_k) \neq 0$, and when w_k is found with $F(w_k, w_k) = 0$ the Gram–Schmidt process comes to a stop.

But before we show that the method just outlined always works, we shall expand our terminology concerning positive definite forms.

Definition 6.2 *A symmetric bilinear form F on a real vector space V is said to be **definite** if*

$$F(v, v) = 0 \Leftrightarrow v = 0$$

*for all $v \in V$; in other words, if $F(v, v)$ can only be zero when v is zero. So F is positive definite if it is definite and $F(v, v) \geqslant 0$ for all $v \in V$; F is said to be **negative definite** if it is definite and $F(v, v) \leqslant 0$ for all $v \in V$.*

In general, for forms F on complex vector spaces, $F(v, v)$ may not always be real, so it doesn't make sense to say whether the form is positive or negative definite. However, for *conjugate-symmetric* sesquilinear forms F on complex vector spaces, Definition 6.2 is valid since if F is conjugate-symmetric and v is in V then $F(v, v) = \overline{F(v, v)}$, so $F(v, v)$ is always real.

The next lemma provides a useful criterion for proving definiteness.

Lemma 6.3 *Suppose F is a symmetric bilinear form on a vector space V over $\mathbb{R}$, or a conjugate-symmetric sesquilinear form on a vector space V over $\mathbb{C}$, and suppose $v_1, \ldots, v_n$ is a basis of V.*

(a) *If $F(v_i, v_i) > 0$ for all i and $F(v_i, v_j) = 0$ for all $i \neq j$, then F is positive definite.*

(b) *If $F(v_i, v_i) < 0$ for all i and $F(v_i, v_j) = 0$ for all $i \neq j$, then F is negative definite.*

Proof For (a), let $v \in V$ be arbitrary. Since the v_i form a basis, $v = \sum_{i=1}^{n} \lambda_i v_i$ for some scalars λ_i. Then

$$F(v, v) = F\left(\sum_{i=1}^{n} \lambda_i v_i, \sum_{j=1}^{n} \lambda_j v_j\right)$$
$$= \sum_{i=1}^{n} \sum_{j=1}^{n} \overline{\lambda_i} \lambda_j F(v_i, v_j)$$
$$= \sum_{i=1}^{n} |\lambda_i|^2 F(v_i, v_i)$$
$$\geqslant 0$$

since each $F(v_i, v_j)$ $(i \neq j)$ is zero, each $F(v_i, v_i) > 0$, and each $|\lambda_i|^2 \geqslant 0$. Also, since each $F(v_i, v_i) > 0$ the only way $F(v, v)$ could be equal to 0 is if $\lambda_1 = \lambda_2 = \cdots = \lambda_n = 0$, i.e. if $v = 0$. This shows that F is positive definite as required.

The argument for (b) is the same, just replacing $>$ with $<$ in the appropriate places. $\qquad\qquad\square$

Consider the Gram–Schmidt process applied to a symmetric bilinear form F on a real vector space V with basis vectors $v_1, v_2, \ldots, v_n$ (or a conjugate-symmetric sesquilinear form on a complex vector space V). We compute vectors

$$w_k = v_k - \sum_{i=1}^{k-1} \frac{F(w_i, v_k)}{F(w_i, w_i)} w_i$$

as far as possible either until we have found some w_k with $F(w_k, w_k) = 0$, or else until we have obtained all n vectors $w_1, w_2, \ldots, w_n$.

Suppose first that $F(w_i, w_i) \neq 0$ for all $i < k$ but $F(w_k, w_k) = 0$. In this case we have

$$w_i \in \mathrm{span}(v_1, v_2, \ldots, v_i)$$

for all $\leqslant k$. But then $w_k \neq 0$ since otherwise

$$v_k = \sum_{i=1}^{k-1} \frac{F(w_i, v_k)}{F(w_i, w_i)} w_i \in \mathrm{span}(v_1, v_2, \ldots, v_{k-1})$$

contradicting our assumption that $v_1, v_2, \ldots, v_k$ is linearly independent. Thus $w_k \neq 0$ and $F(w_k, w_k) = 0$ so F is not definite—in particular neither positive definite nor negative definite.

Now suppose that the Gram–Schmidt process completes, and we are able to compute $w_1, w_2, \ldots, w_n$ with $F(w_i, w_i) \neq 0$ for all $i \leqslant n$. In this case, the proof of Theorem 5.8 shows that $w_1, w_2, \ldots, w_n$ is a basis of V. By Lemma 6.3, if each $F(w_i, w_i)$ is positive then F is positive definite, and if each $F(w_i, w_i)$ is negative then F is negative definite. So in each case the Gram–Schmidt method will determine whether F is positive definite, negative definite, or neither.

Example 6.4 We apply the Gram–Schmidt process to the symmetric bilinear form

$$F(\mathbf{x}, \mathbf{y}) = \mathbf{x}^T \begin{pmatrix} -1 & 1 & 1 \\ 1 & -2 & 0 \\ 1 & 0 & -3 \end{pmatrix} \mathbf{y}$$

on $\mathbb{R}^3$. Take the usual basis $\mathbf{v}_1 = (1,0,0)^T$, $\mathbf{v}_2 = (0,1,0)^T$, $\mathbf{v}_3 = (0,0,1)^T$, and compute

$$\mathbf{w}_1 = \mathbf{v}_1 \qquad\qquad\qquad F(\mathbf{w}_1, \mathbf{w}_1) = -1$$

$$\mathbf{w}_2 = \mathbf{v}_2 - \frac{F(\mathbf{w}_1, \mathbf{v}_2)}{F(\mathbf{w}_1, \mathbf{w}_1)} \mathbf{w}_1 = \begin{pmatrix} 1 \\ 1 \\ 0 \end{pmatrix} \qquad F(\mathbf{w}_2, \mathbf{w}_2) = -1$$

$$\mathbf{w}_3 = \mathbf{v}_3 - \frac{F(\mathbf{w}_1, \mathbf{v}_3)}{F(\mathbf{w}_1, \mathbf{w}_1)} \mathbf{w}_1 - \frac{F(\mathbf{w}_2, \mathbf{v}_3)}{F(\mathbf{w}_2, \mathbf{w}_2)} \mathbf{w}_2 = \begin{pmatrix} 1 \\ 1 \\ 1 \end{pmatrix} \qquad F(\mathbf{w}_3, \mathbf{w}_3) = -2.$$

So, with respect to the ordered basis $(1,0,0)^T, (1,1,0)^T, (1,1,1)^T$, the form F has matrix representation

$$\begin{pmatrix} -1 & 0 & 0 \\ 0 & -1 & 0 \\ 0 & 0 & -2 \end{pmatrix}$$

and hence is negative definite by Lemma 6.3.

The only question left to answer here is whether a form can be definite even if it is neither positive definite nor negative definite. In other words, if the space V has two vectors v, w with $F(v,v) < 0 < F(w,w)$, can F be definite? The answer is no.

Proposition 6.5 *Suppose F is a symmetric bilinear form on a real vector space V, or a conjugate-symmetric sesquilinear form on a complex vector space V, and there are $x, y \in V$ such that $F(x,x) < 0 < F(y,y)$. Then there is $z \in V$ with $z \neq 0$ and $F(z,z) = 0$.*

Proof Note first that $\{x, y\}$ is linearly independent, since if $x = \alpha y$ then $F(x, x) = |\alpha|^2 F(y, y)$ and these would have the same sign.

Suppose $F(x, x) = \mu < 0$, $F(y, y) = \nu > 0$, and $F(x, y) = \lambda$. The strategy is to work in the subspace spanned by x, y, using Gram–Schmidt with normalization to find a basis of this subspace with respect to which the matrix of F is $\begin{pmatrix} -1 & 0 \\ 0 & 1 \end{pmatrix}$.

We start by applying the Gram–Schmidt formula to the vectors x, y, defining

$$x' = x$$

$$y' = y - \frac{F(x, y)}{F(x, x)} x = y - \frac{\lambda}{\mu} x$$

so $F(x', x') = \mu < 0$, $F(y', y') = \nu - |\lambda|^2/\mu$, and $F(x', y') = 0$. (Note that $F(y', y') > 0$ as $\mu < 0 < \nu$.) We now normalize, defining

$$x'' = \frac{1}{\sqrt{-\mu}} x'$$

$$y'' = \frac{1}{\sqrt{\nu - |\lambda|^2/\mu}} y'$$

so $F(x'', x'') = -1$, $F(y'', y'') = 1$, and $F(x'', y'') = 0$. Then $x'' + y'' \neq 0$ (in fact, x'', y'' are linearly independent) and $F(x'' + y'', x'' + y'') = F(x'', x'') + F(y'', y'') = -1 + 1 = 0$. $\qquad\square$

The following corollary is immediate.

Corollary 6.6 *A form F is definite if and only if either it is positive definite or it is negative definite.*

6.2 The leading minor test

There is a well-known and useful test, called the *leading minor test*, which uses determinants to give an alternative to the Gram–Schmidt method to determine if a form is positive or negative definite, which we give now.

First, we fix some notation. For simplicity, for the rest of this section we work in the vector space $V = \mathbb{R}^n$ over the reals, and we suppose $\mathbf{A}$ is an $n \times n$ symmetric matrix defining the symmetric bilinear form $F(\mathbf{x}, \mathbf{y}) = \mathbf{x}^T \mathbf{A} \mathbf{y}$. But if you wish, you can think of $\mathbf{A}$ as being a conjugate-symmetric matrix, $V = \mathbb{C}^n$, and $F(\mathbf{x}, \mathbf{y}) = \overline{\mathbf{x}}^T \mathbf{A} \mathbf{y}$, as all the arguments here work unchanged.

Note also that by the previous chapter, we could have started with any finite dimensional real vector space with a bilinear form F, taken any ordered basis of it, and defined $\mathbf{A}$ to be the matrix representing F with respect to this basis. Then all the results here will also apply to F, since, for example, F is positive definite if and only if the matrix $\mathbf{A}$ is positive definite.

We work throughout with the standard basis $\mathbf{e}_1, \mathbf{e}_2, \ldots, \mathbf{e}_n$ of V. Given scalars (i.e. real numbers) a_{lm} for $1 \leqslant l \leqslant i - 1$ and $1 \leqslant m \leqslant i$, we will use the expression

$$\begin{vmatrix} a_{11} & a_{12} & \dots & a_{1i} \\ a_{21} & a_{22} & \dots & a_{2i} \\ \vdots & \vdots & & \vdots \\ a_{i-1\,1} & a_{i-1\,2} & \dots & a_{i-1\,i} \\ \mathbf{e}_1 & \mathbf{e}_2 & \dots & \mathbf{e}_i \end{vmatrix} \tag{2}$$

with vectors in the bottom row of this 'determinant' as a shorthand for the following expansion of a determinant by its last row:

$$\begin{vmatrix} a_{11} & \dots & a_{1\,i-1} \\ \vdots & & \vdots \\ a_{i-1\,1} & \dots & a_{i-1\,i-1} \end{vmatrix} \mathbf{e}_i - \begin{vmatrix} a_{11} & \dots & a_{1\,i-2} & a_{1i} \\ \vdots & & \vdots & \vdots \\ a_{i-1\,1} & \dots & a_{i-1\,i-2} & a_{i-1\,i} \end{vmatrix} \mathbf{e}_{i-1} + \cdots$$

$$+ (-1)^i \begin{vmatrix} a_{11} & a_{13} & \dots & a_{1i} \\ \vdots & \vdots & & \vdots \\ a_{i-1\,1} & a_{i-1\,3} & \dots & a_{i-1\,i} \end{vmatrix} \mathbf{e}_2 + (-1)^{i+1} \begin{vmatrix} a_{12} & \dots & a_{1i} \\ \vdots & & \vdots \\ a_{i-1\,2} & \dots & a_{i-1\,i} \end{vmatrix} \mathbf{e}_1 .$$

Since $\mathbf{e}_1, \dots, \mathbf{e}_n$ is the standard basis for $\mathbb{R}^n$, this expression denotes a column vector whose first i entries are the $(i-1) \times (i-1)$ determinants indicated, with alternating signs as shown. The remaining $n-i$ entries are all zero.

We have chosen a somewhat unusual way of expanding this determinant to make the work easier later on. Usually the determinant (2) would be expanded by the last row by writing down the $\mathbf{e}_1$ term first. However, the sign of this term has to be adjusted by $(-1)^{i+1}$, and we prefer to start the expansion with a term that does not have its sign adjusted in this way.

You should be aware that strictly *vectors ought not appear as the entries of determinants like this.* (The determinant operation acts on an array of numbers, not a mixture of numbers and vectors.) But as long as we restrict vectors to the bottom row of a determinant and be careful to interpret our determinants as in the expansion of (2) we do get a valid expression. Of course, it is impossible to interpret the expansion of (2) by any row or column other than the last row! Note also that (2) gives a *vector*, i.e. an element of V, and not a scalar, which is what you'd get if you expanded an ordinary determinant.

Example 6.7 In $\mathbb{R}^3$, the determinant

$$\begin{vmatrix} a_1 & a_2 & a_3 \\ b_1 & b_2 & b_3 \\ \mathbf{e}_1 & \mathbf{e}_2 & \mathbf{e}_3 \end{vmatrix}$$

is the *vector product* or *cross product* $\mathbf{a} \times \mathbf{b}$, where $\mathbf{a} = (a_1, a_2, a_3)^T$ and $\mathbf{b} = (b_1, b_2, b_3)^T$ which you may have seen elsewhere. For example,

$$\mathbf{e}_1 \times \mathbf{e}_2 = \begin{vmatrix} 1 & 0 & 0 \\ 0 & 1 & 0 \\ \mathbf{e}_1 & \mathbf{e}_2 & \mathbf{e}_3 \end{vmatrix} = \mathbf{e}_3 .$$

Proposition 6.8 *Let $V = \mathbb{R}^n$, $\mathbf{A} = (a_{ij})$ a symmetric $n \times n$ real matrix, and let $F(\mathbf{x}, \mathbf{y}) = \mathbf{x}^T \mathbf{A} \mathbf{y}$ be the corresponding symmetric bilinear form on V (so $a_{ij} = F(\mathbf{e}_i, \mathbf{e}_j) = F(\mathbf{e}_j, \mathbf{e}_i) = a_{ji}$ for all i, j). For each $i = 1, \ldots, n$ define*

$$
\mathbf{v}_i = \begin{vmatrix}
a_{11} & a_{12} & \cdots & a_{1i} \\
a_{21} & a_{22} & \cdots & a_{2i} \\
\vdots & \vdots & & \vdots \\
a_{i-1\,1} & a_{i-1\,2} & \cdots & a_{i-1\,i} \\
\mathbf{e}_1 & \mathbf{e}_2 & \cdots & \mathbf{e}_i
\end{vmatrix}.
$$

Then the $\mathbf{v}_i$ are orthogonal with respect to F, i.e. $F(\mathbf{v}_i, \mathbf{v}_j) = 0$ for all $i \neq j$. Moreover, $F(\mathbf{v}_1, \mathbf{v}_1) = a_{11}$ and

$$
F(\mathbf{v}_i, \mathbf{v}_i) = \begin{vmatrix}
a_{11} & \cdots & a_{1\,i-1} \\
\vdots & & \vdots \\
a_{i-1\,1} & \cdots & a_{i-1\,i-1}
\end{vmatrix}
\begin{vmatrix}
a_{11} & a_{12} & \cdots & a_{1i} \\
a_{21} & a_{22} & \cdots & a_{2i} \\
\vdots & \vdots & & \vdots \\
a_{i1} & a_{i2} & \cdots & a_{ii}
\end{vmatrix}. \tag{3}
$$

(Note that this proposition does not say whether $F(\mathbf{v}_i, \mathbf{v}_i)$ is positive, negative, or zero, and does not imply that the $\mathbf{v}_i$ form a basis.)

Proof   First, if $j < i$ then by using the expansion of (2), and using bilinearity, $F(\mathbf{e}_j, \mathbf{v}_i)$ equals

$$
F(\mathbf{e}_j, \mathbf{e}_i) \begin{vmatrix}
a_{11} & \cdots & a_{1\,i-1} \\
\vdots & & \vdots \\
a_{i-1\,1} & \cdots & a_{i-1\,i-1}
\end{vmatrix}
$$

$$
- F(\mathbf{e}_j, \mathbf{e}_{i-1}) \begin{vmatrix}
a_{11} & \cdots & a_{1\,i-2} & a_{1i} \\
\vdots & & \vdots & \vdots \\
a_{i-1\,1} & \cdots & a_{i-1\,i-2} & a_{i-1\,i}
\end{vmatrix} \tag{4}
$$

$$
+ \cdots
$$

$$
+ (-1)^{i+1} F(\mathbf{e}_j, \mathbf{e}_1) \begin{vmatrix}
a_{12} & \cdots & a_{1i} \\
\vdots & & \vdots \\
a_{i-1\,2} & \cdots & a_{i-1\,i}
\end{vmatrix}
$$

which (using $F(\mathbf{e}_j, \mathbf{e}_k) = a_{jk}$) equals the determinant

$$
\begin{vmatrix}
a_{11} & a_{12} & \cdots & a_{1i} \\
\vdots & \vdots & & \vdots \\
a_{i-1\,1} & a_{i-1\,2} & \cdots & a_{i-1\,i} \\
a_{j1} & a_{j2} & \cdots & a_{ji}
\end{vmatrix}.
$$

But this determinant has two identical rows, so is zero. Thus $F(\mathbf{e}_j, \mathbf{v}_i) = 0$.

We can use this to prove the orthogonality of the $\mathbf{v}_i$. Here, we must show that $F(\mathbf{v}_j, \mathbf{v}_i) = 0$ if $i \neq j$. Since $F(\mathbf{v}_i, \mathbf{v}_j) = F(\mathbf{v}_j, \mathbf{v}_i)$ we might as well assume $j < i$. By (2), $\mathbf{v}_j = \beta_j \mathbf{e}_j + \cdots + \beta_1 \mathbf{e}_1$ for some scalars β_k (all being plus or minus certain $(j-1) \times (j-1)$ determinants); hence

$$F(\mathbf{v}_j, \mathbf{v}_i) = F(\beta_j \mathbf{e}_j + \cdots + \beta_1 \mathbf{e}_1, \mathbf{v}_i)$$
$$= \beta_j F(\mathbf{e}_j, \mathbf{v}_i) + \cdots + \beta_1 F(\mathbf{e}_1, \mathbf{v}_i)$$
$$= 0$$

as required.

Finally, we must compute $F(\mathbf{v}_i, \mathbf{v}_i)$. First, note that by definition, $\mathbf{v}_1 = \mathbf{e}_1$ so $F(\mathbf{v}_1, \mathbf{v}_1) = F(\mathbf{e}_1, \mathbf{e}_1) = a_{11}$. Now suppose $i > 1$ and note that by (2) again,

$$\mathbf{v}_i = \alpha_i \mathbf{e}_i + \alpha_{i-1} \mathbf{e}_{i-1} \cdots + \alpha_1 \mathbf{e}_1$$

for scalars α_k, where α_i is the determinant

$$\begin{vmatrix} a_{11} & a_{12} & \cdots & a_{1\,i-1} \\ a_{21} & a_{22} & \cdots & a_{2\,i-1} \\ \vdots & \vdots & & \vdots \\ a_{i-1\,1} & a_{i-1\,2} & \cdots & a_{i-1\,i-1} \end{vmatrix}$$

and the other α_k are given by other determinants as in (2). So

$$F(\mathbf{v}_i, \mathbf{v}_i) = F(\alpha_i \mathbf{e}_i + \alpha_{i-1} \mathbf{e}_{i-1} + \cdots + \alpha_1 \mathbf{e}_1, \mathbf{v}_i)$$
$$= \alpha_i F(\mathbf{e}_i, \mathbf{v}_i) + \alpha_{i-1} F(\mathbf{e}_{i-1}, \mathbf{v}_i) + \cdots + \alpha_1 F(\mathbf{e}_1, \mathbf{v}_i)$$
$$= \alpha_i F(\mathbf{e}_i, \mathbf{v}_i) + 0 + \cdots + 0$$

and, by a calculation identical to (4),

$$F(\mathbf{e}_i, \mathbf{v}_i) = \begin{vmatrix} a_{11} & a_{12} & \cdots & a_{1\,i} \\ a_{21} & a_{22} & \cdots & a_{2\,i} \\ \vdots & \vdots & & \vdots \\ a_{i\,1} & a_{i\,2} & \cdots & a_{i\,i} \end{vmatrix}$$

which gives (3). $\qquad \square$

This argument gives an alternative to the Gram–Schmidt algorithm.

Theorem 6.9 *If F is an inner product on the finite dimensional real vector space V, V has basis $\mathbf{e}_1, \ldots, \mathbf{e}_n$, and $a_{ij} = F(\mathbf{e}_i, \mathbf{e}_j)$, then the vectors*

$$\mathbf{v}_i = \begin{vmatrix} a_{11} & a_{12} & \cdots & a_{1i} \\ a_{21} & a_{22} & \cdots & a_{2i} \\ \vdots & \vdots & & \vdots \\ a_{i-1\,1} & a_{i-1\,2} & \cdots & a_{i-1\,i} \\ \mathbf{e}_1 & \mathbf{e}_2 & \cdots & \mathbf{e}_i \end{vmatrix}$$

form an orthogonal basis of V.

Proof Since the $\mathbf{v}_i$ are orthogonal, it suffices to show that they are all nonzero, and to do this it suffices by (3) to show that each determinant

$$D_i = \begin{vmatrix} a_{11} & a_{12} & \cdots & a_{1i} \\ a_{21} & a_{22} & \cdots & a_{2i} \\ \vdots & \vdots & & \vdots \\ a_{i1} & a_{i2} & \cdots & a_{ii} \end{vmatrix}$$

is positive. For $i = 0$, we conventionally define $D_0 = 1$. Now, we use induction on i to show that $D_i > 0$.

Assume inductively that $i > 0$ and $D_{i-1} > 0$, and note that by (2)

$$\mathbf{v}_i = D_{i-1}\mathbf{e}_i + \gamma_{i-1}\mathbf{e}_{i-1} + \cdots + \gamma_1\mathbf{e}_1$$

for some scalars γ_k. If $\mathbf{v}_i = 0$ then

$$0 = D_{i-1}\mathbf{e}_i + \gamma_{i-1}\mathbf{e}_{i-1} + \cdots + \gamma_1\mathbf{e}_1$$

gives a linear dependence between $\mathbf{e}_1, \dots, \mathbf{e}_i$ with not all coefficients zero (since by assumption $D_{i-1} > 0$), which is impossible. So $\mathbf{v}_i \neq 0$ and by (3)

$$F(\mathbf{v}_i, \mathbf{v}_i) = D_{i-1}D_i > 0$$

since F is positive definite. Finally, $D_i = D_{i-1}D_i/D_{i-1}$, the quotient of two positive real numbers, so D_i is also positive, as required. Therefore the induction continues. □

Actually, the vectors $\mathbf{v}_i$ are not very different from the vectors in the Gram–Schmidt process. If F is an inner product on $\mathbb{R}^n$ and

$$\mathbf{w}_k = \mathbf{e}_k - \sum_{i=1}^{k-1} \frac{F(\mathbf{w}_i, \mathbf{e}_k)}{F(\mathbf{w}_i, \mathbf{w}_i)}\mathbf{w}_i, \tag{5}$$

then it turns out that

$$\mathbf{v}_i = D_{i-1}\mathbf{w}_i.$$

See Exercise 6.5.

Our analysis gives necessary and sufficient conditions for a symmetric bilinear form to be positive definite.

Theorem 6.10 (The leading minor test) *If F is a symmetric bilinear form on the finite dimensional real vector space $V = \mathbb{R}^n$, $\mathbf{e}_1, \dots, \mathbf{e}_n$ is the standard basis of V, and $a_{ij} = F(\mathbf{e}_i, \mathbf{e}_j)$, then F is positive definite (i.e. is an inner product) if and only if the determinants*

$$D_i = \begin{vmatrix} a_{11} & a_{12} & \cdots & a_{1i} \\ a_{21} & a_{22} & \cdots & a_{2i} \\ \vdots & \vdots & & \vdots \\ a_{i1} & a_{i-1\,2} & \cdots & a_{ii} \end{vmatrix}$$

are all positive.

Proof If F is positive definite, then the proof of the previous theorem shows that each $D_i > 0$. On the other hand, if each $D_i > 0$, then the vectors $\mathbf{v}_i$ defined in the proposition are orthogonal and have $F(\mathbf{v}_i, \mathbf{v}_i) = D_{i-1} D_i > 0$ (where $D_0 = 1$). Therefore each $\mathbf{v}_i$ is nonzero and hence $\{\mathbf{v}_1, \mathbf{v}_2, \ldots, \mathbf{v}_n\}$ is an orthogonal basis of V. This suffices, by Lemma 6.3. $\qquad\qquad\square$

The determinants D_i here are called the *leading minors* of the matrix $\mathbf{A}$. In calculations, the Gram–Schmidt vectors are usually easier to compute than the leading minors D_i or the vectors $\mathbf{v}_i$ defined here using determinants, but for some matrices $\mathbf{A}$ it may be possible to spot values for the leading minors rather quickly.

Example 6.11 We can apply the leading minor test to the matrix

$$\mathbf{A} = \begin{pmatrix} 2 & 1 & -1 \\ 1 & 2 & 0 \\ -1 & 0 & 2/3 \end{pmatrix}$$

of Example 6.1; $\mathbf{A}$ does not define a positive definite form, since it has determinant $2 \cdot (4/3) - 1 \cdot (2/3) - 1 \cdot 2 = 0$. On the other hand,

$$\mathbf{B} = \begin{pmatrix} 2 & 1 & -1 \\ 1 & 2 & 0 \\ -1 & 0 & 1 \end{pmatrix}$$

has leading minors 2, 3, and 1 as you can check, so does define a positive definite form.

Theorem 6.12 *If F is a symmetric bilinear form on the finite dimensional real vector space $V = \mathbb{R}^n$, V has basis $\mathbf{e}_1, \ldots, \mathbf{e}_n$, and $a_{ij} = F(\mathbf{e}_i, \mathbf{e}_j)$, then F is negative definite if and only if the signs of the determinants*

$$D_i = \begin{vmatrix} a_{11} & a_{12} & \cdots & a_{1i} \\ a_{21} & a_{22} & \cdots & a_{2i} \\ \vdots & \vdots & & \vdots \\ a_{i1} & a_{i-1\,2} & \cdots & a_{ii} \end{vmatrix}$$

alternate, with $D_i < 0$ for odd i and $D_i > 0$ for even i.

Proof If F is negative definite, then for the vectors $\mathbf{v}_i$ in Proposition 6.8 we have $F(v_i, v_i) = D_{i-1} D_i$. By convention, $D_0 = 1 > 0$ and $\mathbf{v}_1 = \mathbf{e}_1 \neq 0$ so $D_1 = a_{11} = F(\mathbf{e}_1, \mathbf{e}_1) < 0$. The proof of Theorem 6.9 can now be modified to show that the signs of the D_i alternate.

Conversely, if the signs of the D_i alternate, then the vectors $\mathbf{v}_i$ defined in the proposition are orthogonal and have $F(\mathbf{v}_i, \mathbf{v}_i) = D_{i-1} D_i < 0$ so F is negative definite, by Lemma 6.3. $\qquad\qquad\square$

Exercises

Exercise 6.1 Which of the following real symmetric matrices $\mathbf{A}$ are positive definite?

(a) $\begin{pmatrix} 2 & -1 & 5 \\ -1 & 1 & -5 \\ 5 & -5 & 6 \end{pmatrix}$
(b) $\begin{pmatrix} 1 & 2 \\ 2 & 1 \end{pmatrix}$

(c) $\begin{pmatrix} 1 & 1 \\ 1 & 1 \end{pmatrix}$
(d) $\begin{pmatrix} 1 & 1 & 1 \\ 1 & 2 & 2 \\ 1 & 2 & 3 \end{pmatrix}$.

Use the Gram–Schmidt process, the leading minor test, and/or row operations. If you find your matrix $\mathbf{A}$ is not positive definite, write down a nonzero vector $\mathbf{x}$ such that $\mathbf{x}^T \mathbf{A} \mathbf{x} \leqslant 0$.

Exercise 6.2 A symmetric bilinear form f on a real vector space V is called *positive semi-definite* if $f(v, v) \geqslant 0$ for all $v \in V$. Similarly, it is called *negative semi-definite* if $f(v, v) \leqslant 0$ for all $v \in V$. Show that if there is a basis of V with respect to which the matrix $\mathbf{A}$ of f is diagonal, then

(a) if all diagonal entries of $\mathbf{A}$ are nonnegative then f is positive semi-definite;
(b) if all diagonal entries of $\mathbf{A}$ are nonpositive then f is negative semi-definite.

Exercise 6.3 For each of the following real symmetric matrices $\mathbf{A}$, determine whether the corresponding symmetric bilinear form is positive definite, negative definite, positive semi-definite, negative semi-definite, or none of these.

(a) $\begin{pmatrix} -6 & 0 & 4 \\ 0 & -3 & -7 \\ 4 & -7 & -19 \end{pmatrix}$
(b) $\begin{pmatrix} 5 & -7 & 3 \\ -7 & 12 & 0 \\ 3 & 0 & 10 \end{pmatrix}$

(c) $\begin{pmatrix} 9 & -1 & -5 \\ -1 & 7 & 1 \\ -5 & 1 & 2 \end{pmatrix}$
(d) $\begin{pmatrix} 1 & -1 & -3 \\ -1 & 3 & 7 \\ -3 & 7 & 17 \end{pmatrix}$.

Exercise 6.4 For each of the following complex conjugate-symmetric matrices $\mathbf{A}$, determine whether the corresponding sesquilinear form is positive definite, negative definite, positive semi-definite, negative semi-definite, or none of these. (The definitions of positive and negative semi-definite are exactly the same as for real symmetric matrices above.)

(a) $\begin{pmatrix} 3 & -i \\ i & 6 \end{pmatrix}$
(b) $\begin{pmatrix} -1 & 2 - 3i \\ 2 + 3i & 4 \end{pmatrix}$
(c) $\begin{pmatrix} -9 & 1 & 4 + i \\ 1 & -15 & 3i \\ 4 - i & -3i & -4 \end{pmatrix}$.

Exercise 6.5 Suppose $\mathbf{v}$ and $\mathbf{w}$ are vectors in $\mathbb{R}^n$, F is an inner product on $\mathbb{R}^n$ given by matrix $\mathbf{A} = (a_{ij})$ with leading minors D_i, and $F(\mathbf{w}, \mathbf{e}_i) = F(\mathbf{v}, \mathbf{e}_i) = 0$

for all $i \leqslant k$, where $k < n$; also, suppose $\mathbf{v} \in \mathrm{span}(\mathbf{e}_1, \dots, \mathbf{e}_k, \mathbf{w})$. Show that $\mathbf{v} = \lambda \mathbf{w}$ for some scalar λ. Hence deduce for

$$\mathbf{w}_k = \mathbf{e}_k - \sum_{i=1}^{k-1} \frac{F(\mathbf{w}_i, \mathbf{e}_k)}{F(\mathbf{w}_i, \mathbf{w}_i)} \mathbf{w}_i$$

and

$$\mathbf{v}_i = \begin{vmatrix} a_{11} & a_{12} & \dots & a_{1i} \\ a_{21} & a_{22} & \dots & a_{2i} \\ \vdots & \vdots & & \vdots \\ a_{i-1\,1} & a_{i-1\,2} & \dots & a_{i-1\,i} \\ \mathbf{e}_1 & \mathbf{e}_2 & \dots & \mathbf{e}_i \end{vmatrix}$$

that $\mathbf{v}_i = D_{i-1}\mathbf{w}_i$. [Hint: for the first part, consider an orthonormal basis of $\mathrm{span}(\mathbf{e}_1, \dots, \mathbf{e}_k)$.]

7

Quadratic forms and Sylvester's law of inertia

Given a symmetric bilinear form F on a finite dimensional real vector space V we attempt to find 'nice' bases where the matrix representation of the form is as simple as possible, in the same sort of way as we did in the Gram–Schmidt orthogonalization process for inner products.

The theorem which says that this is possible and which describes all the possible matrix representations one can get is called *Sylvester's law of inertia*. (Sylvester's law of inertia has nothing to do with inertia: the story goes that Sylvester said, 'If Isaac Newton can have a law of inertia, then so can I', and consequently gave this theorem the name it has been known by ever since.) This 'law' is exactly analogous to Theorem 2.33 which classified all finite dimensional vector spaces using dimension; it turns out that symmetric bilinear forms are classified by the dimension of the underlying vector space and two further numbers, called the *rank* and the *signature* of the form.

7.1 Quadratic forms

Given an inner product $\langle \,|\, \rangle$, we defined a norm $\|\ \|$ by $\|v\|^2 = \langle v|v \rangle$, so $\|v\|$ is the positive square root of $\langle v|v \rangle$. We can also reverse this process, in the sense that, given the norm, we can find the inner product that gave rise to it, as follows.

$$\|v + w\|^2 = \langle v + w|v + w \rangle = \langle v|v \rangle + 2\langle v|w \rangle + \langle w|w \rangle$$

so

$$\langle v|w \rangle = \tfrac{1}{2}(\|v + w\|^2 - \|v\|^2 - \|w\|^2).$$

In the more general context of a symmetric bilinear form F, we find that $F(v, v)$ can be negative, so it does not make sense to take its square root. Apart from this technicality, however, we can do the same thing to an arbitrary symmetric bilinear form as we did to an inner product.

Definition 7.1 *Given a symmetric bilinear form F on a real vector space V, we define a map $Q: V \to \mathbb{R}$ by $Q(v) = F(v, v)$; Q is called the **quadratic form** associated with the symmetric bilinear form F.*

Given Q we can again recover F, since

$$Q(v + w) = F(v + w, v + w)$$
$$= F(v, v) + F(v, w) + F(w, v) + F(w, w)$$
$$= Q(v) + 2F(v, w) + Q(w)$$

so

$$F(v, w) = \tfrac{1}{2}(Q(v + w) - Q(v) - Q(w)).$$

Thus we can pass freely from the symmetric bilinear form to the corresponding quadratic form and back again. We state this formally.

Lemma 7.2 *Let F be a symmetric bilinear form on a vector space V, and Q the associated quadratic form. Then for all $v, w \in V$,*

$$F(v, w) = \tfrac{1}{2}(F(v + w, v + w) - F(v, v) - F(w, w)).$$

(i.e. $F(v, w) = \tfrac{1}{2}(Q(v+w) - Q(v) - Q(w))$ where Q is the corresponding quadratic form.)

Warning. Lemma 7.2 only works for *symmetric* bilinear forms (why?). We will see analogues of this lemma for sesquilinear forms over $\mathbb{C}$ later on. (See Lemma 7.20.)

Example 7.3 If $V = \mathbb{R}^3$ and F is the standard inner product, defined by

$$F\big((x_1, x_2, x_3)^T, (y_1, y_2, y_3)^T\big) = x_1 y_1 + x_2 y_2 + x_3 y_3,$$

then

$$Q\big((x_1, x_2, x_3)^T\big) = F\big((x_1, x_2, x_3)^T, (x_1, x_2, x_3)^T\big) = x_1^2 + x_2^2 + x_3^2.$$

The name 'quadratic form' comes from the fact that Q is given by a homogeneous quadratic function of the coordinates. (Homogeneous means each term has the same degree—in this case degree 2, being x^2 or xy for some variables x and y.) Indeed, this could be taken as the *definition* of a quadratic form (see Proposition 7.6).

Example 7.4 Let $V = \mathbb{R}^3$ and define $Q: V \to \mathbb{R}$ by

$$Q\big((x_1, x_2, x_3)^T\big) = x_1^2 + 2x_1 x_2 - 2x_1 x_3 + 2x_2 x_3 - 3x_3^2.$$

Then

$$F(\mathbf{x}, \mathbf{y}) = x_1 y_1 + x_1 y_2 + x_2 y_1 - x_1 y_3 - x_3 y_1 + x_2 y_3 + x_3 y_2 - 3x_3 y_3,$$

where $\mathbf{x} = (x_1, x_2, x_3)^T$ and $\mathbf{y} = (y_1, y_2, y_3)^T$, and the two forms F and Q can be represented (with respect to the standard basis) by the matrix

$$\mathbf{B} = \begin{pmatrix} 1 & 1 & -1 \\ 1 & 0 & 1 \\ -1 & 1 & -3 \end{pmatrix}.$$

We can work out $F(\mathbf{v}, \mathbf{w})$ as $\mathbf{v}^T \mathbf{B} \mathbf{w}$ and $Q(\mathbf{v})$ as $\mathbf{v}^T \mathbf{B} \mathbf{v}$. For example, if $\mathbf{v} = (1, 2, -1)^T$ then

$$Q(\mathbf{v}) = \begin{pmatrix} 1 & 2 & -1 \end{pmatrix} \begin{pmatrix} 1 & 1 & -1 \\ 1 & 0 & 1 \\ -1 & 1 & -3 \end{pmatrix} \begin{pmatrix} 1 \\ 2 \\ -1 \end{pmatrix}$$

$$= \begin{pmatrix} 4 & 0 & 4 \end{pmatrix} \begin{pmatrix} 1 \\ 2 \\ -1 \end{pmatrix} = 0.$$

Proposition 7.5 *If Q is a quadratic form on a real vector space V, then for all $x \in V$ and $\lambda \in \mathbb{R}$ we have $Q(\lambda x) = \lambda^2 Q(x)$.*

Proof If F is the symmetric bilinear form associated to Q, then $Q(\lambda x) = F(\lambda x, \lambda x) = \lambda^2 F(x, x) = \lambda^2 Q(x)$. □

In fact, *every* homogeneous quadratic function of the coordinates $x_1, \dots, x_n$ in $\mathbb{R}^n$ is a quadratic form, and conversely *every* quadratic form on a finite dimensional space is given by a homogeneous quadratic function of the coordinates.

Proposition 7.6 *Let $V = \mathbb{R}^n$. Then every quadratic form on V is given by a homogeneous function of the coordinates, of degree 2. Conversely, every homogeneous function of degree 2 of the coordinates is a quadratic form.*

Proof As in the above example, we work out $Q(\mathbf{v})$, where $\mathbf{v} = (v_1, \dots, v_n)^T$, by using the formula

$$Q(\mathbf{v}) = \mathbf{v}^T \mathbf{B} \mathbf{v} = \sum_{i=1}^{n} \sum_{j=1}^{n} v_i b_{ij} v_j,$$

and we see that every term in the expansion of the right-hand side is quadratic in the v_i. Conversely, if $q(\mathbf{v})$ is a homogeneous quadratic function of the coordinates of $\mathbf{v}$, then we can write

$$q(\mathbf{v}) = q\big((v_1, \dots, v_n)^T\big) = \sum_{i=1}^{n} b_{ii} v_i^2 + \sum_{i<j} 2 b_{ij} v_i v_j$$

for some constants b_{ij}. By defining $b_{ij} = b_{ji}$ whenever $i > j$, we can rewrite this as

$$q(\mathbf{v}) = \sum_{i=1}^{n} b_{ii} v_i^2 + \sum_{i \neq j} b_{ij} v_i v_j = \sum_{i=1}^{n} \sum_{j=1}^{n} b_{ij} v_i v_j$$

which is, in matrix notation, again in the familiar form $\mathbf{v}^T \mathbf{B} \mathbf{v}$. □

7.2 Sylvester's law of inertia

We return now to considering symmetric bilinear forms, but it should be borne in mind that, because of the previous section, all the results we prove apply equally to the associated quadratic form.

There are two parts to Sylvester's law: the first part says that a 'nice' basis can be found, and the second says that the matrix representation of the form looks essentially the same whichever of these 'nice' bases you choose to take.

Example 7.7 In Example 6.1 we obtained vectors $\mathbf{w}_1, \mathbf{w}_2, \mathbf{w}_3$ using the Gram–Schmidt formula. In fact, $\mathbf{w}_1, \mathbf{w}_2, \mathbf{w}_3$ is a basis of $\mathbb{R}^3$, and with respect to this basis the form F has matrix

$$\begin{pmatrix} 2 & 0 & 0 \\ 0 & 3/2 & 0 \\ 0 & 0 & 0 \end{pmatrix}.$$

The basis can be normalized by setting $\mathbf{w}_1' = \mathbf{w}_1/\sqrt{2}$, $\mathbf{w}_2' = \mathbf{w}_2/\sqrt{3/2}$, and $\mathbf{w}_3' = \mathbf{w}_3$, so that with respect to this normalized basis, the matrix representation of F is

$$\begin{pmatrix} 1 & 0 & 0 \\ 0 & 1 & 0 \\ 0 & 0 & 0 \end{pmatrix}.$$

The following theorem shows that any symmetric bilinear form can be represented by a diagonal matrix (with respect to *some* basis) with 0, 1, or -1 on the diagonal.

Theorem 7.8 (First part of Sylvester's law of inertia) *Let V be an n dimensional real vector space, and let F be a symmetric bilinear form on V. Then there are nonnegative integers k and m and a basis $\{w_1, \ldots, w_n\}$ of V such that $F(w_i, w_j) = 0$ for all $i \neq j$, $F(w_i, w_i) = 1$ for $i \leqslant k$, $F(w_i, w_i) = -1$ for $k < i \leqslant k + m$, and $F(w_i, w_i) = 0$ for $i > k + m$.*

In other words, the matrix representation of F with respect to the ordered basis $w_1, \ldots, w_n$ is

$$\begin{pmatrix} 1 & 0 & & & \cdots & & & 0 \\ 0 & \ddots & & & & & & \\ & & 1 & & & & & \\ & & & -1 & & & & \\ \vdots & & & & \ddots & & & \vdots \\ & & & & & -1 & & \\ & & & & & & 0 & \\ & & & & & & & \ddots \\ 0 & & & \cdots & & & & 0 \end{pmatrix}$$

with k entries equal to '1' and m entries equal to '-1' on the diagonal, and all other entries 0. One or both of k, m may of course be zero.

Before we prove the theorem, we present some simple lemmas. The first says that if a quadratic form is identically zero, then so is the associated symmetric bilinear form.

Lemma 7.9 *Let F be a symmetric bilinear form on a real vector space V, and suppose $F(v, v) = 0$ for all $v \in V$. Then $F(v, w) = 0$ for all $v, w \in V$ (with v not necessarily equal to w).*

Proof From Lemma 7.2,

$$F(v, w) = \tfrac{1}{2}(F(v + w, v + w) - F(v, v) - F(w, w))$$

so if $F(v, v) = F(w, w) = F(v + w, v + w) = 0$ then $F(v, w) = 0$. □

The basic idea of the proof of Theorem 7.8 is to use the Gram–Schmidt method. However, as we have seen, for certain 'bad' choices of initial basis vectors $v_1, \ldots, v_n$ this process may not complete, since some w_k obtained by the formula could have $F(w_k, w_k) = 0$. The trick we will use is not to work with any particular initial basis $v_1, \ldots, v_n$, but to choose these vectors as we go along. To do this, we need two further lemmas, both being simple variations of results we have already seen for inner products, but here proved in more generality.

The following lemma generalizes the fact that, for an inner product, nonzero orthogonal vectors are linearly independent.

Lemma 7.10 *Let F be a bilinear form on a real vector space V and suppose that $w_1, w_2, \ldots, w_n$ are vectors from V which are orthogonal for F, i.e. $F(w_i, w_j) = 0$ for all $i \neq j$. For all scalars λ_i $(i = 1, \ldots, n)$, if*

$$\lambda_1 w_1 + \lambda_2 w_2 + \cdots + \lambda_n w_n = 0$$

then $\lambda_i = 0$ for all i such that $F(w_i, w_i) \neq 0$.

In particular, if $F(w_i, w_i) \neq 0$ for all i then $\{w_1, w_2, \ldots, w_n\}$ is linearly independent.

Proof This is just as in Proposition 5.3. If

$$\lambda_1 w_1 + \lambda_2 w_2 + \cdots + \lambda_n w_n = 0$$

then

$$
\begin{aligned}
0 &= F(w_i, 0) \\
&= F(w_i, \lambda_1 w_1 + \lambda_2 w_2 + \cdots + \lambda_n w_n) \\
&= \sum_{j=1}^{n} \lambda_j F(w_i, w_j) = \lambda_i F(w_i, w_i),
\end{aligned}
$$

since $F(w_j, w_i) = 0$ for $i \neq j$. But if $F(w_i, w_i) \neq 0$ we must have $\lambda_i = 0$. □

The next lemma is just another way of stating the basic idea behind the Gram–Schmidt formula.

Lemma 7.11 *Let F be a bilinear form on a real vector space V, and suppose $w_1, w_2, \ldots, w_k \in V$ is orthogonal for F, i.e. $F(w_i, w_j) = 0$ for all $i \neq j$, and that $F(w_i, w_i) \neq 0$ for all $i \leqslant k$. Then for all $v \in V$ there is $u \in V$ such that $F(w_i, u) = 0$ for all $i \leqslant k$, and v is a linear combination of vectors from $w_1, w_2, \ldots, w_k, u$. In other words, $V = \mathrm{span}(U \cup \{w_1, w_2, \ldots, w_k\})$ where*

$$U = \{u \in V : F(w_i, u) = 0 \text{ for all } i \leqslant k\}.$$

Proof   Define u by the Gram–Schmidt formula,

$$u = v - \sum_{i=1}^{k} \frac{F(w_i, v)}{F(w_i, w_i)} w_i.$$

This expression has a well-defined meaning since each $F(w_i, w_i) \neq 0$. But then

$$F(w_i, u) = F(w_i, v) - \frac{F(w_i, v)}{F(w_i, w_i)} F(w_i, w_i) = 0,$$

for each i, and

$$v = u + \sum_{i=1}^{k} \frac{F(w_i, v)}{F(w_i, w_i)} w_i.$$

So every $v \in V$ is a linear combination of the w_i and some $u \in U$.   □

Proof of Theorem 7.8   Our object first is to find a basis $w_1, \ldots, w_n$ which is orthogonal for F, i.e. one with respect to which the matrix of F is diagonal, and this is done by an inductive argument. We will normalize the basis we get later.

To start things off, choose (if possible) $w_1 \in V$ so that $F(w_1, w_1) \neq 0$. If this is not possible, then it means $F(w, w) = 0$ for all w; hence $F(u, v) = 0$ for all $u, v \in V$, by Lemma 7.9, so the matrix of F with respect to *any* basis of V will be the zero matrix, which is certainly of the form required.

Now suppose inductively that we have found a set $\{w_1, \ldots, w_k\} \subseteq V$, where $k < n$, such that $F(w_i, w_j) = 0$ for all $i \neq j$ and $F(w_i, w_i) \neq 0$ for all i; we show how to obtain w_{k+1}.

Define

$$U = \{u \in V : F(w_i, u) = 0 \text{ for all } i \leqslant k\}.$$

This is a subspace of V, since if $u, v \in U$ then $F(w_i, u) = F(w_i, v) = 0$ for all $i \leqslant k$, so $F(w_i, u + \lambda v) = 0$ for all $i \leqslant k$ and all scalars λ; hence $u + \lambda v \in U$.

Case 1. If there is $u \in U$ with $F(u, u) \neq 0$, let $w_{k+1} \in U$ be such a vector u, i.e. let $w_{k+1} \in U$ with $F(w_{k+1}, w_{k+1}) \neq 0$. Then as $w_{k+1} \in U$, $\{w_1, \ldots, w_k, w_{k+1}\}$ is orthogonal for F, and $F(w_{k+1}, w_{k+1}) \neq 0$ so our induction continues.

Case 2. If $F(u, u) = 0$ for all $u \in U$, we use the Gram–Schmidt idea to get a suitable basis for the whole of V. Lemma 7.11 shows that $V = \mathrm{span}(U \cup$

$\{w_1, \ldots, w_k\}$). Note too that since $F(w, w) = 0$ for all $w \in U$ and U is a subspace of V, then $F(u, v) = 0$ for all $u, v \in U$, by Lemma 7.9.

Now take a basis $w_{k+1}, \ldots, w_l$ of U. Since $V = \text{span}(U \cup \{w_1, \ldots, w_k\})$, we see that $V = \text{span}\{w_1, \ldots, w_k, w_{k+1}, \ldots, w_l\}$. We prove that this is a basis of V (so in particular $l = n$). If

$$\lambda_1 w_1 + \cdots + \lambda_k w_k + \lambda_{k+1} w_{k+1} + \cdots + \lambda_l w_l = 0$$

for some scalars λ_i, then $\lambda_i = 0$ for each $i \leqslant k$ by Lemma 7.10, so

$$\lambda_{k+1} w_{k+1} + \cdots + \lambda_l w_l = 0$$

and hence also $\lambda_i = 0$ for $i > k$ by the linear independence of $\{w_{k+1}, \ldots, w_l\}$, as required. Since $F(w_i, w_j) = 0$ for $i \leqslant k < j$ and $F(w_i, w_j) = 0$ for $i, j \geqslant 0$, it is clear that the matrix of F with respect to $\{w_1, w_2, \ldots, w_n\}$ is diagonal.

Normalization. The construction so far has gone by following 'case 1' above as far as possible, until there are no more vectors u available with $F(u, u) \neq 0$. At this point, 'case 2' completed the basis.

All that remains is to normalize the basis $w_1, \ldots, w_n$ that we have found. This is done by setting

$$w_i' = \begin{cases} \dfrac{1}{\sqrt{F(w_i, w_i)}} w_i & \text{if } F(w_i, w_i) > 0 \\[2mm] \dfrac{1}{\sqrt{-F(w_i, w_i)}} w_i & \text{if } F(w_i, w_i) < 0 \\[2mm] w_i & \text{if } F(w_i, w_i) = 0. \end{cases}$$

By reordering the basis $w_1', \ldots, w_n'$ so that the vectors v with $F(v, v) > 0$ come first, those with $F(v, v) < 0$ next and those with $F(v, v) = 0$ last, we get the required basis. $\square$

There will in general be many different bases with respect to which F has the form given in Theorem 7.8, but the integers k and m are always the same, regardless of the basis.

Theorem 7.12 (Second part of Sylvester's law of inertia) *Suppose F is a symmetric bilinear form on a real vector space V which has two diagonal matrix representations $\mathbf{A}, \mathbf{A}'$ as in Theorem 7.8, with respect to ordered bases $e_1, \ldots, e_n$ and $e_1', \ldots, e_n'$. If $\mathbf{A}$ has k positive diagonal entries and m negative diagonal entries, and $\mathbf{A}'$ has k' positive diagonal entries and m' negative diagonal entries, then $k = k'$ and $m = m'$.*

Proof Let U be the subspace of V of dimension k spanned by $e_1, \ldots, e_k$, and let W be the subspace of dimension $n - k'$ spanned by $e_{k'+1}', \ldots, e_n'$. Then for every vector $u = \sum_{i=1}^{k} u_i e_i$ in U, we have

$$F(u, u) = \sum_{i=1}^{k} u_i^2 \geqslant 0$$

and if $F(u,u) = 0$ then $u = 0$ since each $F(e_i, e_i) > 0$ and so F is positive definite on U, by Lemma 6.3. On the other hand, for every vector $w = \sum_{i=k'+1}^{n} w_i e'_i \in W$, we have

$$F(w,w) = - \sum_{i=k'+1}^{k'+m'} w_i^2 \leqslant 0.$$

Now if $k > k'$, the set

$$B = \{e_1, \ldots, e_k, e'_{k'+1}, \ldots, e'_n\}$$

is a set of size at least $n + 1$, and V has dimension n, so B is linearly dependent. Hence for some scalars $\lambda_1, \ldots, \lambda_k, \mu_{k'+1}, \ldots, \mu_n$ not all zero

$$\lambda_1 e_1 + \cdots + \lambda_k e_k + \mu_{k'+1} e'_{k'+1} + \cdots + \mu_n e'_n = 0.$$

Let

$$v = \lambda_1 e_1 + \cdots + \lambda_k e_k = -(\mu_{k'+1} e'_{k'+1} + \cdots + \mu_n e'_n).$$

Since $v \in \text{span}(e_1, \ldots, e_k) = U$, we have $F(v,v) \geqslant 0$, and similarly, as $v \in \text{span}(e'_{k'+1}, \ldots, e'_n) = W$, we have $F(v,v) \leqslant 0$. Hence $F(v,v) = 0$ and so $v = 0$ since F is positive definite on U. But then

$$0 = \lambda_1 e_1 + \cdots + \lambda_k e_k = \mu_{k'+1} e'_{k'+1} + \cdots + \mu_n e'_n$$

with not all the λ_i and μ_j zero, giving either a linear dependence between the e_i or a linear dependence between the e'_j, a contradiction. So $k \leqslant k'$.

The other cases are proved similarly: if $k < k'$ consider the subspaces spanned by $e'_1, \ldots, e'_{k'}$ and $e_{k+1}, \ldots, e_n$; if $m < m'$ consider $e_1, \ldots, e_k, e_{k+m+1}, \ldots, e_n$ and $e'_{k'+1}, \ldots, e'_{k'+m'}$; and if $m' < m$ consider $e'_1, \ldots, e'_{k'}, e'_{k'+m'+1}, \ldots, e'_n$ and $e_{k+1}, \ldots, e_{k+m}$. In each case, note first whether F is positive definite or negative definite on the spaces spanned by these vectors, using Lemma 6.3. □

This enables us to make the following definition without ambiguity.

Definition 7.13 *With the notation of Theorem 7.8, $k + m$ is the **rank** of F, and $k - m$ is the **signature**.*

Proposition 7.14 *With the same notation, F is positive definite (i.e. is an inner product) if and only if $k = n$ and $m = 0$. Similarly, F is negative definite if and only if $k = 0$ and $m = n$.*

7.3 Examples

Since quadratic forms are nothing but symmetric bilinear forms in disguise, and Sylvester's law of inertia (Section 7.2) classifies symmetric bilinear forms, Sylvester's law tells us something about quadratic forms too.

Example 7.15 We apply the method used to prove Sylvester's law of inertia to Example 7.4. Here we have

$$Q\big((x_1, x_2, x_3)^T\big) = x_1^2 + 2x_1 x_2 - 2x_1 x_3 + 2x_2 x_3 - 3x_3^2$$

and the corresponding symmetric bilinear form F given above. First, choose the vector $\mathbf{w}_1 = (1,0,0)^T$, which has $Q((1,0,0)^T) = 1$. Now let U_1 be the subspace $U_1 = \{\mathbf{u} : F(\mathbf{w}_1, \mathbf{u}) = 0\}$. As $F(\mathbf{w}_1, (y_1, y_2, y_3)^T) = y_1 + y_2 - y_3$ we have

$$U_1 = \{(y_1, y_2, y_3)^T : y_1 + y_2 - y_3 = 0\},$$

which is spanned by $(-1, 1, 0)^T$ and $(1, 0, 1)^T$. Next we need a vector $\mathbf{v} \in U_1$ with $Q(\mathbf{v}) \neq 0$, if there is such a vector. If we take $\mathbf{v} = (1, 0, 1)^T$ then we get $Q(\mathbf{v}) = 0$, so that is no good, but if we take $\mathbf{v} = (-1, 1, 0)^T$ then we get $Q(\mathbf{v}) = -1$, which is all right. So we take $\mathbf{w}_2 = (-1, 1, 0)^T$, and let U_2 be the space $U_2 = \{\mathbf{u} : F(\mathbf{w}_1, \mathbf{u}) = F(\mathbf{w}_2, \mathbf{u}) = 0\}$. Then

$$U_2 = \{(y_1, y_2, y_3)^T : y_1 + y_2 - y_3 = -y_2 + 2y_3 = 0\},$$

which is spanned by $(-1, 2, 1)^T$. Moreover, $Q((-1, 2, 1)^T) = 0$, so we can take $\mathbf{w}_3 = (-1, 2, 1)^T$.

To sum up, with respect to the basis

$$\mathbf{w}_1 = \begin{pmatrix} 1 \\ 0 \\ 0 \end{pmatrix}, \mathbf{w}_2 = \begin{pmatrix} -1 \\ 1 \\ 0 \end{pmatrix}, \mathbf{w}_3 = \begin{pmatrix} -1 \\ 2 \\ 1 \end{pmatrix}$$

the quadratic form Q has the following shape:

$$Q(\lambda_1 \mathbf{w}_1 + \lambda_2 \mathbf{w}_2 + \lambda_3 \mathbf{w}_3) = \lambda_1^2 - \lambda_2^2.$$

In particular, this quadratic form has rank 2 and signature 0.

In practice, there are two principal ways to find the rank and signature of a quadratic form. The first is to 'complete the square'. In the above example, the terms involving x_1 are

$$x_1^2 + 2x_1 x_2 - 2x_1 x_3 = (x_1 + (x_2 - x_3))^2 - (x_2 - x_3)^2,$$

and so $Q(x) = (x_1 + x_2 - x_3)^2 - x_2^2 + 4x_2 x_3 - 4x_3^2$.
Then we collect all remaining terms in x_2, to obtain

$$Q(x) = (x_1 + x_2 - x_3)^2 - (x_2 - 2x_3)^2 + 0x_3^2.$$

To find the corresponding basis for the space in this example, we need to find the three vectors $\mathbf{w}_1, \mathbf{w}_2, \mathbf{w}_3$ such that

$$(x_1, x_2, x_3)^T = (x_1 + x_2 - x_3)\mathbf{w}_1 + (x_2 - 2x_3)\mathbf{w}_2 + x_3 \mathbf{w}_3,$$

for all x_1, x_2, x_3. Putting $x_1 = 1, x_2 = x_3 = 0$ we obtain $(1, 0, 0)^T = \mathbf{w}_1$. Putting $x_2 = 1, x_1 = x_3 = 0$ we obtain $\mathbf{w}_1 + \mathbf{w}_2 = (0, 1, 0)^T$, so $\mathbf{w}_2 = (-1, 1, 0)^T$, and similarly, putting $x_1 = x_2 = 0, x_3 = 1$ we obtain $\mathbf{w}_3 = (-1, 2, 1)^T$.

The second method uses the matrices instead, but is essentially the same calculation. Recall that a quadratic form, and the associated symmetric bilinear form, can be represented by a symmetric matrix. Changing the basis of the underlying space has the effect of changing the matrix $\mathbf{A}$ to $\mathbf{P}^T\mathbf{A}\mathbf{P}$, where $\mathbf{P}$ is the base-change matrix. Now pre-multiplication by $\mathbf{P}^T$ corresponds to performing certain *row operations* on $\mathbf{A}$. Post-multiplication by $\mathbf{P}$ corresponds to performing *the same* operations on the *columns*.

We are trying to put $\mathbf{A}$ into diagonal form $\mathbf{P}^T\mathbf{A}\mathbf{P}$, and to do this, therefore, we must perform certain row operations to clear out each column in turn, simultaneously performing the same column operations to clear out the corresponding row.

Example 7.16 We apply row and column operations to the following matrix $\mathbf{A}$.

$$\mathbf{A} = \begin{pmatrix} 1 & -1 & 0 \\ -1 & 2 & 3 \\ 0 & 3 & -2 \end{pmatrix}$$

$$\xrightarrow{\rho_2 := \rho_2 + \rho_1} \begin{pmatrix} 1 & -1 & 0 \\ 0 & 1 & 3 \\ 0 & 3 & -2 \end{pmatrix} \xrightarrow{\kappa_2 := \kappa_2 + \kappa_1} \begin{pmatrix} 1 & 0 & 0 \\ 0 & 1 & 3 \\ 0 & 3 & -2 \end{pmatrix}$$

$$\xrightarrow{\rho_3 := \rho_3 - 3\rho_2} \begin{pmatrix} 1 & 0 & 0 \\ 0 & 1 & 3 \\ 0 & 0 & -11 \end{pmatrix} \xrightarrow{\kappa_3 := \kappa_3 - 3\kappa_2} \begin{pmatrix} 1 & 0 & 0 \\ 0 & 1 & 0 \\ 0 & 0 & -11 \end{pmatrix}$$

$$\xrightarrow{\rho_3 := \frac{1}{\sqrt{11}}\rho_3} \begin{pmatrix} 1 & 0 & 0 \\ 0 & 1 & 0 \\ 0 & 0 & -\sqrt{11} \end{pmatrix} \xrightarrow{\kappa_3 := \frac{1}{\sqrt{11}}\kappa_3} \begin{pmatrix} 1 & 0 & 0 \\ 0 & 1 & 0 \\ 0 & 0 & -1 \end{pmatrix}.$$

So $\mathbf{A}$ has rank 3 and signature 1. To find the corresponding matrix $\mathbf{P}^T$, we have to keep track of the elementary row operations in the order in which they were carried out. Here we have

$$\mathbf{P}^T = \begin{pmatrix} 1 & 0 & 0 \\ 0 & 1 & 0 \\ 0 & 0 & \sqrt{11} \end{pmatrix} \begin{pmatrix} 1 & 0 & 0 \\ 0 & 1 & 0 \\ 0 & -3 & 1 \end{pmatrix} \begin{pmatrix} 1 & 0 & 0 \\ 1 & 1 & 0 \\ 0 & 0 & 1 \end{pmatrix} = \begin{pmatrix} 1 & 0 & 0 \\ 1 & 1 & 0 \\ \frac{-3}{\sqrt{11}} & \frac{-3}{\sqrt{11}} & \frac{1}{\sqrt{11}} \end{pmatrix}$$

and we can check that

$$\mathbf{P}^T\mathbf{A}\mathbf{P} = \begin{pmatrix} 1 & 0 & 0 \\ 1 & 1 & 0 \\ \frac{-3}{\sqrt{11}} & \frac{-3}{\sqrt{11}} & \frac{1}{\sqrt{11}} \end{pmatrix} \begin{pmatrix} 1 & -1 & 0 \\ -1 & 2 & 3 \\ 0 & 3 & -2 \end{pmatrix} \begin{pmatrix} 1 & 1 & \frac{-3}{\sqrt{11}} \\ 0 & 1 & \frac{-3}{\sqrt{11}} \\ 0 & 0 & \frac{1}{\sqrt{11}} \end{pmatrix}$$

$$= \begin{pmatrix} 1 & 0 & 0 \\ 0 & 1 & 0 \\ 0 & 0 & -1 \end{pmatrix}.$$

This method works fine in most cases, but every so often you will come across a matrix which seems to resist diagonalizing.

Example 7.17 Let $\mathbf{A} = \begin{pmatrix} 0 & 1 \\ 1 & 0 \end{pmatrix}$. An obvious strategy to try is to swap rows 1 and 2, but this doesn't work:

$$\mathbf{A} = \begin{pmatrix} 0 & 1 \\ 1 & 0 \end{pmatrix} \xrightarrow{\text{swap}(\rho_1,\rho_2)} \begin{pmatrix} 1 & 0 \\ 0 & 1 \end{pmatrix} \xrightarrow{\text{swap}(\kappa_1,\kappa_2)} \begin{pmatrix} 0 & 1 \\ 1 & 0 \end{pmatrix}.$$

The problem with this matrix is that for the corresponding quadratic form and the usual basis $\mathbf{e}_1, \mathbf{e}_2$ or $\mathbb{R}^2$ we have $Q(\mathbf{e}_1) = Q(\mathbf{e}_2) = 0$, so the algorithm for Sylvester's law cannot start with either of these two vectors. What's more, the $\text{swap}(\rho_1, \rho_2)$ operation only swaps these two basis vectors around—it doesn't introduce anything new. So we need to find a different vector $\mathbf{v}$ with $Q(\mathbf{v}) \neq 0$. $\mathbf{e}_1 + \mathbf{e}_2$ is one such, so we start off with the operation $\rho_1 := \rho_1 + \rho_2$,

$$\mathbf{A} = \begin{pmatrix} 0 & 1 \\ 1 & 0 \end{pmatrix} \xrightarrow{\rho_1:=\rho_1+\rho_2} \begin{pmatrix} 1 & 1 \\ 1 & 0 \end{pmatrix} \xrightarrow{\kappa_1:=\kappa_1+\kappa_2} \begin{pmatrix} 2 & 1 \\ 1 & 0 \end{pmatrix}$$

$$\begin{pmatrix} 2 & 1 \\ 1 & 0 \end{pmatrix} \xrightarrow{\rho_2:=\rho_2-\frac{1}{2}\rho_1} \begin{pmatrix} 2 & 1 \\ 0 & -1/2 \end{pmatrix} \xrightarrow{\kappa_2:=\kappa_2-\frac{1}{2}\kappa_1} \begin{pmatrix} 2 & 0 \\ 0 & -1/2 \end{pmatrix}.$$

Note that in this case, as in many others, the 'obvious' method of first using row operations to get the matrix into upper triangular form and then doing the corresponding column operations *does not work*.

$$\mathbf{A} = \begin{pmatrix} 0 & 1 \\ 1 & 0 \end{pmatrix} \xrightarrow{\rho_1:=\rho_1+\rho_2} \begin{pmatrix} 1 & 1 \\ 1 & 0 \end{pmatrix} \xrightarrow{\rho_2:=\rho_2-\rho_1} \begin{pmatrix} 1 & 1 \\ 0 & -1 \end{pmatrix}$$

$$\begin{pmatrix} 1 & 1 \\ 0 & -1 \end{pmatrix} \xrightarrow{\kappa_1:=\kappa_1+\kappa_2} \begin{pmatrix} 2 & 1 \\ -1 & -1 \end{pmatrix} \xrightarrow{\kappa_2:=\kappa_2-\kappa_1} \begin{pmatrix} 2 & -1 \\ -1 & 0 \end{pmatrix}.$$

In general, you need to do the column operation corresponding to the last row operation *immediately* after that row operation, or else the method will not work.

Example 7.18 We now do the same example as in Example 7.17, but this time by 'completing the square'.

The quadratic form in question here is $Q(x, y) = 2xy$, and there is no obvious way of 'collecting the x terms'. So we 'change basis' to $(x + y), y$, writing $x = (x + y) - y$.

$$Q(x, y) = 2xy = 2\big((x + y) - y\big)y = -2y^2 + 2y(x + y).$$

We can now collect up the y terms as before,

$$-2y^2 + 2y(x + y) = -2\big(y - \tfrac{1}{2}(x + y)\big)^2 + \tfrac{1}{2}(x + y)^2$$

showing as before that the form has rank 2 and signature 0.

7.4 Applications to surfaces

Sylvester's law, or the leading minor test of the previous chapter, is often useful to determine the nature of stationary points of a function of several variables.

We assume that we have a sufficiently smooth function $F \colon \mathbb{R}^n \to \mathbb{R}$ which is twice differentiable in a neighbourhood of some $\mathbf{a} \in \mathbb{R}^n$. The graph of this function is thought of as a (possibly higher dimensional) surface in $\mathbb{R}^{n+1}$. For example, the function $F(x, y) = x^2 y^2 - x^2 - y^2 + 1$ has as a 'graph' the surface consisting of all points $(x, y, z)^T \in \mathbb{R}^3$ satisfying $z = x^2 y^2 - x^2 - y^2 + 1$.

Just as for a function of a single variable, we say that the function F has a *stationary point* at $(a_1, \dots, a_n)^T \in \mathbb{R}^n$ if all of the partial derivatives of F in the directions of the n coordinate axes vanish. For functions of two variables, these stationary points are the places when the tangent plane to the surface is actually horizontal, and for more variables you have to imagine the analogous thing in higher dimensions. In the case of $F(x, y) = x^2 y^2 - x^2 - y^2 + 1$, we can calculate $\partial F/\partial x = 2xy^2 - 2x$, and $\partial F/\partial x = 2x^2 y - 2y$. Solving $2xy^2 - 2x = 2x^2 y - 2y = 0$ we see that these partial derivatives vanish at the points $(0, 0)^T$, $(1, 1)^T$, $(1, -1)^T$, $(-1, 1)^T$, $(-1, -1)^T$, so there are five stationary points of F. The second derivatives of F, and Sylvester's law, can help us determine the nature of these stationary points.

In the general case of $F(x_1, x_2, \dots, x_n)$, we shall denote by $F_i(x_1, x_2, \dots, x_n)$ the first partial derivative $\partial F/\partial x_i$, and by $F_{ij}(x_1, x_2, \dots, x_n)$ the second partial derivative $\partial^2 F/\partial x_i \partial x_j$. Then the Taylor expansion of the function F about the point $\mathbf{a}$ up to quadratic terms is given by

$$F(\mathbf{a} + \mathbf{x}) \approx F(\mathbf{a}) + (F_1(\mathbf{a}), \dots, F_n(\mathbf{a})) \begin{pmatrix} x_1 \\ \vdots \\ x_n \end{pmatrix}$$

$$+ \frac{1}{2}(x_1, \dots, x_n) \begin{pmatrix} F_{11}(\mathbf{a}) & \dots & F_{1n}(\mathbf{a}) \\ \vdots & \ddots & \vdots \\ F_{n1}(\mathbf{a}) & \dots & F_{nn}(\mathbf{a}) \end{pmatrix} \begin{pmatrix} x_1 \\ \vdots \\ x_n \end{pmatrix} \tag{1}$$

for small values x_i. (If the reader hasn't seen this before, it is a good exercise to verify that the first- and second-order partial derivatives of the function on the right-hand side of (1) agree with those of $F(\mathbf{a} + \mathbf{x})$.)

At a stationary point $\mathbf{a}$ of F, $F_1(\mathbf{a}) = \dots = F_n(\mathbf{a}) = 0$ and the middle term of (1) vanishes, giving

$$F(\mathbf{a} + \mathbf{x}) - F(\mathbf{a}) \approx \frac{1}{2}(x_1, \dots, x_n) \begin{pmatrix} F_{11}(\mathbf{a}) & \dots & F_{1n}(\mathbf{a}) \\ \vdots & \ddots & \vdots \\ F_{n1}(\mathbf{a}) & \dots & F_{nn}(\mathbf{a}) \end{pmatrix} \begin{pmatrix} x_1 \\ \vdots \\ x_n \end{pmatrix}.$$

If the form

$$Q(\mathbf{x}) = (x_1, \dots, x_n) \begin{pmatrix} F_{11}(\mathbf{a}) & \dots & F_{1n}(\mathbf{a}) \\ \vdots & \ddots & \vdots \\ F_{n1}(\mathbf{a}) & \dots & F_{nn}(\mathbf{a}) \end{pmatrix} \begin{pmatrix} x_1 \\ \vdots \\ x_n \end{pmatrix}$$

is positive definite, then for all sufficiently small $\mathbf{x}$, third- and higher order terms in $\mathbf{x}$ in the Taylor expansion of F can be ignored, and

$$F(\mathbf{a} + \mathbf{x}) \approx F(\mathbf{a}) + \tfrac{1}{2}Q(\mathbf{x}) \geqslant F(\mathbf{a})$$

with inequality for all sufficiently small $\mathbf{x} \neq \mathbf{0}$. Thus $\mathbf{a}$ is a *local minimum point* of F. Similarly, if Q were negative definite, then this point $\mathbf{a}$ would be a *local maximum point* of F. If the diagonal form for Q given by Sylvester's law has both positive and negative values on the diagonal, then in some directions (corresponding to a basis vector with positive entry on the diagonal) the function F will increase as we go away from $\mathbf{a}$, and in other directions (corresponding to a negative entry on the diagonal) the function F will decrease, so this point $\mathbf{a}$ is neither a local maximum nor a local minimum—it is called a *saddle point*.

Finally, if the form Q has rank $k < n$ and signature $\pm k$, so its diagonal form consists of all nonnegative entries or all nonpositive entries with at least one zero entry on the diagonal, then the method cannot tell us whether the point is a local maximum, minimum, or saddle. In this case, it is necessary to look at higher derivatives in the directions corresponding to the zeros on the diagonal.

Example 7.19 We analyse the five stationary points $(0, 0)^T$, $(1, 1)^T$, $(1, -1)^T$, $(-1, 1)^T$, and $(-1, -1)^T$ of the function $F(x, y) = x^2 y^2 - x^2 - y^2 + 1$. The second partial derivatives of F are $\partial^2 F / \partial x^2 = 2y^2 - 2$, $\partial^2 F / \partial x \partial y = 4xy$, and $\partial^2 F / \partial y^2 = 2x^2 - 2$. So, if the stationary point is $(a, b)^T$, we must look at the quadratic form

$$Q(x, y)^T = (x, y) \begin{pmatrix} 2b^2 - 2 & 4ab \\ 4ab & 2a^2 - 2 \end{pmatrix} \begin{pmatrix} x \\ y \end{pmatrix}.$$

Substituting $a = \pm 1$, $b = \pm 1$ we get the matrices $\begin{pmatrix} 0 & 4 \\ 4 & 0 \end{pmatrix}$ (twice) and $\begin{pmatrix} 0 & -4 \\ -4 & 0 \end{pmatrix}$ (twice), and each of these have rank 2 and signature 0, so these four stationary points are saddles. On the other hand, substituting $a = b = 0$ gives the negative definite matrix $\begin{pmatrix} -2 & 0 \\ 0 & -2 \end{pmatrix}$, so the origin is a maximum point. Figure 7.1 shows this surface in more detail.

7.5   Sesquilinear and Hermitian forms

The proof of Sylvester's law given in Section 7.2 successfully classifies conjugate-symmetric sesquilinear forms over complex vector spaces too. In this section we describe the results, but do not give full details of the proofs. The vector space V here is a finite dimensional space over the complex numbers throughout.

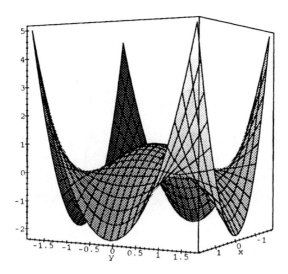

Fig. 7.1 The surface $z = x^2 y^2 - x^2 - y^2 + 1$

The analogue of the quadratic form obtained from a symmetric bilinear form is called the *Hermitian form H* associated with a conjugate-symmetric sesquilinear form F. This is defined in the same way: $H(v) = F(v, v)$. Notice first that $H(v) = F(v, v) = \overline{F(v,v)} = \overline{H(v)}$, since F is conjugate-symmetric, and therefore $H(v)$ is always *real*. Also, if λ is any scalar, then $H(\lambda v) = F(\lambda v, \lambda v) = \overline{\lambda}\lambda F(v, v) = |\lambda|^2 H(v)$, giving a result analogous to Proposition 7.5.

We can still recover F from H, but it is a bit more complicated than in the real case. There are various possible formulae. Here we present two of them.

Lemma 7.20 *If $F \colon V \times V \to \mathbb{C}$ is a conjugate-symmetric sesquilinear form, and H is the associated Hermitian form, then*

(a) $F(v, w) = \frac{1}{2}(H(v + w) + iH(v - iw) - (1 + i)(H(v) + H(w)))$;

(b) $F(v, w) = \frac{1}{4}(H(v + w) - H(v - w) + iH(v - iw) - iH(v + iw))$.

Proof Expand the right-hand sides, and observe that everything cancels out except the terms in $F(v, w)$. □

This gives a lemma analogous to Lemma 7.9.

Lemma 7.21 *Let F be a conjugate-symmetric sesquilinear form on a complex vector space V, and suppose $F(v, v) = 0$ for all $v \in V$. Then $F(v, w) = 0$ for all $v, w \in V$.*

The proof of Sylvester's law now goes through in the complex case exactly as it did in the real case. What's more, the notions of 'rank' and 'signature' make sense in this case, just as they did in the real case.

Theorem 7.22 *Let F be a conjugate-symmetric sesquilinear form on an n dimensional complex vector space V. Then there is a basis $w_1, \ldots, w_n$ of V and $k, m \leqslant n$ such that $F(w_i, w_j) = 0$ for all $i \neq j$, $F(w_i, w_i) = 1$ for $i \leqslant k$, $F(w_i, w_i) = -1$ for $k < i \leqslant k + m$, and $F(w_i, w_i) = 0$ for $i > k + m$. Moreover, k and m are uniquely determined from F (i.e. are independent of the choice of diagonalizing basis $w_1, \ldots, w_n$ as above).*

Example 7.23 We apply row operations to find the rank and signature of the conjugate-symmetric form

$$F(\mathbf{x}, \mathbf{y}) = \overline{\mathbf{x}}^T \begin{pmatrix} 1 & i & i \\ -i & 2 & 0 \\ -i & 0 & 0 \end{pmatrix} \mathbf{y}.$$

The method is as before: apply a row operation and then a corresponding column operation. However, in this case it is necessary to remember that the base-change formula for sesquilinear forms is $\mathbf{B} = \overline{\mathbf{P}}^T \mathbf{AP}$, so each row operation should be followed by the *conjugate* of the same operation applied to columns.

$$\mathbf{A} = \begin{pmatrix} 1 & i & i \\ -i & 2 & 0 \\ -i & 0 & 0 \end{pmatrix}$$

$$\xrightarrow{\rho_2 := \rho_2 + i\rho_1} \begin{pmatrix} 1 & i & i \\ 0 & 1 & -1 \\ -i & 0 & 0 \end{pmatrix} \xrightarrow{\kappa_2 := \kappa_2 - i\kappa_1} \begin{pmatrix} 1 & 0 & i \\ 0 & 1 & -1 \\ -i & -1 & 0 \end{pmatrix}$$

$$\xrightarrow{\rho_3 := \rho_3 + i\rho_1} \begin{pmatrix} 1 & 0 & i \\ 0 & 1 & -1 \\ 0 & -1 & -1 \end{pmatrix} \xrightarrow{\kappa_3 := \kappa_3 - i\kappa_1} \begin{pmatrix} 1 & 0 & 0 \\ 0 & 1 & -1 \\ 0 & -1 & -1 \end{pmatrix}$$

$$\xrightarrow{\rho_3 := \rho_3 + \rho_2} \begin{pmatrix} 1 & 0 & 0 \\ 0 & 1 & -1 \\ 0 & 0 & -2 \end{pmatrix} \xrightarrow{\kappa_3 := \kappa_3 + \kappa_2} \begin{pmatrix} 1 & 0 & 0 \\ 0 & 1 & 0 \\ 0 & 0 & -2 \end{pmatrix}$$

so the form has rank 3 and signature 1.

To determine the base-change matrix, we multiply together the row operation matrices we used,

$$\begin{pmatrix} 1 & 0 & 0 \\ 0 & 1 & 0 \\ 0 & 1 & 1 \end{pmatrix} \begin{pmatrix} 1 & 0 & 0 \\ 0 & 1 & 0 \\ i & 0 & 1 \end{pmatrix} \begin{pmatrix} 1 & 0 & 0 \\ i & 1 & 0 \\ 0 & 0 & 1 \end{pmatrix} = \begin{pmatrix} 1 & 0 & 0 \\ i & 1 & 0 \\ 2i & 1 & 1 \end{pmatrix},$$

and remember that this is $\overline{\mathbf{P}}^T$, the conjugate-transpose of the base-change matrix. So the base-change matrix here is

$$\mathbf{P} = \begin{pmatrix} 1 & -i & -2i \\ 0 & 1 & 1 \\ 0 & 0 & 1 \end{pmatrix}.$$

We conclude this section by linking up Sylvester's law with the leading minor test of the previous chapter. In the proof of the leading minor test, we started with an $n \times n$ matrix $\mathbf{A}$ which was real symmetric or complex and conjugate-symmetric and considered the form $F(\mathbf{v}, \mathbf{w}) = \overline{\mathbf{v}}^T \mathbf{A} \mathbf{w}$. We then defined the leading minors $D_0 = 1, D_1, \ldots, D_n$ of $\mathbf{A}$ by saying that D_i is the determinant of the top left $i \times i$ submatrix of $\mathbf{A}$, and found vectors $\mathbf{f}_1, \ldots, \mathbf{f}_n$ in the underlying vector space V ($\mathbb{R}^n$ or $\mathbb{C}^n$) which were orthogonal for the form F and satisfied $F(\mathbf{f}_i, \mathbf{f}_i) = D_{i-1} D_i$. Note also that, even in the case over the complex numbers, these leading minors D_i are all real.

By Lemma 7.10 (which works over $\mathbb{C}$ in just the same way), if all the D_i are nonzero then $\mathbf{f}_1, \ldots, \mathbf{f}_n$ is a basis of V. In this particular case, we can read off the rank and signature of the form directly.

Proposition 7.24 *Suppose $\mathbf{A}$ is an $n \times n$ symmetric matrix over $\mathbb{R}$ or $\mathbb{C}$ with leading minors $D_0 = 1, D_1, \ldots, D_n$ all nonzero. Then the form F defined by $F(\mathbf{v}, \mathbf{w}) = \overline{\mathbf{v}}^T \mathbf{A} \mathbf{w}$ on $\mathbb{R}^n$ or $\mathbb{C}^n$ has rank n and signature equal to the number of i with $0 \leqslant i < n$ such that D_i and D_{i+1} have the same sign.*

Proof Lemma 7.10 shows the $\mathbf{f}_i$ form a basis, and the signature of F is the number of $i < n$ such that D_i, D_{i+1} have the same sign, minus the number of $i < n$ such that D_i, D_{i+1} have different sign. $\qquad\square$

This proposition says nothing if one or more of the leading minors are zero. For example, the matrix $\mathbf{A} = \begin{pmatrix} 0 & 1 \\ 1 & 0 \end{pmatrix}$ has leading minors $D_1 = 0$ and $D_2 = 1$, so Proposition 7.24 doesn't apply. However, Example 7.17 showed that this matrix has rank 2 and signature 0.

Summary

To sum up, in this part of the book we have studied bilinear forms on both real and complex finite dimensional vector spaces and also sesquilinear forms on finite dimensional complex vector spaces, and went some way to classify them. We showed that they are all given by a square matrix $\mathbf{A}$ and are isomorphic to

$$F(\mathbf{x}, \mathbf{y}) = \mathbf{x}^T \mathbf{A} \mathbf{y}$$

on $\mathbb{R}^n$, in the case of bilinear forms on vector spaces over $\mathbb{R}$, or to

$$F(\mathbf{x}, \mathbf{y}) = \overline{\mathbf{x}}^T \mathbf{A} \mathbf{y}$$

on $\mathbb{C}^n$, in the case of sesquilinear forms on spaces over $\mathbb{C}$.

We then studied three important cases in more detail: firstly, inner products which we showed (using the Gram–Schmidt theorem) are always isomorphic to the usual inner product

$$F(\mathbf{x}, \mathbf{y}) = \overline{\mathbf{x}}^T \mathbf{I} \mathbf{y}$$

on $\mathbb{R}^n$ or $\mathbb{C}^n$; next *symmetric* bilinear forms, which were not all isomorphic, but (by Sylvester's law) could always be diagonalized and could be classified by the dimension of the vector space and the *rank* and *signature* of the form. Sylvester's law can be applied to classify quadratic forms too, because of the close connections between quadratic forms and symmetric bilinear forms in Section 7.1. These results were applied to surface sketching in three or more dimensions in Section 7.4. Finally, we discussed conjugate-symmetric sesquilinear forms and Hermitian forms, and showed that Sylvester's law applies to these also.

In the next part of this book, we shall go on to look at another important application of matrices, that of matrices as representing *linear transformations*, and try to investigate the structure of such linear transformations in detail using matrix methods. There will be many applications too, such as to solving simultaneous differential equations.

Exercises

Exercise 7.1 Which of the following are quadratic forms on $\mathbb{R}^n$ for some n? Write down the matrix representation (with respect to the usual basis) of the corresponding symmetric bilinear forms.

(a) $Q\left((x, y, z)^T\right) = x^2 + 2y^2 + 3z^2 + xyz$

(b) $Q\left((x, y, z)^T\right) = x^2 + xy$

(c) $Q\left((x, y, z)^T\right) = (x - y)^2 + (x - z)^2 + (y - z)^2$.

Exercise 7.2 (a) Let F be the symmetric bilinear form on $\mathbb{R}^3$ defined by

$$F(v, w) = 2v_1 w_1 + v_1 w_2 + w_1 v_2 + 3v_2 w_2 + 2v_1 w_3 + 2w_1 v_3 + 5v_3 w_3$$

where $v = (v_1, v_2, v_3)^T$ and $w = (w_1, w_2, w_3)^T$. By completing the square, or otherwise, find a basis with respect to which the matrix of F is diagonal, and give this matrix.

(b) Do the same for the form

$$F(v, w) = 2v_1 w_1 + 2v_1 w_2 + 2v_2 w_1 - 3v_2 w_2 + 2v_1 w_3 + 2v_3 w_1 + v_3 w_3.$$

Exercise 7.3 Find a basis of $\mathbb{R}^4$ for which the representation of

$$F(\mathbf{x}, \mathbf{y}) = \mathbf{x}^T \begin{pmatrix} 1 & 1 & 2 & 0 \\ 1 & 0 & 1 & -1 \\ 2 & 1 & 3 & -1 \\ 0 & -1 & -1 & -1 \end{pmatrix} \mathbf{y}$$

is diagonal, and write the matrix of F with respect to your basis. (Use row and column operations.)

Exercise 7.4 Find the rank and signature of each of the following quadratic forms. Use both the methods (i) completing the square, and (ii) row and column operations, and verify that they give the same answers.

(a) $Q\left((a, b, c, d)^T\right) = 2a^2 - 6ab + 2ac - 8bc - c^2 + 4bd - 2d^2$

(b) $Q \begin{pmatrix} x \\ y \\ z \end{pmatrix} = (x \quad y \quad z) \begin{pmatrix} 3 & 2 & 4 \\ 2 & 0 & 2 \\ 4 & 2 & 3 \end{pmatrix} \begin{pmatrix} x \\ y \\ z \end{pmatrix}$

(c) $Q\left((x,y,z)^T\right) = 14x^2 + y^2 + 5z^2 - 10xy + 6yz - 54xz$

(d) $Q\left((x,y,z)^T\right) = x^2 + y^2 + z^2 - 2xy + 2xz.$

In each case, find a basis with respect to which the matrix of Q is diagonal.

Exercise 7.5 Check that the proof of Theorem 7.22 goes through exactly as in the real case.

Exercise 7.6 Determine the rank and signature of the following conjugate-symmetric matrices.

(a) $\begin{pmatrix} -1 & i & i \\ -i & -1 & -1 \\ -i & -1 & -1 \end{pmatrix}$ (b) $\begin{pmatrix} 2 & 2 & 1 \\ 2 & 3 & 1+2i \\ 1 & 1-2i & 2 \end{pmatrix}$ (c) $\begin{pmatrix} 0 & i & 0 \\ -i & -2 & 1 \\ 0 & 1 & 0 \end{pmatrix}.$

Exercise 7.7 For each of the following matrices $\mathbf{A}$, find an invertible matrix $\mathbf{P}$ so that $\overline{\mathbf{P}}^T \mathbf{A} \mathbf{P}$ is diagonal.

(a) $\begin{pmatrix} 1 & 1+i \\ 1-i & -1 \end{pmatrix}$ (b) $\begin{pmatrix} -7 & 1+3i \\ 1-3i & -1 \end{pmatrix}$

(c) $\begin{pmatrix} 1 & -i & 1+i \\ i & 5 & -1+3i \\ 1-i & -1-3i & 3 \end{pmatrix}$ (d) $\begin{pmatrix} 3 & 2 & 2+i & 2 \\ 2 & 6 & 1+i & 4 \\ 2-i & 1-i & 3 & 2 \\ 2 & 4 & 2 & 5 \end{pmatrix}.$

Exercise 7.8 Suppose $\mathbf{A}$ is an $n \times n$ complex conjugate-symmetric matrix, and $F(\mathbf{x},\mathbf{y}) = \overline{\mathbf{x}}^T \mathbf{A}\mathbf{y}$ for all $\mathbf{x},\mathbf{y} \in \mathbb{C}^n$. Show that if $\det \mathbf{A} \neq 0$ then F has rank n.

Exercise 7.9 Find all stationary points of the following functions and investigate their nature.

(a) $f(x,y) = x^2y^2 - 4x^2y + 3x^2 - y^2 + 4y$

(b) $f(x,y) = x^3 - x^2y - x^2 + y^2$

(c) $f(x,y) = x^3 + y^3 + 9(x^2 + y^2) + 12xy$

(d) $f(x,y) = x^3 - 15x^2 - 20y^2 + 5$

(e) $f(x,y) = x^2 + y^2 + (2/x) + (2/y)$

(f) $f(x,y) = x^3 + y^3 - 2(x^2 + y^2) + 3xy$

(g) $f(x,y) = y^2 \cos x - \cos x + y$.

The next few exercises discuss *skew-symmetric* or *alternating* bilinear forms F on a vector space V over the reals. These forms are defined to be bilinear forms F satisfying $F(v,v) = 0$ for all $v \in V$, or equivalently (see Exercise 4.8) bilinear forms F satisfying $F(x,y) = -F(y,x)$ for all $x,y \in V$.

Exercise 7.10 Let V be a finite dimensional real vector space with a skew-symmetric bilinear form F. Suppose $v_1, v_2 \in V$ satisfy $F(v_1,v_2) \neq 0$.

(a) Show that $\{v_1, v_2\}$ is linearly independent, and hence extends to a basis $\{v_1, v_2, \ldots, v_n\}$ of V.

(b) Show that the vectors $w_1, w_2, \ldots, w_n$ defined by

$$w_1 = v_1, \qquad w_2 = (1/F(v_1, v_2))v_2$$

and, for each $i \geqslant 3$,

$$w_i = v_i + F(w_2, v_i)w_1 - F(w_1, v_i)w_2.$$

form a basis of V.

(c) Compute the first two rows (and first two columns) of the matrix of F with respect to $w_1, w_2, \ldots, w_n$.

Exercise 7.11 Let F be a skew-symmetric bilinear form on a finite dimensional real vector space V. Show that there is a basis $v_1, \ldots, v_n$ of V such that, with respect to this basis, F has matrix

$$\begin{pmatrix} \mathbf{B}_1 & \mathbf{0} & \ldots & \mathbf{0} \\ \mathbf{0} & \mathbf{B}_2 & & \vdots \\ \vdots & & \ddots & \\ \mathbf{0} & \ldots & & \mathbf{B}_r \end{pmatrix}$$

where each submatrix $\mathbf{B}_i$ is either the 1×1 zero matrix (0) or is the 2×2 matrix

$$\begin{pmatrix} 0 & 1 \\ -1 & 0 \end{pmatrix}.$$

[Hint: construct a basis inductively using the previous exercise. There are two cases. Given nonzero v_1, if $F(v_1, w) = 0$ for all $w \in V$ then we get $\mathbf{B}_1 = (0)$; else there is v_2 with $F(v_1, v_2) \neq 0$ and the previous exercise applies.]

Part III

Linear transformations

8

Linear transformations

In this part of the book, we will study *linear transformations* in the same spirit as we studied inner products and quadratic forms. In this chapter we will see how a linear transformation can be represented by a matrix, with respect to a particular basis, and in later chapters we will discuss how to find the 'best' basis, so that the corresponding matrix is as simple as possible, with applications to the solution of many kinds of differential and difference equations, and to quadratic forms.

Our vector spaces will be over a field F. For this chapter, F can be any field whatsoever. If it helps, it is possible to think of F as either $\mathbb{R}$ or $\mathbb{C}$ without much being lost. In later chapters, we will need some further properties of the field, taking the form that certain polynomials have roots; these properties are always true of the field $\mathbb{C}$, so it is safe to think of F as $\mathbb{C}$ throughout the rest of the book.

8.1 Basics

We start with the definition of the basic objects of study in this part of the book.

Definition 8.1 *If V and W are two vector spaces over the same field F, then a **linear transformation** from V to W (also called a **linear map**, or **homomorphism**) is a map $f\colon V \to W$ satisfying the condition*

$$f(\lambda u + \mu v) = \lambda f(u) + \mu f(v) \tag{1}$$

*for all $u, v \in V$ and all scalars $\lambda, \mu \in F$. The space V is called the **domain** of f, and W is called the **codomain** of f.*

Thus a linear transformation is the same as an isomorphism of vector spaces (Definition 2.32), except that we drop the requirements that it be injective and surjective.

Lemma 8.2 *A linear transformation $f\colon V \to W$ satisfies*

(a) $f(0) = 0$,
(b) $f(\lambda u) = \lambda f(u)$,
(c) $f(-u) = -f(u)$,
(d) $f(u + v) = f(u) + f(v)$, *and*

(e) $f(\sum_{i=1}^{n} \lambda_i u_i) = \sum_{i=1}^{n} \lambda_i f(u_i)$,

for all $u, v \in V$ and all scalars λ, λ_i.

Proof For (a) use $f(0) = f(0u + 0v) = 0f(u) + 0f(v) = 0$ for all vectors $u, v \in V$. For (b), apply Definition 8.1 for $\mu = 0$, and for (c), apply part (b) with $\lambda = -1$. Part (d) is just Definition 8.1 where $\lambda = \mu = 1$, and (e) is by repeated applications of Definition 8.1. $\qquad\square$

Example 8.3 Let V be the real vector space $\mathbb{R}^3$ and let W be $\mathbb{R}^2$. We define linear transformations $f, g \colon V \to W$ by

$$f\begin{pmatrix} x \\ y \\ z \end{pmatrix} = \begin{pmatrix} 2x + y \\ y + z \end{pmatrix}$$

and

$$g\begin{pmatrix} x \\ y \\ z \end{pmatrix} = \begin{pmatrix} x \\ x \end{pmatrix}.$$

It is easy to check that Definition 8.1 is satisfied. For example,

$$f\left(\lambda \begin{pmatrix} x_1 \\ y_1 \\ z_1 \end{pmatrix} + \mu \begin{pmatrix} x_1 \\ y_1 \\ z_1 \end{pmatrix} \right) = f\begin{pmatrix} \lambda x_1 + \mu x_2 \\ \lambda x_1 + \mu x_2 \\ \lambda x_1 + \mu x_2 \end{pmatrix}$$

$$= \begin{pmatrix} 2\lambda x_1 + 2\mu x_2 + \lambda y_1 + \mu y_2 \\ \lambda y_1 + \mu y_2 + \lambda z_1 + \mu z_2 \end{pmatrix}$$

$$= \lambda \begin{pmatrix} 2x_1 + y_1 \\ y_1 + z_1 \end{pmatrix} + \mu \begin{pmatrix} 2x_2 + y_2 \\ y_2 + z_2 \end{pmatrix}$$

$$= \lambda f\begin{pmatrix} x_1 \\ y_1 \\ z_1 \end{pmatrix} + \mu f\begin{pmatrix} x_2 \\ y_2 \\ z_2 \end{pmatrix}.$$

Definition 8.4 *Given $f \colon V \to W$ as in Definition 8.1, the **image** (or **range**) of f is $\{f(v) : v \in V\}$ (written $f(V)$ or $\mathrm{im}(f)$). The **kernel** (or **nullspace**) of f is $\{v \in V : f(v) = 0\}$ (written $\ker(f)$).*

Proposition 8.5 *If $f \colon V \to W$ is a linear transformation, then $\mathrm{im}(f)$ is a subspace of W and $\ker(f)$ is a subspace of V.*

Proof We verify the conditions in Lemma 2.9.

If $v, w \in \ker f$ then $f(v + \lambda w) = f(v) + \lambda f(w) = 0 + \lambda 0 = 0$, so $v + \lambda w \in \ker f$. If $v, w \in \mathrm{im} f$ then $v = f(x)$ and $w = f(y)$ for some $x, y \in V$, so $f(x + \lambda y) = f(x) + \lambda f(y) = v + \lambda w \in \mathrm{im} f$. $\qquad\square$

Example 8.6 For f and g as in Example 8.3, f is surjective since given $(u, v)^T \in \mathbb{R}^2$ we have $f(u, -u, u + v)^T = (u, v)^T$, but g is not surjective as for example $(1, -1)^T$ is not equal to any $(x, x)^T$. The kernel of f is

$$\ker f = \{(x, y, z)^T : 2x + y = y + z = 0\}$$

which as you can check is spanned by the vector $(1, -2, 2)^T$. The kernel of g is spanned by $(0, 1, 0)^T, (0, 0, 1)^T$.

The following is a particularly useful criterion for testing if a linear transformation is injective.

Proposition 8.7 *A linear transformation* $f : V \to W$ *is injective if and only if* $\ker f$ *is the zero subspace* $\{0\}$ *of* V.

Proof If $v \neq 0$ is in $\ker f$ then $f(v) = f(0) = 0$ so f is not injective. Conversely, if $f(v) = f(w)$ for some $v \neq w$, then $v - w \neq 0$ and $f(v - w) = f(v) - f(w) = 0$ so $v - w \in \ker f$ and hence $\ker f \neq \{0\}$. $\square$

Proposition 8.5 allows us to make the following definition.

Definition 8.8 *The* **rank** *of* f *is the dimension of* $\operatorname{im}(f)$, *and the* **nullity** *of* f *is the dimension of* $\ker(f)$. *We write* $r(f)$ *for the rank of* f, *and* $n(f)$ *for its nullity.*

Example 8.9 In Examples 8.3 and 8.6 we have $n(f) = 1$ and $n(g) = 2$. We can calculate the ranks of f and g as follows: firstly, $r(f) = 2$ as f maps $\mathbb{R}^3$ to $\mathbb{R}^2$ surjectively; secondly, $r(g) = 1$ as $(1, 1)^T$ forms a one-element basis of $\operatorname{im} g$. In both cases we check that $r(f) + n(f) = 3 = \dim \mathbb{R}^3$ and $r(g) + n(g) = 3 = \dim \mathbb{R}^3$. This is no accident; in fact, these are just particular cases of the rank–nullity formula.

Theorem 8.10 (The rank–nullity formula) *If* $f : V \to W$ *is a linear map, then* $r(f) + n(f) = \dim(V)$.

Proof Choose a basis $\{v_1, \dots, v_k\}$ for $\ker(f)$, and extend this basis to a basis $\{v_1, \dots, v_m\}$ of V, so that $k = n(f)$ and $m = \dim(V)$. Now any vector v in V is of the form $v = \sum_{i=1}^{m} \lambda_i v_i$, so

$$f(v) = f\left(\sum_{i=1}^{m} \lambda_i v_i\right)$$

$$= \sum_{i=1}^{m} \lambda_i f(v_i), \qquad \text{since } f \text{ is linear,}$$

$$= \sum_{i=k+1}^{m} \lambda_i f(v_i)$$

since $f(v_1) = \cdots = f(v_k) = 0$. Thus the vectors $f(v_{k+1}), \dots, f(v_m)$ span the image of f. On the other hand if $\sum_{i=k+1}^{m} \lambda_i f(v_i) = 0$ then

$$f\left(\sum_{i=k+1}^{m} \lambda_i v_i\right) = 0,$$

so

$$\sum_{i=k+1}^{m} \lambda_i v_i \in \ker(f),$$

and hence there are scalars μ_j such that

$$\sum_{i=k+1}^{m} \lambda_i v_i = \sum_{j=1}^{k} \mu_j v_j.$$

But $\{v_1, \ldots, v_m\}$ is a linearly independent set and

$$\sum_{j=1}^{k} \mu_j v_j + \sum_{i=k+1}^{m} (-\lambda_i) v_i = 0$$

so $\lambda_i = 0$ for all i. Therefore $\{f(v_{k+1}), \ldots, f(v_m)\}$ is a linearly independent set, and hence a basis for $\mathrm{im}(f)$. So

$$r(f) = m - k = \dim(V) - n(f)$$

as required.   □

Rank and nullity give useful ways of determining if a transformation is injective or surjective.

Proposition 8.11 *If $f\colon V \to W$ is a linear transformation of finite dimensional vector spaces V, W (over the same field F) then (a) f is injective if and only if $n(f) = 0$, and (b) f is surjective if and only if $r(f) = \dim W$.*

Proof The map f is injective if and only if $\ker f = \{0\}$, if and only if $n(f) = 0$ by Proposition 8.7, and f is surjective if and only if $\dim(\mathrm{im}\, f) = \dim W$ by Corollary 2.28.   □

Corollary 8.12 *If $f\colon V \to W$ is a linear transformation of vector spaces V, W, then f is injective if and only if $r(f) = \dim V$ and f is surjective if and only if $n(f) = \dim V - \dim W$.*

Proof By the previous proposition and the rank–nullity formula.   □

Just as matrices provided a canonical family of examples of bilinear forms in the previous part of this book, they provide examples of linear transformations too.

Example 8.13 Let $V = \mathbb{R}^n$ and $W = \mathbb{R}^m$, and let $\mathbf{A}$ be an $m \times n$ matrix with real entries. Then we may define the linear transformation $f_{\mathbf{A}} : V \to W$ by

$$f_{\mathbf{A}}(\mathbf{v}) = \mathbf{A}\mathbf{v}.$$

(Note that if $\mathbf{v}$ is an $n \times 1$ matrix, $\mathbf{A}\mathbf{v}$ is an $m \times 1$ matrix, so the matrix multiplication in this definition makes sense.) The transformation $f_{\mathbf{A}}$ is linear by distributivity of matrix multiplication:

$$\mathbf{A}(\lambda \mathbf{u} + \mu \mathbf{v}) = \lambda \mathbf{A}\mathbf{u} + \mu \mathbf{A}\mathbf{v}.$$

We can do the same over any field F. If $\mathbf{A}$ is an $m \times n$ matrix with entries from F, we can define the linear transformation $f_{\mathbf{A}} : F^n \to F^m$ by $f_{\mathbf{A}}(\mathbf{v}) = \mathbf{A}\mathbf{v}$.

It is convenient to extend the terminology and notation for image, kernel, rank, and nullity to matrices in this way: if $\mathbf{A}$ is an $m \times n$ matrix over the field F, then im $\mathbf{A}$ denotes the image of $f_{\mathbf{A}}$,

$$\operatorname{im} \mathbf{A} = \{\mathbf{A}\mathbf{x} : \mathbf{x} \in F^n\},$$

ker $\mathbf{A}$ denotes the kernel

$$\ker \mathbf{A} = \{\mathbf{x} \in F^n : \mathbf{A}\mathbf{x} = \mathbf{0}\},$$

and $r(\mathbf{A})$, $n(\mathbf{A})$ denote the rank and nullity of $\mathbf{A}$, i.e. the dimensions of im $\mathbf{A}$ and ker $\mathbf{A}$ respectively.

Exercise 8.1 Show that, for an $m \times n$ matrix $\mathbf{A}$ over a field F, the subspace im $\mathbf{A}$ of F^m is the subspace spanned by the columns of $\mathbf{A}$.

It is interesting to note that the rank of an $m \times n$ matrix $\mathbf{A}$ over $\mathbb{R}$, as defined using echelon form in Section 1.5, is the same as the rank $r(\mathbf{A})$ of the linear transformation $f_{\mathbf{A}}$ just defined. To see this, consider a sequence of row operations converting $\mathbf{A}$ to echelon form $\mathbf{B}$. Since each such row operation corresponds to an $m \times m$ matrix $\mathbf{R}_i$ over $\mathbb{R}$, we have

$$\mathbf{R}_k \ldots \mathbf{R}_2 \mathbf{R}_1 \mathbf{A} = \mathbf{B}.$$

Now, each row operation matrix $\mathbf{R}_i$ is invertible, so

$$\mathbf{A} = \mathbf{R}_1^{-1} \mathbf{R}_2^{-1} \ldots \mathbf{R}_k^{-1} \mathbf{B}.$$

This means that the linear transformation $f_{\mathbf{A}}$ is equal to the composition of the transformations $f_{\mathbf{R}_1^{-1}}, f_{\mathbf{R}_2^{-1}}, \ldots, f_{\mathbf{R}_k^{-1}}, f_{\mathbf{B}}$. You can verify that each $\mathbf{R}_i^{-1}$ is actually a bijection $\mathbb{R}^m \to \mathbb{R}^m$. It follows that $\dim(\operatorname{im} f_{\mathbf{A}}) = \dim(\operatorname{im} f_{\mathbf{B}})$. But since $\mathbf{B}$ is in echelon form we can spot a basis of $\operatorname{im} f_{\mathbf{B}}$ immediately. If

$$\mathbf{B} = \begin{pmatrix} 0 & \cdots & 0 & b_{1n_1} & \cdots & & & \\ 0 & \cdots & 0 & 0 & \cdots & 0 & b_{2n_2} & \cdots \\ \vdots & & \vdots & \vdots & & \vdots & & \\ 0 & \cdots & 0 & 0 & \cdots & 0 & & \cdots \end{pmatrix}$$

with each $b_{i\,n_i} \neq 0$ then the subspace $\operatorname{im} f_{\mathbf{B}}$ which is spanned by the columns of $\mathbf{B}$ has basis given by

$$\begin{pmatrix} 1 \\ 0 \\ 0 \\ \vdots \\ 0 \end{pmatrix}, \begin{pmatrix} 0 \\ 1 \\ 0 \\ \vdots \\ 0 \end{pmatrix}, \begin{pmatrix} 0 \\ 0 \\ 1 \\ \vdots \\ 0 \end{pmatrix}, \dots.$$

(There is one of these vectors for each nonzero row in $\mathbf{B}$.) So $\operatorname{im} f_{\mathbf{B}}$ has dimension equal to the number of nonzero rows in $\mathbf{B}$, i.e. equal to rk $\mathbf{A}$ as defined in Section 1.5.

Example 8.14 Let $\mathbf{A}$ be the matrix

$$\begin{pmatrix} 1 & 1 & 2 & 3 \\ 1 & 2 & -1 & 1 \\ 1 & 3 & -4 & -1 \end{pmatrix}.$$

This is a 3×4 matrix so represents a linear transformation $\mathbb{R}^3 \to \mathbb{R}^4$.

We can perform row operations to get $\mathbf{A}$ into echelon form as follows.

$$\mathbf{A} \xrightarrow{\rho_2 := \rho_2 - \rho_1} \begin{pmatrix} 1 & 1 & 2 & 3 \\ 0 & 1 & -3 & -2 \\ 1 & 3 & -4 & -1 \end{pmatrix} \xrightarrow{\rho_3 := \rho_3 - \rho_1} \begin{pmatrix} 1 & 1 & 2 & 3 \\ 0 & 1 & -3 & -2 \\ 0 & 2 & -6 & -4 \end{pmatrix}$$

$$\xrightarrow{\rho_3 := \rho_3 - 2\rho_1} \begin{pmatrix} 1 & 1 & 2 & 3 \\ 0 & 1 & -3 & -2 \\ 0 & 0 & 0 & 0 \end{pmatrix} = \mathbf{B}.$$

So $\mathbf{A}$ has rank 2, and hence nullity $\dim(\mathbb{R}^4) - 2 = 2$.

By multiplying the elementary row operation matrices together we find that $\mathbf{R}\mathbf{A} = \mathbf{B}$, where $\mathbf{R}$ is the matrix

$$\begin{pmatrix} 1 & 0 & 0 \\ -1 & 1 & 0 \\ 1 & -2 & 1 \end{pmatrix}.$$

Now, $\operatorname{im} \mathbf{B}$ is spanned by the columns of $\mathbf{B}$, so a basis of this space is $(1,0,0)^T$ and $(0,1,0)^T$. It follows that $\operatorname{im} \mathbf{A}$ is spanned by $\mathbf{R}^{-1}(1,0,0)^T$ and $\mathbf{R}^{-1}(0,1,0)^T$, and on calculating we find that

$$\mathbf{R}^{-1} = \begin{pmatrix} 1 & 0 & 0 \\ 1 & 1 & 0 \\ 1 & 2 & 1 \end{pmatrix}$$

so a basis of im $\mathbf{A}$ is formed by $\mathbf{R}^{-1}(1,0,0)^T = (1,1,1)^T$ and $\mathbf{R}^{-1}(0,1,0)^T = (0,1,2)^T$.

To work out the kernels, note that

$$\ker \mathbf{B} = \left\{ \begin{pmatrix} x \\ y \\ z \\ w \end{pmatrix} : \begin{matrix} x + y + 2z + 3w = 0 \\ y - 3z - 2w = 0 \end{matrix} \right\}$$

$$= \left\{ \begin{pmatrix} x \\ y \\ z \\ w \end{pmatrix} : \begin{matrix} x = -2z - 3w - y \\ y = 3z + 2w \end{matrix} \right\}$$

$$= \left\{ \begin{pmatrix} -5z - 5w \\ 3z + 2w \\ z \\ w \end{pmatrix} : z, w \in \mathbb{R} \right\}$$

so is spanned by $(-5, 3, 1, 0)^T$ and $(-5, 2, 0, 1)^T$. Also, $\ker \mathbf{A} = \ker \mathbf{B}$ since for any vector $\mathbf{v}$,

$$\mathbf{Av} = \mathbf{0} \Leftrightarrow \mathbf{RAv} = \mathbf{0} \Leftrightarrow \mathbf{Bv} = \mathbf{0}$$

because $\mathbf{R}$ is invertible.

8.2 Arithmetic operations on linear transformations

You will be used to the idea of adding functions together, defining the sum of two functions $f + g$ by

$$(f + g)(x) = f(x) + g(x),$$

and of scaling functions, defining the function λf by

$$(\lambda f)(x) = \lambda(f(x)).$$

For this to make sense, we simply need the codomain of the functions to have an addition and a scalar multiplication defined on it, and this is true in the case of linear transformations since this codomain here is a vector space W. It should come as no surprise that we expect these operations on functions to satisfy the vector space axioms.

In fact, if we take a set S of functions from any set X to a vector space W, such that S is closed under addition and scalar multiplication, then S is automatically a vector space. We do not need any other conditions on the functions, or on the domain X of these functions. To prove this, we simply have to check the vector space axioms: for example, for any $f, g \in S$ and any scalar λ we have $\lambda(f + g) = (\lambda f) + (\lambda g)$, since

$$(\lambda(f + g))(x) = \lambda((f + g)(x))$$
$$= \lambda(f(x) + g(x))$$
$$= \lambda(f(x)) + \lambda(g(x))$$

since W is a vector space, and so

$$(\lambda(f + g))(x) = (\lambda f)(x) + (\lambda g)(x)$$
$$= ((\lambda f) + (\lambda g))(x).$$

The other axioms are equally easy to check.

Example 8.15 Let $\mathscr{D}[0, 1]$ be the set of all differentiable functions from $[0, 1]$ to $\mathbb{R}$. The codomain of these functions is $\mathbb{R}$, which may be regarded as a one-dimensional vector space over itself. Then $\mathscr{D}[0, 1]$ is a vector space with the operations defined above, since it is closed under addition and scalar multiplication.

Example 8.16 Let $\mathscr{L}(V, W)$ be the set of all linear transformations from a vector space V to a vector space W over the same field F. Then $\mathscr{L}(V, W)$ is itself a vector space, over the same field F. The zero element in $\mathscr{L}(V, W)$ is the map that takes every vector in V to zero in W.

The special case $\mathscr{L}(V, V)$ will be especially important for the study of linear transformations and applications. This space may be given an additional operation, namely *composition of functions*, by setting $f \circ g$ to be the function defined by

$$(f \circ g)(x) = f(g(x)).$$

Composition is a sort of 'multiplication' of linear transformations from V to V, but you should be warned that not all the laws you might expect for multiplication hold in this case. For example, the commutativity law $f \circ g = g \circ f$ is false in general, although the associativity law $(f \circ g) \circ h = f \circ (g \circ h)$ is always true.

For the record, we list all the properties of the arithmetic operations on linear transformations defined that hold for all $f, g, h \in \mathscr{L}(V, V)$ and all scalars λ, μ.

1. (Associativity.) $(f + g) + h = f + (g + h)$.
2. (Commutativity.) $f + g = g + f$.
3. (Zero.) $0 + f = f + 0 = f$, where 0 is the zero map defined by $0(x) = 0$.
4. $\lambda(\mu f) = (\lambda \mu) f$.
5. $(\lambda + \mu)f = \lambda f + \mu f$.
6. $0f = 0$.
7. $f + (-1)f = 0$.

8. $(f \circ g) \circ h = f \circ (g \circ h)$.
9. $I \circ f = f \circ I = f$, where I is the identity map defined by $I(x) = x$.
10. $f \circ (g + h) = (f \circ g) + (f \circ h)$.
11. $(g + h) \circ f = (g \circ f) + (h \circ f)$.

A diligent reader will stop to check these properties at this point, but for others, the main idea of the next section provides an alternative proof.

8.3 Representation by matrices

If $f: V \to W$ is a linear map, $v_1, \ldots, v_n$ is a basis for V, and $w_1, \ldots, w_m$ is a basis for W, then each $f(v_j)$ is in W, so can be written as a linear combination of the basis vectors. Thus we have

$$f(v_j) = \sum_{i=1}^{m} a_{ij} w_i$$

for some scalars a_{ij}. (These scalars are uniquely determined by Proposition 2.18.) The matrix $\mathbf{A} = (a_{ij})$ is called *the matrix of f with respect to the ordered bases* $v_1, \ldots, v_n$ *of V and* $w_1, \ldots, w_m$ *of W.*

Example 8.17 (a) The matrix of the zero map $0: V \to W$ taking any $v \in V$ to 0 is just the zero matrix $\mathbf{0}$. To see this, note that

$$\sum_{i=1}^{m} 0 w_i = 0 = f(v_j) = \sum_{i=1}^{m} a_{ij} w_i$$

so each $a_{ij} = 0$ since these coefficients are unique.

(b) The matrix of the identity map $I: V \to V$ with $I(v) = v$, *with respect to the same basis* $v_1, \ldots, v_n$ *in both domain and codomain*, is the identity matrix $\mathbf{I}$. Again, to see this we just need to note that

$$f(v_j) = v_j = \sum_{i=1}^{n} \delta_{ij} v_i$$

where $\delta_{ij} = 0$ if $i \neq j$ and $\delta_{ij} = 1$ if $i = j$.

Given the matrix $\mathbf{A}$ of a linear map f with respect to basis $v_1, \ldots, v_n$, we can work out $f(v)$ for any vector $v \in V$, as follows. First express v as a linear combination of the basis vectors of V, say $v = \sum_{j=1}^{n} \lambda_j v_j$, so that

$$f(v) = f\left(\sum_{j=1}^{n} \lambda_j v_j \right) = \sum_{j=1}^{n} \lambda_j f(v_j) \tag{2}$$

since f is linear, and therefore

$$f(v) = \sum_{j=1}^{n} \lambda_j \sum_{i=1}^{m} a_{ij} w_i.$$

We can view this matrix in terms of the coordinates $\lambda_1, \ldots, \lambda_n$ of the vector v with respect to the ordered basis $v_1, \ldots, v_n$ of V. (See Section 2.5 for a discussion of coordinates.) Equation (2) shows that the ith coordinate μ_i of $f(v)$ with respect to the ordered basis $w_1, \ldots, w_m$ of W is

$$\mu_i = \sum_{j=1}^{n} a_{ij}\lambda_j,$$

or, in other words, the column vector $\boldsymbol{\mu} = (\mu_1, \ldots, \mu_m)^T$ is related to the column vector $\boldsymbol{\lambda} = (\lambda_1, \ldots, \lambda_n)^T$ and the matrix $\mathbf{A} = (a_{ij})$ by matrix multiplication,

$$\boldsymbol{\mu} = \mathbf{A}\boldsymbol{\lambda}.$$

Example 8.18 Consider the linear transformations $f, g \colon \mathbb{R}^3 \to \mathbb{R}^2$ given by

$$f \colon \begin{pmatrix} x \\ y \\ z \end{pmatrix} \mapsto \begin{pmatrix} 2x + y \\ y + z \end{pmatrix}, \qquad g \colon \begin{pmatrix} x \\ y \\ z \end{pmatrix} \mapsto \begin{pmatrix} x \\ x \end{pmatrix}.$$

The matrices representing these with respect to the usual bases of $\mathbb{R}^3$ and $\mathbb{R}^2$ are

$$\begin{pmatrix} 2 & 1 & 0 \\ 0 & 1 & 1 \end{pmatrix} \qquad \text{and} \qquad \begin{pmatrix} 1 & 0 & 0 \\ 1 & 0 & 0 \end{pmatrix}$$

respectively. To see this note that

$$f\begin{pmatrix} 1 \\ 0 \\ 0 \end{pmatrix} = \begin{pmatrix} 2 \\ 0 \end{pmatrix} = 2\begin{pmatrix} 1 \\ 0 \end{pmatrix} + 0\begin{pmatrix} 0 \\ 1 \end{pmatrix},$$

giving the first column of the matrix for f, and so on.

Alternatively, note that $(x, y, z)^T$ is the coordinate form for this vector with respect to the usual basis, and

$$f\begin{pmatrix} x \\ y \\ z \end{pmatrix} = \begin{pmatrix} 2 & 1 & 0 \\ 0 & 1 & 1 \end{pmatrix}\begin{pmatrix} x \\ y \\ z \end{pmatrix} \qquad \text{and} \qquad g\begin{pmatrix} x \\ y \\ z \end{pmatrix} = \begin{pmatrix} 1 & 0 & 0 \\ 1 & 0 & 0 \end{pmatrix}\begin{pmatrix} x \\ y \\ z \end{pmatrix}.$$

The next important question is: what happens to the matrix $\mathbf{A}$ if we change either the basis of V or the basis of W? First, let us replace the basis $v_1, \ldots, v_n$ by $v'_1, \ldots, v'_n$, related by the base-change matrix $\mathbf{P} = (p_{ij})$, so that

$$v'_j = \sum_{i=1}^{n} p_{ij}v_i.$$

Then

$$f(v_j') = f\left(\sum_{i=1}^{n} p_{ij}v_i\right)$$

$$= \sum_{i=1}^{n} p_{ij}f(v_i)$$

$$= \sum_{i=1}^{n} p_{ij} \sum_{k=1}^{m} a_{ki}w_k$$

$$= \sum_{k=1}^{m}\left(\sum_{i=1}^{n} a_{ki}p_{ij}\right) w_k.$$

Now the sum in brackets in this last expression is simply the (k,j)th entry of the matrix $\mathbf{AP}$, so the matrix of f with respect to the pair of bases $v_1', \ldots, v_n'$ and $w_1, \ldots, w_m$ is $\mathbf{AP}$.

Similarly, we can replace the basis $w_1, \ldots, w_m$ by $w_1', \ldots, w_m'$, a new basis related to the old one by the base-change matrix $\mathbf{Q} = (q_{ij})$ so that

$$w_j' = \sum_{i=1}^{m} q_{ij}w_i.$$

Let $\mathbf{Q}^{-1} = (r_{ij})$, so that

$$w_i = \sum_{j=1}^{m} r_{ji}w_j'.$$

Then

$$f(v_j) = \sum_{i=1}^{m} a_{ij}w_i = \sum_{i=1}^{m} a_{ij} \sum_{k=1}^{m} r_{ki}w_k' = \sum_{k=1}^{m}\left(\sum_{i=1}^{m} r_{ki}a_{ij}\right) w_k',$$

so that the matrix of f with respect to $v_1, \ldots, v_n$ and $w_1', \ldots, w_m'$ is $\mathbf{Q}^{-1}\mathbf{A}$. Putting the two together we obtain the general form $\mathbf{Q}^{-1}\mathbf{AP}$, where $\mathbf{P}$ is the base-change matrix for V and $\mathbf{Q}$ is the base-change matrix for W.

We are most interested in the case when $V = W$, where it is reasonable to suppose that we will use the same basis for both the domain and codomain.

Thus if $f: V \to V$ is a linear map, and $v_1, \ldots, v_n$ is a basis for V, we can write

$$f(v_j) = \sum_{i=1}^{n} a_{ij}v_i,$$

and say that $\mathbf{A} = (a_{ij})$ is the matrix of f with respect to the ordered basis $v_1, \ldots, v_n$. In this case, if we change basis by $\mathbf{P} = (p_{ij})$, so that

$$v_j' = \sum_{i=1}^{n} p_{ij}v_i,$$

then we change *both* occurrences of the basis, so the matrix of f with respect to $v'_1, \ldots, v'_n$ is $\mathbf{P}^{-1}\mathbf{A}\mathbf{P}$. Let us state this formally for future reference.

Proposition 8.19 *Let V be a vector space with ordered bases B given by $v_1, \ldots, v_n$ and B' given by $v'_1, \ldots, v'_n$. Let $\mathbf{P}$ be the base-change matrix from B to B', so*

$$v'_j = \sum_{i=1}^{n} p_{ij} v_i.$$

Suppose that $f: V \to V$ is a linear transformation which has matrix $\mathbf{A}$ with respect to the basis B and matrix $\mathbf{B}$ with respect to the basis B'. Then $\mathbf{B} = \mathbf{P}^{-1}\mathbf{A}\mathbf{P}$.

Warning. A matrix in isolation can represent many different things. It does not make sense to talk about changing the basis unless you know what the matrix in question represents. Thus changing the basis for a quadratic form has the effect of changing the representing matrix from $\mathbf{A}$ to $\mathbf{P}^T\mathbf{A}\mathbf{P}$, whereas changing the basis for a linear transformation has the effect of changing $\mathbf{A}$ to $\mathbf{P}^{-1}\mathbf{A}\mathbf{P}$. These are only the same under the very special circumstances when $\mathbf{P}^{-1} = \mathbf{P}^T$, circumstances which are examined in more detail in Chapter 13.

Definition 8.20 *If $\mathbf{B} = \mathbf{P}^{-1}\mathbf{A}\mathbf{P}$, then $\mathbf{A}$ and $\mathbf{B}$ are called **similar** matrices. If $\mathbf{B} = \mathbf{P}^T\mathbf{A}\mathbf{P}$, for some invertible matrix $\mathbf{P}$, then $\mathbf{A}$ and $\mathbf{B}$ are called **congruent**.*

The final thing that we need to consider with matrix representation of linear transformations is how the matrix representation of the product λf of a scalar λ and a linear transformation f or the representation of the sum $f + g$ or composition $f \circ g$ of two linear transformations can be obtained from the matrix representations of f, g.

For scalar multiplication and addition, this is quite straightforward.

Proposition 8.21 *Let $f, g \in \mathscr{L}(V, W)$ where V, W are finite dimensional vector spaces, and let f, g have matrix representations $\mathbf{A}, \mathbf{B}$ respectively, with respect to some ordered bases A and B of V, W respectively. Then*

(a) *The matrix representation of λf with respect to A, B is the scalar product $\lambda \mathbf{A}$ of the matrix $\mathbf{A}$.*

(b) *The matrix representation of the sum $f + g$ with respect to A, B is the matrix sum $\mathbf{A} + \mathbf{B}$.*

Proof This is almost immediate from the definition. The matrix of f is given by

$$f(v_j) = \sum_{j=1}^{m} a_{ij} w_i$$

where A is the ordered basis $v_1, \ldots, v_n$ and B is $w_1, \ldots, w_m$. Then

$$(\lambda f)(v_j) = \lambda \sum_{j=1}^{m} a_{ij}w_i = \sum_{j=1}^{m}(\lambda a_{ij})w_i$$

so the matrix for λf is $\lambda \mathbf{A}$. Similarly if

$$g(v_j) = \sum_{j=1}^{m} b_{ij}w_i$$

then

$$(f + g)(v_j) = \sum_{j=1}^{m} a_{ij}w_i + \sum_{j=1}^{m} b_{ij}w_i = \sum_{j=1}^{m}(a_{ij} + b_{ij})w_i$$

and hence the matrix for $f + g$ is $\mathbf{A} + \mathbf{B}$. $\qquad\square$

Note also that the matrix for the zero transformation $0 \in \mathscr{L}(V, W)$ given by $0(x) = 0$ is the zero matrix of the appropriate size.

The other arithmetic operation on linear transformations is composition, and this corresponds to matrix multiplication. Indeed, this perhaps explains why matrix multiplication is defined in the way it is.

Proposition 8.22 (Composition of functions) *Let U, V, W be finite dimensional vector spaces, with ordered bases A, B, C respectively, and let $g: U \to V$ and $f: V \to W$ be linear maps. If f and g are represented by the matrices $\mathbf{A}$ and $\mathbf{B}$ with respect to A, B, C, then $f \circ g$ is represented with respect to the same bases by the matrix product $\mathbf{AB}$.*

Proof If A is the ordered basis $u_1, \dots, u_l$, B is $v_1, \dots, v_m$, and C is $w_1, \dots, w_n$, then

$$(f \circ g)(u_k) = f(g(u_k))$$

$$= f\left(\sum_{j=1}^{m} b_{jk}v_j\right)$$

$$= \sum_{j=1}^{m} b_{jk} f(v_j)$$

$$= \sum_{j=1}^{m} b_{jk} \sum_{i=1}^{n} a_{ij}w_i$$

$$= \sum_{i=1}^{n}\left(\sum_{j=1}^{m} a_{ij}b_{jk}\right) v_i.$$

But $\sum_{j=1}^{m} a_{ij}b_{jk}$ is precisely the (i, k)th element of the matrix product $\mathbf{AB}$, so the matrix representing $f \circ g$ is $\mathbf{AB}$. $\qquad\square$

In the special case of $\mathscr{L}(V,V)$, where V is an n-dimensional vector space over a field F, what we have proved is this. Given an ordered basis B of the vector space V, we have a map from $\mathscr{L}(V,V)$ to the set $M_{n,n}(F)$ of $n \times n$ matrices with entries from F, taking f to the matrix representing f. Every linear transformation corresponds to some matrix (which is uniquely determined once we have fixed B), and every matrix is the matrix of *some* transformation f. What's more, the zero transformation corresponds to the zero matrix, the identity transformation $\mathrm{I}(x) = x$ corresponds to the identity matrix $\mathbf{I}_n$, and the operations of scalar multiplication and addition of linear transformations correspond to scalar multiplication and addition of matrices. In other words, the vector spaces $\mathscr{L}(V,V)$ and $M_{n,n}(F)$ over F are isomorphic. But we can say a little bit more: these vector spaces have 'multiplication' operations—for $M_{n,n}(F)$ this is just multiplication of matrices, and for $\mathscr{L}(V,V)$ it is composition $f \circ g$ of functions, and Proposition 8.22 shows that $M_{n,n}(F)$ and $\mathscr{L}(V,V)$ are isomorphic with this operation too.

Exercises

Exercise 8.2 For each of the following $n \times m$ matrices $\mathbf{A}$, give bases in $\mathbb{R}^n$ and $\mathbb{R}^m$ for im $f_{\mathbf{A}}$ and ker $f_{\mathbf{A}}$, where $f_{\mathbf{A}} : \mathbb{R}^m \to \mathbb{R}^n$ is left-multiplication by $\mathbf{A}$.

(a) $\begin{pmatrix} 1 & 2 & 1 & 1 \\ 2 & 1 & 1 & 1 \end{pmatrix}$ (b) $\begin{pmatrix} 1 & 1 \\ 2 & 1 \\ -1 & 0 \\ 1 & 1 \end{pmatrix}$ (c) $\begin{pmatrix} 1 & 2 & -1 \\ -1 & 1 & 1 \\ -1 & 4 & 1 \end{pmatrix}$.

Exercise 8.3 For each of the following sets S of vectors in $\mathbb{R}^n$, find a linear map $\mathbb{R}^n \to \mathbb{R}^3$ whose kernel is spanned by S.

(a) $S = \{(1,1,0),(1,0,-1)\}$ in $\mathbb{R}^3$.
(b) $S = \{(1,-2,1,0)\}$ in $\mathbb{R}^4$.
(c) $S = \{(-2,1)\}$ in $\mathbb{R}^2$.
(d) $S = \{(-1,1,2,-1),(0,1,2,3)\}$ in $\mathbb{R}^4$.

Exercise 8.4 Let $f : \mathbb{R}^3 \to \mathbb{R}^3$ be the linear transformation defined by

$$f((x,y,z)^T) = (-x+y-z, x+2y, -y+3z)^T.$$

Calculate the images of the vectors

$$\mathbf{v}_1 = (0,1,1)$$
$$\mathbf{v}_2 = (1,-1,1)$$
$$\mathbf{v}_3 = (2,1,0).$$

Verify that $f(\mathbf{v}_1) = -\mathbf{v}_1 - 2\mathbf{v}_2 + \mathbf{v}_3$, and derive similar expressions for $f(\mathbf{v}_2)$ and $f(\mathbf{v}_3)$. Hence write down the matrix of f with respect to the basis $\mathbf{v}_1, \mathbf{v}_2, \mathbf{v}_3$ of $\mathbb{R}^3$.

Exercise 8.5 Do the same for the map g defined by

$$g((x, y, z)^T) = (y - 2z, -x + 2y - z, x + y + z)^T.$$

Exercise 8.6 Define $f \colon \mathbb{C}^3 \to \mathbb{C}^3$ by $f((a, b, c)^T) = (a + 3b - c, 2a - b, b + 2c)^T$. Write down the matrix of f with respect to

(a) the standard basis,
(b) the basis $(1, -1, 0)^T, (0, 1, 2)^T, (1, 0, 1)^T$,
(c) the basis $(0, 1, -1)^T, (1, 1, 1)^T, (2, 0, -1)^T$.

Verify the identity $\mathbf{B} = \mathbf{P}^{-1}\mathbf{A}\mathbf{P}$ for the base change from (a) to (b), and for the base change from (b) to (c).

Exercise 8.7 Let e_{ij} be the $n \times m$ matrix with 1 in the (i, j)th position and 0 elsewhere. Show that $\{e_{ij} : 1 \leqslant i \leqslant n, 1 \leqslant j \leqslant m\}$ is a basis for $M_{n,m}(\mathbb{R})$ and deduce the dimension of $M_{n,m}(\mathbb{R})$ as a real vector space.

Exercise 8.8 If V, W are vector spaces of dimensions m, n respectively, what is the dimension of the vector space $\mathscr{L}(V, W)$?

9

Polynomials

In the last chapter we saw how we can add and multiply linear transformations in $\mathscr{L}(V,V)$, i.e. how $\mathscr{L}(V,V)$ is given arithmetic operations analogous to the matrix operations of addition and multiplication. Much of the study of linear transformations $f\colon V \to V$ concerns their properties under these arithmetic operations, and before we can take this further we must review some material on polynomials and discuss how polynomials can be applied to linear transformations such as f.

9.1 Polynomials

We consider polynomials over a field F to be expressions of the form

$$a_n x^n + a_{n-1} x^{n-1} + \cdots + a_1 x + a_0,$$

where the a_r are from F (these are called the *coefficients* of the polynomial) and x is an 'indeterminate'. This polynomial will also be written as $\sum_{i=0}^{n} a_n x^n$. The *degree* of this polynomial is the largest r such that $a_r \neq 0$. The polynomial is *monic*, or is a *monomial* if $a_r = 1$ for this particular r, i.e. if the first term of the polynomial is x^r for some r. (If all the a_r are 0, we have the *zero polynomial*, whose degree may be defined as -1, $-\infty$, or not defined at all, according to taste.)

Polynomials can be added and multiplied in the familiar way. Moreover, you can divide one polynomial $f(x)$ by another nonzero polynomial $g(x)$ to get a *quotient* $q(x)$ and a *remainder* $r(x)$. The essential property of the remainder is that it is 'smaller' than $g(x)$ in the sense that it has smaller degree. Since this is the crux of what follows, we state and prove this formally.

Proposition 9.1 (Division algorithm for polynomials) *If $f(x)$ and $g(x)$ are two polynomials, and $g(x)$ is not the zero polynomial, then there exist polynomials $q(x)$ and $r(x)$ such that $f(x) = g(x)q(x) + r(x)$, and either $r(x)$ is the zero polynomial or else $\deg(r) < \deg(g)$.*

Proof The proof is by induction on the degree of $f(x)$. If $\deg(f) < \deg(g)$, we take $q(x)$ to be the zero polynomial, and $r(x) = f(x)$. Thus the induction starts.

If $\deg(f) \geqslant \deg(g)$, say $f(x) = \lambda x^n + h(x)$ and $g(x) = \mu x^m + k(x)$, where $\deg(h) < \deg(f) = n$, $\deg(k) < \deg(g) = m$, and $m \leqslant n$, then

$$f(x) - (\lambda/\mu)x^{n-m}g(x) = \lambda x^n + h(x) - (\lambda/\mu)x^{n-m}\mu x^m - (\lambda/\mu)x^{n-m}k(x)$$
$$= h(x) - (\lambda/\mu)x^{n-m}k(x),$$

which has degree less than n. Therefore, by induction, this polynomial can be written in the form $g(x)s(x) + r(x)$ for some polynomials $r(x)$ and $s(x)$ with either $\deg(r) < \deg(g)$ or $r(x)$ identically zero. It follows immediately that

$$f(x) = \big((\lambda/\mu)x^{n-m} + s(x)\big)g(x) + r(x),$$

as required. □

In practice, you find the quotient and remainder by long division of polynomials. At each stage you just divide the highest terms into each other, as we did above, and let the smaller terms look after themselves.

Example 9.2 Divide $2x^3 + 4x^2 + 9x + 7$ by $x^2 - 2x$.

$$
\begin{array}{r}
2x + 8 \\
x^2 - 2x \,\big|\, \overline{\,2x^3 + 4x^2 + 9x + 7} \\
2x^3 - 4x^2 \\
\hline
8x^2 + 9x + 7 \\
8x^2 - 16x \\
\hline
25x + 7
\end{array}
$$

so we get a quotient of $2x + 8$ and a remainder of $25x + 7$.

9.2 Evaluating polynomials

If $p(x) = \sum_{r=0}^{n} a_r x^r$, then you know what it means to evaluate $p(\lambda)$ (or to evaluate $p(x)$ at $x = \lambda$), when λ is in the same field as the coefficients a_r. You simply form the expression $\sum_{r=0}^{n} a_r \lambda^r$ and evaluate it in the field. Now if $\mathbf{A}$ is a *matrix*, what can $p(\mathbf{A})$ mean? Presumably we want $p(\mathbf{A}) = \sum_{r=0}^{n} a_r \mathbf{A}^r$; can we make sense of this? Well, $\mathbf{A}^r$ just means the product $\mathbf{A}\mathbf{A}\ldots\mathbf{A}$ of r copies of $\mathbf{A}$, and you can certainly add matrices. Multiplying a matrix by a scalar simply means multiplying all entries by that scalar. Finally, since x^0 is interpreted as 1, presumably $\mathbf{A}^0$ should be interpreted as the identity matrix of the appropriate size.

Example 9.3 Suppose $\mathbf{A} = \begin{pmatrix} 1 & 2 \\ -1 & 0 \end{pmatrix}$ and $p(x) = -x^2 + 2x + 3$. Then

$$p(\mathbf{A}) = -\mathbf{A}^2 + 2\mathbf{A} + 3\mathbf{I}_2$$
$$= -\begin{pmatrix} -1 & 2 \\ -1 & -2 \end{pmatrix} + 2\begin{pmatrix} 1 & 2 \\ -1 & 0 \end{pmatrix} + \begin{pmatrix} 3 & 0 \\ 0 & 3 \end{pmatrix}$$
$$= \begin{pmatrix} 6 & 2 \\ -1 & 5 \end{pmatrix}.$$

But there appears to be a problem with evaluations like this, which we illustrate with a simple example.

Given the polynomial $p(x) = x^2 + 3x + 2$, we know that $p(x) = (x+2)(x+1)$ which is to say that $p(x)$ equals the product of the polynomials $x + 2$ and $x + 1$. But $p(x) = (x+1)(x+2)$ and $p(x) = (x+1)^2 + (x+1)$ also. Now if $\mathbf{A}$ is an $n \times n$ square matrix, $p(\mathbf{A})$ is *defined* above to mean

$$\mathbf{A}^2 + 3\mathbf{A} + 2\mathbf{I}_n,$$

but what are the values of the following?

$$(\mathbf{A} + 2\mathbf{I}_n)(\mathbf{A} + \mathbf{I}_n)$$
$$(\mathbf{A} + \mathbf{I}_n)(\mathbf{A} + 2\mathbf{I}_n)$$
$$(\mathbf{A} + \mathbf{I}_n)^2 + (\mathbf{A} + \mathbf{I}_n).$$

Since matrix multiplication is not commutative, it is not at all obvious whether these are all equal.

Example 9.4 We evaluate the above expressions for the matrix

$$\mathbf{A} = \begin{pmatrix} 1 & 2 \\ -1 & 0 \end{pmatrix}.$$

First,

$$p(\mathbf{A}) = \mathbf{A}^2 + 3\mathbf{A} + 2\mathbf{I}_2 = \begin{pmatrix} -1 & 2 \\ -1 & -2 \end{pmatrix} + \begin{pmatrix} 3 & 6 \\ -3 & 0 \end{pmatrix} + \begin{pmatrix} 2 & 0 \\ 0 & 2 \end{pmatrix} = \begin{pmatrix} 4 & 8 \\ -4 & 0 \end{pmatrix}.$$

Then

$$(\mathbf{A} + 2\mathbf{I}_2)(\mathbf{A} + \mathbf{I}_2) = \begin{pmatrix} 3 & 2 \\ -1 & 2 \end{pmatrix} \begin{pmatrix} 2 & 2 \\ -1 & 1 \end{pmatrix} = \begin{pmatrix} 4 & 8 \\ -4 & 0 \end{pmatrix},$$

$$(\mathbf{A} + \mathbf{I}_2)(\mathbf{A} + 2\mathbf{I}_2) = \begin{pmatrix} 2 & 2 \\ -1 & 1 \end{pmatrix} \begin{pmatrix} 3 & 2 \\ -1 & 2 \end{pmatrix} = \begin{pmatrix} 4 & 8 \\ -4 & 0 \end{pmatrix},$$

$$(\mathbf{A} + \mathbf{I}_2)^2 + (\mathbf{A} + \mathbf{I}_2) = \begin{pmatrix} 2 & 2 \\ -1 & 1 \end{pmatrix}^2 + \begin{pmatrix} 2 & 2 \\ -1 & 1 \end{pmatrix} = \begin{pmatrix} 4 & 8 \\ -4 & 0 \end{pmatrix},$$

so they are all the same.

In fact, it is generally true for *polynomials* $p(x), q(x)$ *in a single free variable* x that whenever we have a polynomial identity $p(x) = q(x)$, such as $x^2 + 3x + 2 = (x+2)(x+1)$, and an $n \times n$ matrix $\mathbf{A}$, the matrix equation $p(\mathbf{A}) = q(\mathbf{A})$ is true. On the other hand, for polynomials in more than one variable the corresponding statement is definitely false, since the equality $xy = yx$ holds for polynomials of two variables, but it is not true that $\mathbf{AB} = \mathbf{BA}$ for all $n \times n$ matrices $\mathbf{A}, \mathbf{B}$.

The proof is rather technical, and relies on a precise technical definition of 'polynomial', which we have avoided giving. In fact, the details are not difficult, just rather unenlightening. It's rather more interesting to outline the basic idea, and this is as follows.

First, note that the key properties of addition and multiplication of polynomials are the associativity, commutativity, and distributivity laws. That is,

$$(p(x) + q(x)) + r(x) = q(x) + (p(x) + r(x))$$
$$p(x) + q(x) = q(x) + p(x)$$
$$(p(x)q(x))r(x) = q(x)(p(x)r(x))$$
$$p(x)q(x) = q(x)p(x)$$
$$p(x)(q(x) + r(x)) = p(x)q(x) + p(x)r(x)$$
$$(p(x) + q(x))r(x) = p(x)r(x) + q(x)r(x).$$

All of these laws hold equally well for matrix addition and multiplication with the exception of the commutativity of multiplication. However, for the special case we are dealing with here—polynomials $p(\mathbf{A}), q(\mathbf{A})$ where $\mathbf{A}$ is a fixed matrix—we find that even matrix multiplication commutes. The essential point to note is that for $n, m \geqslant 0$ and any scalars λ, μ, we have

$$\lambda \mathbf{A}^n \mu \mathbf{A}^m = \mu \mathbf{A}^m \lambda \mathbf{A}^n$$

since each side equals $(\lambda\mu)\mathbf{A}^{n+m}$. This together with an induction on degree gives the following result, which will be used in several places in the rest of the book.

Proposition 9.5 *If $p(x), q(x)$ are polynomials and* $\mathbf{A}$ *is an $n \times n$ matrix, then* $p(\mathbf{A})q(\mathbf{A}) = q(\mathbf{A})p(\mathbf{A})$.

It is because of this that any valid polynomial identity in a single variable x, such as $x^2 + 3x + 2 = (x + 2)(x + 1)$, holds equally well when you substitute a square matrix $\mathbf{A}$ for x.

So far, we have stated everything here in terms of matrices, but it makes sense to evaluate polynomials at linear transformations $f \in \mathcal{L}(V, V)$ too, by using composition of functions in place of multiplication. Thus f^2 is interpreted as $f \circ f$, f^0 is interpreted as the identity transformation, and so on. Then since $\mathcal{L}(V, V)$ with the operations of addition, scalar multiplication, and composition is isomorphic to the set $M_{n,n}(F)$ of $n \times n$ square matrices (where $n = \dim V$) over the appropriate field $F = \mathbb{R}$ or $\mathbb{C}$, all the results here apply immediately to the case of linear transformations too. In particular,

Proposition 9.6 *If $f \in \mathcal{L}(V, V)$ where V is a finite dimensional vector space, and if $p(x), q(x)$ are polynomials, then $p(f)q(f) = q(f)p(f)$.*

9.3   Roots of polynomials over $\mathbb{C}$

Suppose $p(x)$ is a polynomial with coefficients from $\mathbb{R}$ or $\mathbb{C}$. A *root* of $p(x)$ is a number α (from $\mathbb{R}$ or $\mathbb{C}$) such that $p(\alpha) = 0$. The following theorem concerning roots of polynomials is very important.

Theorem 9.7 (Remainder theorem) *Suppose $p(x)$ is a polynomial of degree at least 1 with coefficients from $\mathbb{R}$ or $\mathbb{C}$, and $\alpha \in \mathbb{C}$. Then α is a root of $p(x)$ if and only if $(x - \alpha)$ divides $p(x)$ exactly.*

Proof If $(x - \alpha)$ divides $p(x)$, then $p(x) = (x - \alpha)q(x)$ for some polynomial $q(x)$, so $p(\alpha) = (\alpha - \alpha)q(\alpha) = 0$. Conversely, suppose $p(\alpha) = 0$. By the division algorithm there are polynomials $q(x)$ and $r(x)$ such that

$$p(x) = (x - \alpha)q(x) + r(x)$$

and the degree of $r(x)$ is less than 1, i.e. $r(x)$ is a constant, r. Substituting $x = \alpha$ we have

$$p(\alpha) = (\alpha - \alpha)q(\alpha) + r = r$$

so if α is a root of $p(x)$, then $r = 0$ and $(x - \alpha)$ divides $p(x)$ exactly.   □

From this a very important fact follows.

Corollary 9.8 *A polynomial $p(x)$ of degree $d \geqslant 1$ has at most d roots.*

Proof By induction on the degree d of $p(x)$.

By the preceding theorem, if α is a root of a polynomial $p(x)$ and $p(x)$ has degree d, then $p(x) = (x - \alpha)q(x)$ for some polynomial $q(x)$ of degree $d - 1$. By induction, $q(x)$ has at most $d - 1$ roots, and if β is a root of $p(x)$ then $0 = p(\beta) = (\beta - \alpha)q(\beta)$ so β is a root of $q(x)$ or else equals α. Thus $p(x)$ has at most d roots.   □

It may be that a polynomial of degree 2 has only one root. For example, $p(x) = x^2 + 4x + 4$ has only the root -2, since $x^2 + 4x + 4 = (x + 2)^2$. In cases like this, the root -2 is called a *repeated root*, in this case of *multiplicity* 2. On the other hand, the polynomial $x^2 + 1$ has no roots at all in $\mathbb{R}$ since -1 does not have a square root in $\mathbb{R}$. But this polynomial does, however, have its maximum possible number of roots in $\mathbb{C}$ as i and $-i$ are both roots, so $x^2 + 1 = (x - i)(x + i)$.

It is a beautiful and significant fact about the complex numbers (usually called the fundamental theorem of algebra, even though it is really a theorem of analysis) that every polynomial over $\mathbb{C}$ has its maximum possible number of roots (counting multiplicities) in $\mathbb{C}$. In other words, for any polynomial $p(x)$ of degree $d \geqslant 1$, there are $\alpha_1, \ldots, \alpha_d \in \mathbb{C}$ and $c \in \mathbb{C}$ such that $p(x) = c \prod_{j=1}^{d} (x - \alpha_j)$. (We will often describe this situation by saying that $p(x)$ factorizes into *linear factors*.)

The situation is not quite so nice over the reals, since some polynomials such as $x^2 + 1$ may fail to have roots there, as we have seen. The terminology used is that the field of complex numbers $\mathbb{C}$ is *algebraically closed*, i.e. every polynomial $p(x)$ with coefficients from $\mathbb{C}$ has its maximum possible number of roots in $\mathbb{C}$, whereas the field of real numbers $\mathbb{R}$ is not algebraically closed.

In fact, $\mathbb{C}$ and $\mathbb{R}$ are very closely related as fields. Every complex number ζ satisfies a polynomial equation $p(\zeta) = 0$, since, obviously, ζ is a root of the polynomial $x - \zeta$ over $\mathbb{C}$. However, we can improve on this, showing ζ is a root of some polynomial $p(x)$ *with coefficients from* $\mathbb{R}$. To see this just consider the polynomial $p(x) = (x - \zeta)(x - \overline{\zeta})$, where $\overline{\zeta}$ is the complex conjugate of ζ. On multiplying this out, we have $p(x) = x^2 - (\zeta + \overline{\zeta})x + \zeta\overline{\zeta} = x^2 - 2\operatorname{Re}\zeta\, x + |\zeta|^2$

which has real coefficients. We say that $\mathbb{C}$ is the *algebraic closure* of $\mathbb{R}$ since it has the two properties that (1) $\mathbb{C}$ is algebraically closed, and (2) every element $\zeta \in \mathbb{C}$ satisfies a polynomial equation over $\mathbb{R}$.

We finish this discussion of roots with an application—a special case of Euclid's algorithm—that will be needed in the proof of the 'primary decomposition theorem' in Section 14.4. If the reader prefers, he or she may skip the proof here until it is needed later.

Proposition 9.9 *If $a(x)$ and $b(x)$ are nonzero polynomials over $\mathbb{C}$ which have no root in common, then there are polynomials $s(x)$ and $t(x)$ such that*

$$a(x)s(x) + b(x)t(x) = 1.$$

Proof By induction on $\deg(a) + \deg(b)$.

Without loss of generality we may assume $\deg(a) \geqslant \deg(b)$, so by the division algorithm we can write

$$a(x) = b(x)q(x) + r(x)$$

with $\deg(r) < \deg(b)$. We then have $\deg(r) + \deg(b) < \deg(a) + \deg(b)$, and if r and b have a root in common, then $a(x) = b(x)q(x) + r(x)$ has that root in common with them both. This would contradict our assumption, so $b(x)$ and $r(x)$ have no root in common.

If $r(x)$ is identically zero, then $a(x) = b(x)q(x)$, so $b(x)$ is a constant polynomial (for otherwise $a(x)$ and $b(x)$ would have a root in common), $b(x) = \lambda$, say, and then

$$a(x) + b(x)(1/\lambda)(1 - a(x)) = a(x) + \lambda(1/\lambda)(1 - a(x)) = 1$$

as required. Otherwise $r(x)$ is nonzero and we can apply the inductive hypothesis. This gives polynomials $f(x)$ and $g(x)$ such that

$$r(x)f(x) + b(x)g(x) = 1.$$

Substituting back for $r(x)$ we obtain

$$(a(x) - b(x)q(x))f(x) + b(x)g(x) = 1$$

and hence

$$a(x)f(x) + b(x)(q(x)f(x) + g(x)) = 1$$

as required. □

9.4 Roots of polynomials over other fields

Most of the rest of the book can be applied to vector spaces over arbitrary fields. This (optional) section is provided for readers who would like to see how the theory of linear transformations as we set it out applies to vector spaces over fields F other than $\mathbb{C}$ and $\mathbb{R}$. In the chapters that follow, there is essentially very little that needs to be changed when we work over an arbitrary field F, except that we often need our polynomials to have a root in the field, or to factorize into linear factors over that field. The simplest way to ensure this is to restrict attention to *algebraically closed* fields, which are defined to satisfy these two (equivalent) conditions.

Definition 9.10 *A field F is **algebraically closed** if every polynomial with coefficients from F factorizes into linear factors which themselves have coefficients from F.*

However, many of the results below are true even for fields which are not algebraically closed, although the proofs sometimes require us to work in a larger field than we started with (just as results about $\mathbb{R}$ often require us to work in $\mathbb{C}$). To do this in general, we need the existence of an 'algebraic closure' of our field F, which is a 'smallest' algebraically closed field containing F. This idea is given formally in the next definition, but a proof that such a field exists is beyond the scope of this book.

Definition 9.11 *If F is any field, then $\overline{F}$ is an **algebraic closure** of F if*

(a) *every polynomial with coefficients in $\overline{F}$ has all its roots in $\overline{F}$, and*
(b) *every element of $\overline{F}$ is a root of some polynomial with coefficients from F.*

Theorem 9.12 *If F is any field, then F has an algebraic closure $\overline{F}$. Moreover, $\overline{F}$ is unique (up to isomorphism).*

In the case of the field of real numbers, $\mathbb{R}$, its algebraic closure is just the complex number field, $\mathbb{C}$, which is formed from $\mathbb{R}$ by 'adding a square root of -1' (normally called i), i.e. by adding a root of the polynomial equation $x^2 + 1 = 0$.

This may suggest a method for constructing the algebraic closure of an arbitrary field F: simply adjoin roots of polynomials one after another until we can go no further. In general there are infinitely many polynomials to consider, and so this is an infinite process. It is not obvious that it can be 'completed' in a sensible way, or that the result is uniquely determined by the original field F.

Example 9.13 As indicated in Example 2.41, the field of order 9 may be constructed by adjoining a root of $x^2 + 1$ to the field of order 3.

In general, we may as well assume that we adjoin a root of an *irreducible* polynomial $f(x)$ (i.e. one which cannot be factorized in any nontrivial way over the original field). It can then be shown that by taking all polynomials *modulo* $f(x)$, we obtain a field. If the original field had order q, and the degree of $f(x)$ is n, then the new field will have order q^n. (See Exercises 9.6 and 9.7.)

The existence part of Theorem 9.12 is proved by an infinite process as follows. Given a field $F' \supseteq F$, either F' is the algebraic closure of F, so there is nothing else to do, or else there is a polynomial equation $f(x) = 0$ over F' with not all its roots in F. By factorizing $f(x)$ if possible, we may assume $f(x)$ is irreducible. Then we may add a root of $f(x)$ to F' by the method of Exercise 9.6, obtaining a new field $F'' \supseteq F'$, and the process continues. It may be that this process finishes rather quickly, as would be the case in constructing $\overline{\mathbb{R}}$, or it may take infinitely many stages. However, even if infinitely many stages are required, we may take the union of all fields constructed at some stage as our algebraic closure.

The results on the fields $\mathbb{R}$ and $\mathbb{C}$ and polynomials earlier in this chapter apply to other fields too. For instance, we can generalize Euclid's algorithm (Proposition 9.9) immediately to algebraically closed fields, but for general fields it needs to be stated slightly differently.

Proposition 9.14 *If $a(x)$ and $b(x)$ are nonzero polynomials with coefficients from an arbitrary field F, and $a(x)$ and $b(x)$ have no common factors other than constants, then there are polynomials $s(x)$ and $t(x)$ such that*

$$a(x)s(x) + b(x)t(x) = 1.$$

Proof As before, but replacing 'root' by 'nonconstant factor' everywhere in the previous proof. □

Exercises

Exercise 9.1 Evaluate each of the polynomials $p(x) = x^3 - 2x^2 + x + 1$, $q(x) = x^2 - 5x - 2$, and $r(x) = x^2 + 6x + 9$ at each of the matrices

$$\mathbf{A} = \begin{pmatrix} 1 & 2 \\ 3 & 4 \end{pmatrix} \quad \text{and} \quad \mathbf{B} = \begin{pmatrix} -3 & 0 \\ 1 & -3 \end{pmatrix}.$$

Exercise 9.2 Evaluate the same polynomials as in the last exercise at the linear transformations $f((a, b)^T) = (2a - b, a + b)^T$ and $g((a, b, c)^T) = (a + b, b + c, c - a)^T$ from $\mathbb{R}^2 \to \mathbb{R}^2$ and $\mathbb{R}^3 \to \mathbb{R}^3$ respectively.

Exercise 9.3 Expand $(\mathbf{A} - 3\mathbf{I})(\mathbf{B}^2 + 4\mathbf{B} + 2\mathbf{I})$ and $(\mathbf{A} - 3\mathbf{B})(\mathbf{B}^2 + 4\mathbf{B} + 3\mathbf{A} + 2\mathbf{I})$ where $\mathbf{A}$ and $\mathbf{B}$ are unknown $n \times n$ real matrices and $\mathbf{I}$ is the $n \times n$ identity matrix.

Exercise 9.4 Show that every polynomial of *odd* degree with coefficients from $\mathbb{R}$ has at least one real root. [Hint: show that if ζ is a complex root of a polynomial $p(x)$ over $\mathbb{R}$ then $\overline{\zeta}$ is also a root, so nonreal roots occur in pairs.]

Exercise 9.5 Show that every polynomial over $\mathbb{R}$ can be factorized into linear and quadratic factors over $\mathbb{R}$, i.e. written as a product of polynomials over $\mathbb{R}$,

$$\prod_{j=1}^{r} q_j(x),$$

where each q_j has degree at most 2.

Exercise 9.6 Let F be a field, and let $f(x)$ be an irreducible polynomial of degree k over F. Use Euclid's algorithm to show that any nonzero polynomial

$g(x)$ of degree less than k has a multiplicative inverse modulo $f(x)$; that is, there exists $h(x)$ such that

$$g(x)h(x) \equiv 1 \pmod{f(x)}.$$

Deduce that the set of polynomials modulo $f(x)$ is a field.

Exercise 9.7 Let F be a field, $f(x)$ an irreducible polynomial of degree k over F, and G be the field defined by adjoining a root of $f(x)$ to F. Prove that G is a vector space of dimension k over F.

Exercise 9.8 Let F' be a field and let F be a subfield of F' such that the dimension of F' as a vector space over F is finite. (Such field extensions $F' \supseteq F$ are often called *finite extensions*.) By considering $1, a, a^2, a^3, \ldots, a^n$, or otherwise, show that every $a \in F'$ satisfies a polynomial equation $p(x) = 0$ for some polynomial $p(x)$ with coefficients from F.

10

Eigenvalues and eigenvectors

10.1 An example

Rather than giving the formal definition of eigenvalues and eigenvectors—the subject of this chapter, indeed of the rest of the book—straight away, we shall give a hypothetical example of their use to motivate their study.

We imagine a team of biologists studying a single-celled organism which reproduces by cell division. They have identified two different types of the organism, X and Y, and have noticed that on cell division a type X sometimes mutates into type Y, and sometimes a type Y mutates into type X.

Their measurements indicate that both types of the organism reproduce at the same speed, on average doubling in number in unit time. Moreover, starting from a population of 100 type X organisms, after unit time they observe a population of 180 type X and 20 type Y. Similarly, starting from 100 type Y organisms, after unit time 190 type Y and 10 type X are observed.

They want to use this data to predict the development of a mixed population over several units of time. If x_n represents the number of type X at time n and y_n represents the number of type Y at time n they suggest these numbers should be related by

$$x_{n+1} = 1.8x_n + 0.2y_n \qquad y_{n+1} = 0.1x_n + 1.9y_n.$$

Equations like these are called *simultaneous difference equations*. In matrix form, they are written as

$$\begin{pmatrix} x_{n+1} \\ y_{n+1} \end{pmatrix} = \begin{pmatrix} 1.8 & 0.2 \\ 0.1 & 1.9 \end{pmatrix} \begin{pmatrix} x_n \\ y_n \end{pmatrix}. \tag{1}$$

We shall solve these equations by 'pulling a rabbit out of a hat', considering the vectors $(1, 1)^T$ and $(-2, 1)^T$. These vectors are chosen because they have the nice property that

$$\begin{pmatrix} 1.8 & 0.2 \\ 0.1 & 1.9 \end{pmatrix} \begin{pmatrix} 1 \\ 1 \end{pmatrix} = 2 \begin{pmatrix} 1 \\ 1 \end{pmatrix}, \qquad \begin{pmatrix} 1.8 & 0.2 \\ 0.1 & 1.9 \end{pmatrix} \begin{pmatrix} -2 \\ 1 \end{pmatrix} = 1.7 \begin{pmatrix} -2 \\ 1 \end{pmatrix}. \tag{2}$$

These two vectors are linearly independent so form a basis of $\mathbb{R}^2$. We shall also consider the base-change matrix formed from these two vectors,

$$\mathbf{P} = \begin{pmatrix} 1 & -2 \\ 1 & 1 \end{pmatrix}.$$

This matrix has inverse

$$\mathbf{P}^{-1} = \tfrac{1}{3} \begin{pmatrix} 1 & 2 \\ -1 & 1 \end{pmatrix}$$

and we define numbers u_n, v_n by

$$\begin{pmatrix} u_n \\ v_n \end{pmatrix} = \mathbf{P}^{-1} \begin{pmatrix} x_n \\ y_n \end{pmatrix}.$$

Now, from (1) and (2) we have

$$\begin{pmatrix} u_{n+1} \\ v_{n+1} \end{pmatrix} = \mathbf{P}^{-1} \begin{pmatrix} x_{n+1} \\ y_{n+1} \end{pmatrix} = \mathbf{P}^{-1} \begin{pmatrix} 1.8 & 0.2 \\ 0.1 & 1.9 \end{pmatrix} \mathbf{P} \mathbf{P}^{-1} \begin{pmatrix} x_n \\ y_n \end{pmatrix}$$

or

$$\begin{pmatrix} u_{n+1} \\ v_{n+1} \end{pmatrix} = \mathbf{P}^{-1} \begin{pmatrix} 1.8 & 0.2 \\ 0.1 & 1.9 \end{pmatrix} \mathbf{P} \begin{pmatrix} u_n \\ v_n \end{pmatrix}.$$

But it follows from (2) that

$$\mathbf{P}^{-1} \begin{pmatrix} 1.8 & 0.2 \\ 0.1 & 1.9 \end{pmatrix} \mathbf{P} = \begin{pmatrix} 2 & 0 \\ 0 & 1.7 \end{pmatrix}.$$

(Alternatively, this matrix multiplication can be checked directly.) We deduce

$$\begin{pmatrix} u_{n+1} \\ v_{n+1} \end{pmatrix} = \begin{pmatrix} 2 & 0 \\ 0 & 1.7 \end{pmatrix} \begin{pmatrix} u_n \\ v_n \end{pmatrix}$$

and hence

$$u_n = 2^n u_0, \quad v_n = (1.7)^n v_0.$$

Also, $(u_0, v_0)^T = \mathbf{P}^{-1}(x_0, y_0)^T = \tfrac{1}{3}(x_0 + 2y_0, y_0 - x_0)^T$ and

$$\begin{pmatrix} x_n \\ y_n \end{pmatrix} = \mathbf{P} \begin{pmatrix} u_n \\ v_n \end{pmatrix} = \mathbf{P} \begin{pmatrix} 2^n u_0 \\ (1.7)^n v_0 \end{pmatrix} = \tfrac{1}{3} \mathbf{P} \begin{pmatrix} 2^n(x_0 + 2y_0) \\ (1.7)^n(y_0 - x_0) \end{pmatrix},$$

which can be expanded to give

$$x_n = \tfrac{1}{3}(2^n + 2(1.7)^n)x_0 + \tfrac{1}{3}(2^{n+1} - 2(1.7)^n)y_0$$

and

$$y_n = \tfrac{1}{3}(2^n - (1.7)^n)x_0 + \tfrac{1}{3}(2^{n+1} + (1.7)^n)y_0,$$

being formulae for the numbers x_n, y_n of organisms of type X, Y in terms of the initial populations x_0, y_0 of types X and Y.

10.2 Eigenvalues and eigenvectors

Examples like the one given in the previous section show the importance of vectors such as $(1, 1)^T$ and $(-2, 1)^T$ satisfying properties like those in (2) above. We start by making this into a formal definition.

Definition 10.1 *Let* $\mathbf{A}$ *be an* $n \times n$ *matrix over a field* F. *Then a column vector* $\mathbf{x}$ *in* F^n *is called an* **eigenvector** *of* $\mathbf{A}$, *with* **eigenvalue** $\lambda \in F$, *if* $\mathbf{x} \neq \mathbf{0}$ *and* $\mathbf{A}\mathbf{x} = \lambda\mathbf{x}$.

Theorem 10.2 *A scalar,* λ, *is an eigenvalue of an* $n \times n$ *matrix,* $\mathbf{A}$, *if and only if the matrix* $\mathbf{A} - \lambda\mathbf{I}$ *has nullity* $n(\mathbf{A} - \lambda\mathbf{I}) > 0$.

Proof If λ is an eigenvalue of $\mathbf{A}$ with eigenvector $\mathbf{x} \neq \mathbf{0}$ then $(\mathbf{A} - \lambda\mathbf{I})\mathbf{x} = \mathbf{A}\mathbf{x} - \lambda\mathbf{I}\mathbf{x} = \lambda\mathbf{x} - \lambda\mathbf{x} = \mathbf{0}$. Since $\mathbf{x} \neq \mathbf{0}$ and lies in the kernel of $\mathbf{A} - \lambda\mathbf{I}$, $n(\mathbf{A} - \lambda\mathbf{I}) > 0$.

Conversely, if $n(\mathbf{A} - \lambda\mathbf{I}) > 0$ then there is a nonzero vector $\mathbf{x}$ in the kernel of $\mathbf{A} - \lambda\mathbf{I}$, so $(\mathbf{A} - \lambda\mathbf{I})\mathbf{x} = \mathbf{0}$ or $\mathbf{A}\mathbf{x} = \lambda\mathbf{x}$. Hence λ is an eigenvalue. $\square$

The proof of the preceding theorem also shows the following.

Theorem 10.3 *Suppose* λ *is an eigenvalue of an* $n \times n$ *matrix* $\mathbf{A}$. *Then the eigenvectors of* $\mathbf{A}$ *having eigenvalue* λ *are precisely the nonzero vectors in*

$$\ker(\mathbf{A} - \lambda\mathbf{I}) = \{\mathbf{x} : (\mathbf{A} - \lambda\mathbf{I})\mathbf{x} = \mathbf{0}\}.$$

Theorem 10.4 *Every* $n \times n$ *matrix* $\mathbf{A}$ *over* $F = \mathbb{R}$ *or* $\mathbb{C}$ *has an eigenvalue* λ *in* $\mathbb{C}$ *and an eigenvector* $\mathbf{x}$ *in* $\mathbb{C}^n$ *with eigenvalue* λ.

Proof Fix any nonzero vector $\mathbf{v} \in \mathbb{C}^n$. The vectors

$$\mathbf{v}, \mathbf{A}\mathbf{v}, \ldots, \mathbf{A}^n\mathbf{v},$$

form a set of $n + 1$ vectors in a n-dimensional vector space $\mathbb{C}^n$, so are linearly dependent. In other words there are scalars $a_0, a_1, \ldots, a_n \in \mathbb{C}$ with

$$a_n\mathbf{A}^n\mathbf{v} + \cdots + a_1\mathbf{A}\mathbf{v} + a_0\mathbf{v} = \mathbf{0}.$$

Now consider the polynomial $a_n z^n + \cdots + a_1 z + a_0$. Since we are working over $\mathbb{C}$ and every polynomial over $\mathbb{C}$ has all its roots in $\mathbb{C}$, there are $r_1, \ldots, r_n \in \mathbb{C}$ and $c \in \mathbb{C}$ with

$$a_n z^n + \cdots + a_1 z + a_0 = c(z - r_1) \ldots (z - r_n).$$

Then

$$\begin{aligned} \mathbf{0} &= (a_n\mathbf{A}^n + \cdots + a_1\mathbf{A} + a_0\mathbf{I})\mathbf{v} \\ &= c(\mathbf{A} - r_1\mathbf{I}) \ldots (\mathbf{A} - r_n\mathbf{I})\mathbf{v}. \end{aligned}$$

Since $\mathbf{v} \neq \mathbf{0}$, there is some i such that $(\mathbf{A} - r_i\mathbf{I}) \ldots (\mathbf{A} - r_n\mathbf{I})\mathbf{v} = \mathbf{0}$ but $(\mathbf{A} - r_{i+1}\mathbf{I}) \ldots (\mathbf{A} - r_n\mathbf{I})\mathbf{v} \neq \mathbf{0}$. It follows from this that at least one of the matrices $(\mathbf{A} - r_i\mathbf{I})$ has nonzero nullity, so at least one of the r_i is an eigenvalue for $\mathbf{A}$. $\square$

Remark 10.5 *In general, the same result holds for a matrix* **A** *over an arbitrary field F if we replace* $\mathbb{C}$ *by the algebraic closure* $\overline{F}$ *of F throughout.*

Example 10.6 The above proof shows one way we might try to find eigenvalues and eigenvectors. For example, consider the matrix

$$\mathbf{A} = \begin{pmatrix} 1 & -1 & -1 \\ -1 & 1 & -1 \\ -1 & -1 & 1 \end{pmatrix}.$$

Taking the nonzero vector $\mathbf{e}_1 = (1, 0, 0)^T$ we have

$$\mathbf{A}^0 \mathbf{e}_1 = \begin{pmatrix} 1 \\ 0 \\ 0 \end{pmatrix}, \quad \mathbf{A}^1 \mathbf{e}_1 = \begin{pmatrix} 1 \\ -1 \\ -1 \end{pmatrix}, \quad \mathbf{A}^2 \mathbf{e}_1 = \begin{pmatrix} 3 \\ -1 \\ -1 \end{pmatrix}, \quad \mathbf{A}^3 \mathbf{e}_1 = \begin{pmatrix} 5 \\ -3 \\ -3 \end{pmatrix}.$$

In this case, we have the linear dependence

$$-\mathbf{A}^2 \mathbf{e}_1 + \mathbf{A}\mathbf{e}_1 + 2\mathbf{A}^0 \mathbf{e}_1 = \mathbf{0}$$

which yields the polynomial $p(z) = -z^2 + z + 2 = -(z - 2)(z + 1)$. The proof of the theorem now says that one of the roots of $p(z) = 0$, i.e. one of 2 or -1, is an eigenvalue of **A**. In fact in this case both are, for here

$$\mathbf{A} \begin{pmatrix} 1 \\ -1 \\ 0 \end{pmatrix} = 2 \begin{pmatrix} 1 \\ -1 \\ 0 \end{pmatrix}, \quad \mathbf{A} \begin{pmatrix} 1 \\ 0 \\ -1 \end{pmatrix} = 2 \begin{pmatrix} 1 \\ 0 \\ -1 \end{pmatrix}, \quad \mathbf{A} \begin{pmatrix} 1 \\ 1 \\ 1 \end{pmatrix} = -1 \begin{pmatrix} 1 \\ 1 \\ 1 \end{pmatrix},$$

which gives a *basis* $\{(1, -1, 0)^T, (1, 0, -1)^T, (1, 1, 1)^T\}$ of $\mathbb{R}^3$ of eigenvectors of **A**.

We can also define the concepts of eigenvector and eigenvalue in the more abstract context of vector spaces and linear transformations. There is a very close relationship between the two situations.

Definition 10.7 *If* $f: V \to V$ *is a linear map, where V is a vector space over a field F, and* $0 \neq v \in V$ *with* $f(v) = \lambda v$ *for some* $\lambda \in F$, *then v is an* **eigenvector** *of f, with* **eigenvalue** λ.

Suppose that $\mathbf{A} = (a_{ij})$ is the matrix of the linear map $f: V \to V$ with respect to a basis $v_1, \dots, v_n$ of V. Suppose that v is an eigenvector of f, with eigenvalue λ, and write v in terms of the basis vectors as $v = \sum_{i=1}^n \mu_i v_i$. Then we have $f(v) = \lambda v$. Expressing both sides of this equation in terms of the basis vectors, we have

$$\lambda v = \lambda \sum_{j=1}^n \mu_j v_j = \sum_{j=1}^n (\lambda \mu_j) v_j.$$

On the other hand,

$$f(v) = f\left(\sum_{i=1}^{n} \mu_i v_i\right) = \sum_{i=1}^{n} \mu_i f(v_i)$$

$$= \sum_{i=1}^{n} \mu_i \sum_{j=1}^{n} a_{ji} v_j = \sum_{j=1}^{n} \left(\sum_{i=1}^{n} \mu_i a_{ji}\right) v_j.$$

So, comparing coefficients of each basis vector, we have $\lambda \mu_j = \sum_{i=1}^{n} \mu_i a_{ji}$ for each j, and therefore

$$\begin{pmatrix} a_{11} & a_{12} & \cdots & a_{1n} \\ a_{21} & a_{22} & \cdots & a_{2n} \\ \vdots & \vdots & \ddots & \vdots \\ a_{n1} & a_{n2} & \cdots & a_{nn} \end{pmatrix} \begin{pmatrix} \mu_1 \\ \mu_2 \\ \vdots \\ \mu_n \end{pmatrix} = \lambda \begin{pmatrix} \mu_1 \\ \mu_2 \\ \vdots \\ \mu_n \end{pmatrix},$$

which means that $(\mu_1, \dots, \mu_n)^T$ is an eigenvector of $\mathbf{A}$ with eigenvalue λ.

Conversely, if $(\mu_1, \dots, \mu_n)^T$ is an eigenvector of $\mathbf{A}$ with eigenvalue λ, then the corresponding vector

$$v = \sum_{i=1}^{n} \mu_i v_i$$

in V is an eigenvector of f with eigenvalue λ.

In particular, this result implies that the eigenvalues of f are the same as the eigenvalues of a matrix representing f with respect to *any* basis. Therefore any two representing matrices have the same eigenvalues. Using Proposition 8.19 we can restate this as follows.

Proposition 10.8 *If* $\mathbf{A}$, $\mathbf{B}$, *and* $\mathbf{P}$ *are* $n \times n$ *matrices related by* $\mathbf{B} = \mathbf{P}^{-1} \mathbf{A} \mathbf{P}$, *then* $\mathbf{B}$ *and* $\mathbf{A}$ *have the same eigenvalues.*

10.3 Upper triangular matrices

As an important application of eigenvalues and eigenvectors, we aim to show here that any square matrix over $\mathbb{R}$ or $\mathbb{C}$ is similar to an *upper triangular* matrix over $\mathbb{C}$. In other words, for any square matrix $\mathbf{A}$ over $\mathbb{R}$ or $\mathbb{C}$ there is an invertible matrix $\mathbf{P}$ over $\mathbb{C}$ with

$$\mathbf{P}^{-1} \mathbf{A} \mathbf{P} = \begin{pmatrix} b_{11} & b_{12} & b_{13} & \cdots & b_{1n} \\ 0 & b_{22} & b_{23} & \cdots & b_{2n} \\ 0 & 0 & b_{33} & \cdots & b_{3n} \\ \vdots & \vdots & \ddots & \ddots & \vdots \\ 0 & 0 & \cdots & 0 & b_{nn} \end{pmatrix}.$$

This theorem isn't quite as simple as it may seem. Firstly, it will turn out that all the diagonal entries in the form above will necessarily be eigenvalues of $\mathbf{A}$. Secondly, it isn't always possible to put a *real* matrix $\mathbf{A}$ into a *real* upper triangular form; in other words, the diagonal entries may turn out to be complex.

The same result is true more generally for a matrix $\mathbf{A}$ over a field F. It turns out that we can always find an upper triangular matrix B which is similar to $\mathbf{A}$, but once again the entries from $\mathbf{B}$ may have to be from the algebraic closure $\overline{F}$ of F rather than F itself.

The first lemma we need is quite straightforward, and has nothing to do with eigenvectors.

Lemma 10.9 *Suppose $f\colon V \to V$ is a linear transformation of an n-dimensional vector space V over a field F. If f has nullity at least 1 then there is a basis $v_1, v_2, \ldots, v_n$ such that*

$$f(v_j) \in \mathrm{span}(v_1, \ldots, v_{n-1})$$

for all $j = 1, \ldots, n$. In other words, the matrix of f with respect to $v_1, v_2, \ldots, v_n$ is

$$\begin{pmatrix} a_{11} & a_{12} & \cdots & a_{1n} \\ \vdots & \vdots & \ddots & \vdots \\ a_{n-1\,1} & a_{n-1\,2} & \cdots & a_{n-1\,n} \\ 0 & 0 & \cdots & 0 \end{pmatrix}.$$

Proof Let $r = r(f)$. By the rank–nullity formula, $r(f) < n$. So suppose $v_1, \ldots, v_r$ is a basis for $\mathrm{im}\, f$, and extend this to a basis $v_1, \ldots, v_n$ for the whole of V. Then $f(v_j) \in \mathrm{span}(v_1, \ldots, v_r) \subseteq \mathrm{span}(v_1, \ldots, v_{n-1})$, as required. □

Proposition 10.10 *Let $V = \mathbb{C}^n$ be the n-dimensional vector space over $\mathbb{C}$, and suppose f is a linear transformation from V to V. Then there is a basis $v_1, \ldots, v_n$ of V such that, with respect to this basis, the matrix of f is upper triangular.*

Proof We use induction on n; assume the result is true for all spaces V of dimension $n - 1$ over $\mathbb{C}$.

Given V of dimension n and $f\colon V \to V$, let λ be an eigenvalue of f. Such λ exists by Theorem 10.4. Note that $f - \lambda I$ has nullity at least 1, so by the previous lemma, there is a basis $u_1, u_2, \ldots, u_n$ such that $(f - \lambda I)(u_i) \in \mathrm{span}(u_1, \ldots, u_{n-1})$ for all i. Then the matrix of f with respect to this basis is of the form

$$\begin{pmatrix} a_{11} & a_{12} & \cdots & a_{1n} \\ \vdots & \vdots & \ddots & \vdots \\ a_{n-1\,1} & a_{n-1\,2} & \cdots & a_{n-1\,n} \\ 0 & \cdots & 0 & \lambda \end{pmatrix}.$$

Let W be the subspace spanned by $u_1, \ldots, u_{n-1}$ and note that $f(w) \in W$ for all $w \in W$, so that the restriction of f to W is a linear transformation of W. By the induction hypothesis there is a basis $v_1, v_2, \ldots, v_{n-1}$ of W such that the matrix of the restriction of f to W with respect to this is upper triangular. Put $v_n = u_n$; then $\{v_1, v_2, \ldots, v_n\}$ is a basis of V, since $v_n \notin \mathrm{span}(v_1, \ldots, v_{n-1}) = \mathrm{span}(u_1, \ldots, u_{n-1})$ and hence $\{v_1, \ldots, v_{n-1}, v_n\}$ is a linearly independent set of size $n = \dim V$. Also, $f(v_n) = \lambda v_n + w$ for some $w \in W$; hence the matrix of f with respect to the basis $v_1, v_2, \ldots, v_n$ is upper triangular too. □

Remark 10.11 *Once again, the proposition remains valid for any algebraically closed field $\overline{F}$ in place of $\mathbb{C}$, by the same proof.*

Example 10.12 Consider the matrix

$$\mathbf{A} = \begin{pmatrix} -1 & -1 & -1 \\ 0 & 0 & 1 \\ 0 & -1 & -2 \end{pmatrix}.$$

To put $\mathbf{A}$ into upper triangular form we first need an eigenvector, and from the first column of $\mathbf{A}$ it is obvious that $(1,0,0)^T$ is an eigenvector with eigenvalue -1. Then

$$\mathbf{A} + \mathbf{I} = \begin{pmatrix} 0 & -1 & -1 \\ 0 & 1 & 1 \\ 0 & -1 & -1 \end{pmatrix}$$

so the image of $\mathbf{A} + \mathbf{I}$ (considered as a linear transformation on $\mathbb{R}^3$ by left-multiplication) has basis $(-1, 1, -1)^T$, which as it happens in this particular case is also an eigenvector of $\mathbf{A}$. We extend the vector we have so far to a basis $(-1, 1, -1)^T, (1, 0, 0)^T, (0, 1, 0)^T$ of $\mathbb{R}^3$. Then the base-change matrix is

$$\mathbf{P} = \begin{pmatrix} -1 & 1 & 0 \\ 1 & 0 & 1 \\ -1 & 0 & 0 \end{pmatrix}.$$

This has inverse

$$\mathbf{P}^{-1} = \begin{pmatrix} 0 & 0 & -1 \\ -1 & 0 & 1 \\ 0 & 1 & 1 \end{pmatrix}$$

and

$$\mathbf{P}^{-1}\mathbf{A}\mathbf{P} = \begin{pmatrix} -1 & 0 & 1 \\ 0 & -1 & 0 \\ 0 & 0 & -1 \end{pmatrix},$$

which is in upper triangular form.

Example 10.13 By good fortune, we found the new basis rather easily in the last example. For a slightly more typical example, consider

$$\mathbf{A} = \begin{pmatrix} 3 & 0 & -1 \\ -1 & 4 & -3 \\ -1 & 0 & 5 \end{pmatrix}.$$

Here $(0, 1, 0)^T$ is obviously an eigenvector with eigenvalue 4. The subspace which is the image of the linear transformation

$$\mathbf{A} - 4\mathbf{I} = \begin{pmatrix} -1 & 0 & -1 \\ -1 & 0 & -3 \\ -1 & 0 & 1 \end{pmatrix}$$

has basis formed from $\mathbf{f}_1 = (-1, -1, -1)^T$, and $\mathbf{f}_2 = (1, -3, 1)^T$. We extend this to a basis of the whole space by adjoining $\mathbf{f}_3 = (1, 0, 0)^T$, and so we have base-change matrix

$$\mathbf{P} = \begin{pmatrix} -1 & 1 & 1 \\ -1 & -3 & 0 \\ -1 & 1 & 0 \end{pmatrix}.$$

On calculating, we find that

$$\mathbf{P}^{-1}\mathbf{A}\mathbf{P} = \begin{pmatrix} 3 & 1 & 1 \\ -1 & 5 & 0 \\ 0 & 0 & 4 \end{pmatrix}.$$

We now look at $\mathbf{B} = \begin{pmatrix} 3 & 1 \\ -1 & 5 \end{pmatrix}$ which has eigenvalue 4 and eigenvector $(1, 1)^T$. Also, $\mathbf{B} - 4\mathbf{I} = \begin{pmatrix} -1 & 1 \\ -1 & 1 \end{pmatrix}$, so a basis for the image of this is $(1, 1)^T$. We extend this to the basis $(1, 1)^T, (1, 0)^T$ of $\mathbb{R}^2$. Going back to $\mathbb{R}^3$, what we have done is to replace the basis $\mathbf{f}_1, \mathbf{f}_2, \mathbf{f}_3$ with $\mathbf{f}_1 + \mathbf{f}_2, \mathbf{f}_1, \mathbf{f}_3$, and the base-change matrix for this operation is

$$\mathbf{Q} = \begin{pmatrix} 1 & 1 & 0 \\ 1 & 0 & 0 \\ 0 & 0 & 1 \end{pmatrix}.$$

On calculating, we have

$$\mathbf{Q}^{-1}\mathbf{P}^{-1}\mathbf{A}\mathbf{P}\mathbf{Q} = \begin{pmatrix} 4 & -1 & 0 \\ 0 & 4 & 1 \\ 0 & 0 & 4 \end{pmatrix}$$

in upper triangular form.

Example 10.14 Proposition 10.10 may fail for vector spaces over other fields. For example, let $V = \mathbb{R}^2$ over $\mathbb{R}$ and f be the linear transformation $f(x, y)^T = (y, -x)^T$. The matrix of f with respect to the usual basis is $\begin{pmatrix} 0 & 1 \\ -1 & 0 \end{pmatrix}$. Suppose for the sake of obtaining a contradiction that $\mathbf{P} = \begin{pmatrix} a & b \\ c & d \end{pmatrix}$ is a real base-change matrix putting this into upper triangular form,

$$\mathbf{P}^{-1}\begin{pmatrix} 0 & 1 \\ -1 & 0 \end{pmatrix}\mathbf{P} = \frac{1}{ad - bc}\begin{pmatrix} d & -b \\ -c & a \end{pmatrix}\begin{pmatrix} 0 & 1 \\ -1 & 0 \end{pmatrix}\begin{pmatrix} a & b \\ c & d \end{pmatrix} = \begin{pmatrix} x & y \\ 0 & z \end{pmatrix}.$$

Multiplying out gives

$$\frac{1}{ad - bc} \begin{pmatrix} cd + ab & d^2 + b^2 \\ -c^2 - a^2 & -cd - ab \end{pmatrix} = \begin{pmatrix} x & y \\ 0 & z \end{pmatrix}$$

so $-c^2 - a^2 = 0$ giving $a = c = 0$ (since a and c were supposed to be real) and hence **P** is singular. We conclude that the original matrix is not similar to any upper triangular matrix over $\mathbb{R}$.

Note, however, that

$$\begin{pmatrix} 1 & 1 \\ i & 0 \end{pmatrix}^{-1} \begin{pmatrix} 0 & 1 \\ -1 & 0 \end{pmatrix} \begin{pmatrix} 1 & 1 \\ i & 0 \end{pmatrix} = \begin{pmatrix} i & i \\ 0 & -i \end{pmatrix}$$

so the original matrix can be put in upper triangular form over $\mathbb{C}$. The eigenvalues of this particular example are i and $-i$ which are imaginary and not real.

In general, the diagonal entries of any upper triangular form for a matrix **A** are the eigenvalues of **A**, as shown by the next proposition.

Proposition 10.15 *If A is an $n \times n$ upper triangular matrix, then the diagonal entries in **A** are precisely the eigenvalues of **A**.*

Proof Suppose $\mathbf{A} = (a_{ij})$ and $\lambda = a_{ii}$ is a diagonal entry of **A**. Then

$$\mathbf{A}\mathbf{e}_j = a_{j1}\mathbf{e}_1 + \cdots + a_{j-1\,j}\mathbf{e}_{j-1} + a_{jj}\mathbf{e}_j$$

since **A** is upper triangular. Using $a_{ii} = \lambda$, we obtain

$$(\mathbf{A} - \lambda\mathbf{I})\mathbf{e}_i = a_{i1}\mathbf{e}_1 + \cdots + a_{i-1\,i}\mathbf{e}_{i-1} \in \text{span}(\mathbf{e}_1, \ldots, \mathbf{e}_{i-1}).$$

Thus $\mathbf{A} - \lambda\mathbf{I}$ defines a linear transformation T from an i-dimensional space, $\text{span}(\mathbf{e}_1, \ldots, \mathbf{e}_i)$, into an $(i-1)$-dimensional space, $\text{span}(\mathbf{e}_1, \ldots, \mathbf{e}_{i-1})$. By the rank–nullity formula, T has nullity at least 1, so there is $\mathbf{v} \in \text{span}(\mathbf{e}_1, \ldots, \mathbf{e}_i)$ with $T(\mathbf{v}) = (\mathbf{A} - \lambda\mathbf{I})\mathbf{v} = \mathbf{0}$, i.e. $\mathbf{v}$ is an eigenvector of **A** with eigenvalue λ.

To see that all the eigenvalues of **A** appear along the diagonal of this representation, suppose λ is not equal to any diagonal entry a_{ii}. Then $\mathbf{A} - \lambda\mathbf{I}$ is in upper triangular form and has each diagonal entry nonzero. In other words, $\mathbf{A} - \lambda\mathbf{I}$ is in echelon form and has rank n, and hence nullity 0. Therefore λ is not an eigenvalue. $\square$

It would be possible to *define* the determinant of a matrix **A** to be the product of the diagonal entries in any upper triangular form $\mathbf{P}^{-1}\mathbf{A}\mathbf{P}$ for **A**. The problems are (1) that it is not immediately obvious that this definition agrees with the usual definition, and (2) in showing that this definition does not depend on the choice of **P** or the choice of the upper triangular form **A**. In fact these problems can be got round, and the definition can be made sound, but doing so would take us too far off track.[1]

We conclude with a particularly useful observation concerning upper triangular matrices.

[1] The interested reader can follow up the details in 'Down with determinants', by Sheldon Axler. *American Math Monthly*, February 1995, pp. 139–145.

Theorem 10.16 *If* **A** *is any upper triangular* $n \times n$ *matrix with entries from* $\mathbb{R}$ *or* $\mathbb{C}$, *and* $\lambda_1, \lambda_2, \ldots, \lambda_n$ *are the diagonal entries of* **A** *including repetitions, then the matrix*

$$(\mathbf{A} - \lambda_1 \mathbf{I})(\mathbf{A} - \lambda_2 \mathbf{I}) \ldots (\mathbf{A} - \lambda_n \mathbf{I})$$

is the zero matrix.

Proof If $n = 1$, this is obvious. We prove the general statement using induction on n.

Given an upper triangular $n \times n$ matrix, take the standard basis $\mathbf{e}_1, \ldots, \mathbf{e}_n$ of the underlying vector space $\mathbb{R}^n$ or $\mathbb{C}^n$. Observe first that

$$(\mathbf{A} - \lambda_1 \mathbf{I}) \ldots (\mathbf{A} - \lambda_{n-1}\mathbf{I})(\mathbf{A} - \lambda_n\mathbf{I}) = (\mathbf{A} - \lambda_n\mathbf{I})(\mathbf{A} - \lambda_1\mathbf{I}) \ldots (\mathbf{A} - \lambda_{n-1}\mathbf{I})$$

from Section 9.2 in the last chapter.

Now

$$(\mathbf{A} - \lambda_1 \mathbf{I}) \ldots (\mathbf{A} - \lambda_{n-1}\mathbf{I})(\mathbf{A} - \lambda_n\mathbf{I})\mathbf{e}_n = \mathbf{0}$$

since $(\mathbf{A} - \lambda_n\mathbf{I})\mathbf{e}_n = \mathbf{0}$ since **A** is upper triangular with last row equal to $(0, 0, \ldots, 0, \lambda_n)$. Also, the linear transformation given by **A** on the subspace $\text{span}(\mathbf{e}_1, \ldots, \mathbf{e}_{n-1})$ has upper triangular matrix with respect to this basis, with diagonal entries $\lambda_1, \ldots, \lambda_{n-1}$, so by the induction hypothesis we have

$$(\mathbf{A} - \lambda_1 \mathbf{I}) \ldots (\mathbf{A} - \lambda_{n-1}\mathbf{I})\mathbf{e}_i = \mathbf{0}$$

for all $i < n$. Thus

$$(\mathbf{A} - \lambda_n\mathbf{I})(\mathbf{A} - \lambda_1\mathbf{I}) \ldots (\mathbf{A} - \lambda_{n-1}\mathbf{I})\mathbf{e}_i = \mathbf{0}$$

for $i < n$ and therefore

$$(\mathbf{A} - \lambda_1 \mathbf{I}) \ldots (\mathbf{A} - \lambda_{n-1}\mathbf{I})(\mathbf{A} - \lambda_n\mathbf{I})\mathbf{e}_i = \mathbf{0}$$

for all $i \leqslant n$.

We have shown that the matrix

$$(\mathbf{A} - \lambda_1 \mathbf{I}) \ldots (\mathbf{A} - \lambda_{n-1}\mathbf{I})(\mathbf{A} - \lambda_n\mathbf{I})$$

represents the zero transformation, since multiplying $\mathbf{e}_i$ by it gives $\mathbf{0}$ for all $i \leqslant n$. Therefore this matrix is zero, as required. $\qquad\qquad \square$

Exercises

Exercise 10.1 Let

$$\mathbf{A} = \begin{pmatrix} 2 & -3 & 9 \\ -1 & 0 & -1 \\ 0 & 1 & -1 \end{pmatrix}.$$

Given that **A** has eigenvalues 1, 2, and -2, find all the eigenvectors of **A**, and hence write down a basis of $\mathbb{R}^3$ consisting of eigenvectors of **A**.

Exercise 10.2 Let

$$\mathbf{B} = \begin{pmatrix} 2 & 1 & 4 \\ 0 & 2 & -1 \\ 0 & 0 & -3 \end{pmatrix}.$$

Find all the eigenvalues and eigenvectors of **B**, and hence show that there is no basis of $\mathbb{R}^3$ consisting of eigenvectors of **B**.

Exercise 10.3 Find eigenvectors and eigenvalues of the following. (Try to guess the eigenvectors and verify your guess, calculating the eigenvalues, or use the ideas in this chapter.)

(a) $\begin{pmatrix} 0 & 3 \\ 3 & 0 \end{pmatrix}$ (b) $\begin{pmatrix} 1 & -3 \\ -3 & 1 \end{pmatrix}$ (c) $\begin{pmatrix} 2 & 0 \\ 0 & 1/2 \end{pmatrix}$ (d) $\begin{pmatrix} 4 & 2 \\ 2 & 1 \end{pmatrix}$.

Exercise 10.4 (a) Let T be the rotation on the plane $\mathbb{R}^2$ by $\pi/4$ in the clockwise direction about the origin. Explain why T doesn't have any real eigenvectors.

(b) What are the real eigenvalues and eigenvectors in $\mathbb{R}^3$ of a rotation of $\mathbb{R}^3$ by $\pi/4$ about an axis through the origin perpendicular to the plane given by $2x + y - 5z = 0$?

Exercise 10.5 For each of the following matrices, **A**, find an invertible matrix **P** over $\mathbb{C}$ such that $\mathbf{P}^{-1}\mathbf{AP}$ is upper triangular. (You don't have to compute $\mathbf{P}^{-1}$.)

(a) $\begin{pmatrix} 0 & -1 \\ 1 & 0 \end{pmatrix}$ (b) $\begin{pmatrix} -1 & 1 & 1 \\ -1-i & i & 2 \\ -1-i & 2i & 2-i \end{pmatrix}$ (c) $\begin{pmatrix} 1 & 1 \\ -1 & 1 \end{pmatrix}$.

11

The minimum polynomial

The central result of this chapter is the remarkable and important fact that for any $n \times n$ matrix $\mathbf{A}$ there is a polynomial $p(x)$ of degree at most n such that $p(\mathbf{A})$ is the zero matrix. This can be proved either from results in the previous chapter on upper triangular matrices, or more directly. Because of this, it turns out that we can define a special *minimum* polynomial $m_{\mathbf{A}}(x)$ (i.e. of least degree) such that $m_{\mathbf{A}}(\mathbf{A}) = \mathbf{0}$. The properties of this minimum polynomial will be central to all the material in the remainder of this book.

11.1   The minimum polynomial

This section is devoted to the *minimum polynomial* and its properties, and this theory underpins the remainder of the book. The rest of this chapter provides further illustrations and methods of calculation which will be chiefly of use in applying these ideas.

We start straight off with the key theoretical observation.

Theorem 11.1 *Let $f \colon V \to V$ be a linear transformation of a finite dimensional vector space V over a field F. Then there is a polynomial $q(x)$ with coefficients from F such that $q(f) = 0$.*

Proof   Let $\{e_1, e_2, \ldots, e_n\}$ be a basis of V. For each basis vector e_j we consider the $n + 1$ vectors

$$e_j, f(e_j), f^2(e_j), \ldots, f^n(e_j).$$

Since these are $n + 1$ vectors in an n-dimensional space V, they are linearly dependent, i.e.

$$\lambda_0 e_j + \lambda_1 f(e_j) + \lambda_2 f^2(e_j) + \cdots + \lambda_n f^n(e_j) = 0$$

for some scalars λ_i, not all zero. Alternatively,

$$(\lambda_0 + \lambda_1 f + \lambda_2 f^2 + \cdots + \lambda_n f^n)(e_j) = 0.$$

This shows that there is a nonzero polynomial $q_j(x)$ of degree at most n such that $q_j(f)(e_j) = 0$. We find such a polynomial for each j, and multiply them together to give the polynomial

$$q(x) = q_1(x)q_2(x)\ldots q_n(x).$$

Then

$$q(f)(e_j) = q_1(f)\ldots q_{j-1}(f)q_{j+1}(f)\ldots q_n(f)q_j(f)(e_j) = 0$$

for every basis element e_j, so $q(f)$ is the zero transformation $q(f) = 0$, as required. □

Next, we ask about other polynomials $p(x)$ with $p(f) = 0$. First note that if $p(f) = 0$ and $r(x)$ is any other polynomial then $p(f)r(f) = 0$, so we should look for polynomials $p(x)$ of the smallest possible degree such that $p(f) = 0$. Note too that if $p(f) = 0$ then we can divide the polynomial $p(x)$ by its leading coefficient (i.e. the coefficient of the highest power of x) to get another polynomial $q(x)$ with $q(f) = 0$, such that $q(x)$ is *monic*, i.e. has leading coefficient equal to 1.

Definition 11.2 *Given a linear transformation $f\colon V \to V$, the **minimum polynomial** of f is the monic polynomial $p(x)$ of least degree which has $p(f) = 0$. We write $m_f(x)$ for this minimum polynomial. Similarly, the minimum polynomial of a square matrix $\mathbf{A}$, $m_{\mathbf{A}}(x)$, is the monic polynomial $p(x)$ of least degree such that $p(\mathbf{A}) = \mathbf{0}$.*

We need to show that $m_f(x)$ is well-defined, i.e. that there cannot be two different monic polynomials $p(x), q(x)$ of minimum degree such that $p(f) = q(f) = 0$. The following proposition is stated for linear transformations, but the proof applies equally well to matrices instead.

Proposition 11.3 *Let $f\colon V \to V$ be a linear transformation of a finite dimensional vector space V over a field F. Suppose $p(x)$ is any polynomial with $p(f) = 0$ and $m(x)$ is any monic polynomial of minimal degree with $m(f) = 0$. Then $p(x)$ is a multiple of $m(x)$, in the sense that there is a polynomial $q(x)$ such that $p(x) = m(x)q(x)$ as polynomials. Moreover, $m_f(x)$ is well-defined.*

Proof If $p(f) = 0$, and $m(x)$ is of minimal degree such that $m(f) = 0$, then by the division algorithm we can write $p(x) = m(x)q(x) + r(x)$, where the degree of $r(x)$ is less than the degree of $m(x)$. Then $r(f) = p(f) - m(f)q(f) = 0$. But $m(x)$ is of minimal degree with the property that $m(f) = 0$, so $r(x)$ must be the zero polynomial. Therefore $p(x) = m(x)q(x)$ as required.

Now suppose that polynomials $m_f(x)$ and $k_f(x)$ both satisfy Definition 11.2. Then $k_f(f) = 0$, so by what we have just proved $k_f(x) = m_f(x)q(x)$ for some $q(x)$. Similarly, $m_f(x) = k_f(x)s(x)$, for some polynomial $s(x)$. Therefore both $q(x)$ and $s(x)$ are constant polynomials, and since $m_f(x)$ and $k_f(x)$ were defined to be monic, they must be equal. □

Theorem 11.4 *Given a linear transformation $f\colon V \to V$ where V is a vector space over F, and given a scalar $\lambda \in F$, then λ is an eigenvalue of f if and only if λ is a root of the minimum polynomial $m_f(\lambda) = 0$.*

Proof Suppose λ is a root of $m_f(x)$. Then $m_f(x) = (x - \lambda)p(x)$ for some polynomial $p(x)$, by the remainder theorem. Now $p(f) \neq 0$, since $p(x)$ has smaller degree than $m_f(x)$, so there is $v \in V$ with $w = p(f)(v) \neq 0$. But $m_f(f)$ is the transformation taking every vector to 0, so

$$0 = m_f(f)(v) = (f - \lambda\mathrm{I})p(f)(v) = f(w) - \lambda w$$

so $f(w) = \lambda w$, and hence λ is an eigenvalue of f.

Conversely, suppose that $v \in V$ is an eigenvector of f with eigenvalue λ, so that $f(v) = \lambda v$ and $v \neq 0$. Then $m_f(f) = 0$, so $m_f(f)(v)$ is the zero vector, and

$$\begin{aligned}
0 = m_f(f)(v) &= (f^r + a_{r-1}f^{r-1} + \cdots + a_0\mathrm{I})v \\
&= f^r(v) + a_{r-1}f^{r-1}(v) + \cdots + a_0 v \\
&= \lambda^r v + a_{r-1}\lambda^{r-1}v + \cdots + a_0 v \\
&= (\lambda^r + a_{r-1}\lambda^{r-1} + \cdots + a_0)v \\
&= m_f(\lambda)v.
\end{aligned}$$

But $v \neq 0$, so the scalar $m_f(\lambda)$ equals 0.   □

Example 11.5 Consider the two matrices,

$$\mathbf{A} = \begin{pmatrix} 2 & 1 & -1 \\ 0 & 1 & 1 \\ 0 & 0 & 1 \end{pmatrix}, \qquad \mathbf{B} = \begin{pmatrix} 2 & 1 & -1 \\ 0 & 1 & 0 \\ 0 & 0 & 1 \end{pmatrix}.$$

Both are upper triangular, with $2, 1, 1$ on the diagonal, so 2 and 1 are the eigenvalues of both $\mathbf{A}$ and $\mathbf{B}$. By Theorem 10.16, we know that $p(\mathbf{A}) = p(\mathbf{B}) = \mathbf{0}$, where $p(x) = (x - 1)^2(x - 2)$, and by Theorem 11.4 both $(x - 1)$ and $(x - 2)$ must divide the minimum polynomials $m_{\mathbf{A}}(x)$ and $m_{\mathbf{B}}(x)$. So we compute

$$(\mathbf{A} - \mathbf{I})(\mathbf{A} - 2\mathbf{I}) = \begin{pmatrix} 1 & 1 & -1 \\ 0 & 0 & 1 \\ 0 & 0 & 0 \end{pmatrix}\begin{pmatrix} 0 & 1 & -1 \\ 0 & -1 & 1 \\ 0 & 0 & -1 \end{pmatrix} = \begin{pmatrix} 0 & 0 & 1 \\ 0 & 0 & -1 \\ 0 & 0 & 0 \end{pmatrix},$$

and

$$(\mathbf{B} - \mathbf{I})(\mathbf{B} - 2\mathbf{I}) = \begin{pmatrix} 1 & 1 & -1 \\ 0 & 0 & 0 \\ 0 & 0 & 0 \end{pmatrix}\begin{pmatrix} 0 & 1 & -1 \\ 0 & -1 & 0 \\ 0 & 0 & -1 \end{pmatrix} = \begin{pmatrix} 0 & 0 & 0 \\ 0 & 0 & 0 \\ 0 & 0 & 0 \end{pmatrix}.$$

So $m_{\mathbf{A}}(x) = (x - 1)^2(x - 2)$ and $m_{\mathbf{B}}(x) = (x - 1)(x - 2)$.

11.2   The characteristic polynomial

By Theorem 11.4, the minimum polynomial is an example of a polynomial whose roots are exactly the eigenvalues of a square matrix $\mathbf{A}$. Other polynomials with this property can be obtained using the idea of the determinant of a matrix.

Consider Theorem 10.2, which states that a scalar λ is an eigenvalue of a matrix $\mathbf{A}$ if and only if the nullity of the matrix $\mathbf{A} - \lambda\mathbf{I}$ is at least 1. Since a matrix has nonzero nullity if and only if it has determinant equal to zero, we can rephrase this result in terms of determinants as follows.

Theorem 11.6 *Let $\mathbf{A}$ be an $n \times n$ matrix. Then λ is an eigenvalue of $\mathbf{A}$ if and only if $\det(\mathbf{A} - \lambda\mathbf{I}) = 0$.*

Note too that, since $\det(\mathbf{B}) = \det(\mathbf{B}^T)$ and $(\mathbf{A} - \lambda\mathbf{I})^T = (\mathbf{A}^T - \lambda\mathbf{I})$, λ is an eigenvalue of the matrix $\mathbf{A}$ if and only if λ is an eigenvalue of its transpose $\mathbf{A}^T$. (But the reader should note that although the eigen*values* of $\mathbf{A}$ and $\mathbf{A}^T$ are the same, the eigen*vectors* are in general completely different.)

Definition 11.7 *The **characteristic polynomial** of a square matrix $\mathbf{A}$ is the polynomial $\chi_\mathbf{A}(x)$ defined by $\chi_\mathbf{A}(x) = \det(\mathbf{A} - x\mathbf{I})$. The **characteristic equation** is the equation $\chi_\mathbf{A}(x) = 0$.*

Thus the eigenvalues are exactly the solutions of the characteristic equation (i.e. the roots of the characteristic polynomial). Once again, the characteristic polynomial of $\mathbf{A}^T$ is identical to that of $\mathbf{A}$.

Remark 11.8 *Many people define $\chi_\mathbf{A}(x)$ to be the polynomial $\det(x\mathbf{I} - \mathbf{A})$. Since $\det(-\mathbf{B}) = (-1)^n \det \mathbf{B}$ for any $n \times n$ matrix $\mathbf{B}$, we have that*

$$\det(x\mathbf{I} - \mathbf{A}) = (-1)^n \det(\mathbf{A} - x\mathbf{I}),$$

so this alternative definition agrees with ours when n is even, but is the negative of ours for odd n. Unfortunately, there doesn't seem to be any convention here, and the reader will need to be aware of this potential source of difficulty.

The main points of interest in the characteristic polynomial $\chi_\mathbf{A}(x)$ are its coefficients. The constant term turns out to be the determinant of $\mathbf{A}$, and perhaps more surprisingly the trace of $\mathbf{A}$ appears in the second coefficient.

Proposition 11.9 *For a square matrix $\mathbf{A}$,*

$$\chi_\mathbf{A}(x) = (-1)^n x^n + (-1)^{n-1} \operatorname{tr}(\mathbf{A})x^{n-1} + \cdots + \det(\mathbf{A}).$$

Proof $\chi_\mathbf{A}(x) = \det(\mathbf{A} - x\mathbf{I})$ is equal to the determinant

$$\begin{vmatrix} a_{11} - x & a_{12} & \cdots & a_{1n} \\ a_{21} & a_{22} - x & \cdots & a_{2n} \\ \vdots & \vdots & \ddots & \vdots \\ a_{n1} & a_{n2} & \cdots & a_{nn} - x \end{vmatrix}.$$

The only term in x^n comes from taking $-x$ in each diagonal entry, and so the coefficient of x^n is $(-1)^n$. The terms in x^{n-1} must be obtained by taking $n - 1$ terms of the form $-x$ from the diagonal, so the remaining term must be one of the diagonal constant terms a_{ii}. Therefore the term in x^{n-1} is $\sum_{i=1}^n a_{ii}(-x)^{n-1} = (-1)^{n-1} \operatorname{tr}(\mathbf{A})x^{n-1}$. The constant term is obtained by putting $x = 0$, so is $\chi_\mathbf{A}(0) = \det(\mathbf{A})$. $\square$

Of course, the other coefficients of the characteristic polynomial are of interest too. These coefficients provide useful *invariants* of the matrix **A**, with many applications and more advanced material outside the scope of this book.

Example 11.10 We can write down the characteristic polynomial of an upper triangular matrix immediately. For example, if

$$\mathbf{A} = \begin{pmatrix} 2 & 1 & 1 \\ 0 & 1 & 2 \\ 0 & 0 & -1 \end{pmatrix}, \qquad \mathbf{B} = \begin{pmatrix} 3 & 1 & 2 \\ 0 & 3 & -1 \\ 0 & 0 & -2 \end{pmatrix}$$

then $\chi_{\mathbf{A}}(x) = (2-x)(1-x)(-1-x)$ and $\chi_{\mathbf{B}}(x) = (3-x)^2(-2-x)$.

Example 11.11 For the matrix

$$\mathbf{A} = \begin{pmatrix} 1 & 2 \\ 3 & 4 \end{pmatrix},$$

we have

$$\mathbf{A} - x\mathbf{I} = \begin{pmatrix} 1-x & 2 \\ 3 & 4-x \end{pmatrix},$$

and

$$\chi_{\mathbf{A}}(x) = \begin{vmatrix} 1-x & 2 \\ 3 & 4-x \end{vmatrix} = (1-x)(4-x) - 6 = x^2 - 5x - 2.$$

Example 11.12 Sometimes it is possible to find the characteristic polynomial without evaluating determinants at all, but using the remainder theorem instead. Take for example

$$\mathbf{A} = \begin{pmatrix} -1 & 2 & 1 \\ 2 & -2 & 2 \\ 1 & 2 & -1 \end{pmatrix},$$

so

$$\chi_{\mathbf{A}}(x) = \det(\mathbf{A} - x\mathbf{I}) = \begin{vmatrix} -1-x & 2 & 1 \\ 2 & -2-x & 2 \\ 1 & 2 & -1-x \end{vmatrix}. \qquad (1)$$

This is a polynomial of degree 3 with the x^3 term having coefficient -1. So if we can find three roots $\lambda_1, \lambda_2, \lambda_3$ of it, then $\chi_{\mathbf{A}}(x) = (\lambda_1 - x)(\lambda_2 - x)(\lambda_3 - x)$. Now if $x = -2$, the determinant in (1) is zero since the first and the third rows would both be equal to $(1 \quad 2 \quad 1)$. If $x = 2$ then the sum of all three rows would be zero, so the determinant is zero here too, and finally, if $x = -4$ then the first and third rows add to give -2 times the second row, so again the determinant is zero. Thus

$$\chi_{\mathbf{A}}(x) = (2-x)(-2-x)(-4-x).$$

11.3 The Cayley–Hamilton theorem

We would like now to define the characteristic polynomial of a linear transformation $f: V \to V$. In fact, it *is* possible to define the determinant of a linear map and its characteristic polynomial in a completely direct (and rather abstract) way, but the following is entirely adequate for our purposes here.

Definition 11.13 *If $f: V \to V$ is a linear map on a finite dimensional vector space, then the **determinant** of f (written $\det(f)$) is the determinant of any matrix $\mathbf{A}$ representing f. The characteristic polynomial of f is the polynomial $\chi_f(x) = \det(\mathbf{A} - x\mathbf{I})$.*

Of course, we now have to prove that these are well-defined. In other words, it doesn't matter which basis we choose. If $\mathbf{A}$ and $\mathbf{B}$ are two matrices representing f with respect to different bases, then by Proposition 8.19 we have $\mathbf{B} = \mathbf{P}^{-1}\mathbf{A}\mathbf{P}$ for some invertible matrix $\mathbf{P}$. Therefore

$$
\begin{aligned}
\det(\mathbf{B}) &= \det(\mathbf{P}^{-1}\mathbf{A}\mathbf{P}) \\
&= \det(\mathbf{P}^{-1})\det(\mathbf{A})\det(\mathbf{P}) \\
&= (\det(\mathbf{P}))^{-1}\det(\mathbf{A})\det(\mathbf{P}) \\
&= \det(\mathbf{A}).
\end{aligned}
$$

Thus $\det(f)$ is well-defined.

Similarly,

$$
\begin{aligned}
\det(\mathbf{A} - x\mathbf{I}) &= (\det \mathbf{P})^{-1}\det(\mathbf{A} - x\mathbf{I})(\det \mathbf{P}) \\
&= \det(\mathbf{P}^{-1}(\mathbf{A} - x\mathbf{I})\mathbf{P}) \\
&= \det(\mathbf{P}^{-1}\mathbf{A}\mathbf{P} - x\mathbf{I}) \\
&= \det(\mathbf{B} - x\mathbf{I}) = \chi_\mathbf{B}(x)
\end{aligned}
$$

which shows that the characteristic polynomial of a linear transformation f does not depend on choice of basis either.

These observations have some very important consequences. Firstly, that the characteristic polynomial of a matrix is *invariant* in the sense that if the matrices $\mathbf{A}$ and $\mathbf{B}$ are similar then $\chi_\mathbf{A}(x) = \chi_\mathbf{B}(x)$. This combined with Proposition 11.9 shows also that the trace of a matrix is invariant in the same way: $\text{tr}(\mathbf{A}) = \text{tr}(\mathbf{P}^{-1}\mathbf{A}\mathbf{P})$ for any square matrix $\mathbf{A}$ and any invertible $\mathbf{P}$. Alternatively, this can be proved from first principles—see Exercise 11.10.

The invariance of the characteristic polynomial also enables us to prove the following well-known theorem.

Theorem 11.14 (The Cayley–Hamilton theorem) *If $\mathbf{A}$ is any square matrix over a field $F = \mathbb{R}$ or $\mathbb{C}$, and $\chi_\mathbf{A}(x)$ is its characteristic polynomial, then $\chi_\mathbf{A}(\mathbf{A}) = \mathbf{0}$.*

Proof By Proposition 10.10 there is an invertible matrix $\mathbf{P}$ over $\mathbb{C}$ such that $\mathbf{B} = \mathbf{P}^{-1}\mathbf{A}\mathbf{P}$ is upper triangular, and Theorem 10.16 says that if $\lambda_1, \lambda_2, \ldots, \lambda_n \in \mathbb{C}$ are the diagonal entries in $\mathbf{B}$ then

$$(\mathbf{B} - \lambda_1\mathbf{I})(\mathbf{B} - \lambda_2\mathbf{I}) \ldots (\mathbf{B} - \lambda_n\mathbf{I}) = \mathbf{0}.$$

It follows that

$$\begin{aligned}
(\mathbf{A} &- \lambda_1\mathbf{I})(\mathbf{A} - \lambda_2\mathbf{I}) \ldots (\mathbf{A} - \lambda_n\mathbf{I}) \\
&= \mathbf{P}\mathbf{P}^{-1}(\mathbf{A} - \lambda_1\mathbf{I})\mathbf{P}\mathbf{P}^{-1}(\mathbf{A} - \lambda_2\mathbf{I}) \ldots \mathbf{P}\mathbf{P}^{-1}(\mathbf{A} - \lambda_n\mathbf{I})\mathbf{P}\mathbf{P}^{-1} \\
&= \mathbf{P}(\mathbf{P}^{-1}\mathbf{A}\mathbf{P} - \lambda_1\mathbf{I})(\mathbf{P}^{-1}\mathbf{A}\mathbf{P} - \lambda_2\mathbf{I}) \ldots (\mathbf{P}^{-1}\mathbf{A}\mathbf{P} - \lambda_n\mathbf{I})\mathbf{P}^{-1} \\
&= \mathbf{P}(\mathbf{B} - \lambda_1\mathbf{I})(\mathbf{B} - \lambda_2\mathbf{I}) \ldots (\mathbf{B} - \lambda_n\mathbf{I})\mathbf{P}^{-1} \\
&= \mathbf{0}.
\end{aligned}$$

Now, since $\mathbf{B}$ is upper triangular, we have

$$\chi_{\mathbf{B}}(x) = (x - \lambda_1)(x - \lambda_2) \ldots (x - \lambda_n).$$

Since $\mathbf{P}^{-1}\mathbf{A}\mathbf{P} = \mathbf{B}$, then $\chi_{\mathbf{A}}(x) = \chi_{\mathbf{B}}(x)$; hence $\chi_{\mathbf{A}}(\mathbf{A}) = \mathbf{0}$ as required. $\square$

As usual, the proof just given works over an arbitrary field F in place of $\mathbb{R}$ or $\mathbb{C}$ by simply replacing $\mathbb{C}$ by the algebraic closure $\overline{F}$ of F.

Example 11.15 The matrix

$$\mathbf{A} = \begin{pmatrix} 1 & 2 \\ -1 & 0 \end{pmatrix}$$

has characteristic polynomial defined by

$$\chi_{\mathbf{A}}(x) = \det(\mathbf{A} - x\mathbf{I}) = \begin{vmatrix} 1 - x & 2 \\ -1 & -x \end{vmatrix} = x^2 - x + 2.$$

We have

$$\chi_{\mathbf{A}}(\mathbf{A}) = \begin{pmatrix} -1 & 2 \\ -1 & -2 \end{pmatrix} - \begin{pmatrix} 1 & 2 \\ -1 & 0 \end{pmatrix} + \begin{pmatrix} 2 & 0 \\ 0 & 2 \end{pmatrix} = \begin{pmatrix} 0 & 0 \\ 0 & 0 \end{pmatrix}.$$

The Cayley–Hamilton theorem has the following immediate corollary.

Corollary 11.16 *For a square matrix* $\mathbf{A}$, $m_{\mathbf{A}}(x)$ *divides* $\chi_{\mathbf{A}}(x)$.

Proof Immediate from Theorem 11.14 and Proposition 11.3. $\square$

What this means is that the characteristic polynomial gives a rather direct way of computing minimum polynomials. Rather than finding all the eigenvalues of a matrix $\mathbf{A}$ by the rather long-winded method we have used so far and looking at the product of various polynomials of the form $(x - \lambda)$ where λ is an eigenvalue, we now see that it suffices to compute the characteristic polynomial and look at polynomials dividing that.

Example 11.17 To find the minimum polynomial of

$$A = \begin{pmatrix} 5 & 3 & 1 \\ -1 & 1 & -1 \\ 2 & 6 & 6 \end{pmatrix},$$

first find the characteristic polynomial $\chi_A(x) = \det(A - xI)$. This is

$$\begin{vmatrix} 5-x & 3 & 1 \\ -1 & 1-x & -1 \\ 2 & 6 & 6-x \end{vmatrix} = (5-x)(6-7x+x^2+6)$$
$$- 3(-6+x+2) + (-6-2+2x)$$
$$= 60 - 35x + 5x^2 - 12x + 7x^2 - x^3 + 12 - 3x - 8 + 2x$$
$$= 64 - 48x + 12x^2 - x^3$$
$$= -(x-4)^3,$$

so $m_A(x)$ is $(x-4)$, $(x-4)^2$, or $(x-4)^3$. Now

$$(A - 4I) = \begin{pmatrix} 1 & 3 & 1 \\ -1 & -3 & -1 \\ 2 & 6 & 2 \end{pmatrix} \neq 0$$

so $m_A(x)$ is not $(x-4)$. On the other hand,

$$(A - 4I)^2 = \begin{pmatrix} 1 & 3 & 1 \\ -1 & -3 & -1 \\ 2 & 6 & 2 \end{pmatrix} \begin{pmatrix} 1 & 3 & 1 \\ -1 & -3 & -1 \\ 2 & 6 & 2 \end{pmatrix} = \begin{pmatrix} 0 & 0 & 0 \\ 0 & 0 & 0 \\ 0 & 0 & 0 \end{pmatrix}$$

so $m_A(x) = (x-4)^2$.

In doing calculations of this sort, it is useful to remember Theorem 11.4 which says that *every eigenvalue of a matrix* **A** *appears as a root of the minimum polynomial*. This limits the number of possible polynomials considerably, and can sometimes enable you to identify the minimum polynomial directly without any matrix multiplications at all.

Example 11.18 Consider the matrix

$$A = \begin{pmatrix} 3 & 0 & 2 \\ -4 & 2 & -5 \\ -4 & 0 & -3 \end{pmatrix}.$$

We compute

$$\chi_A = (2-x) \begin{vmatrix} 3-x & 2 \\ -4 & -3-x \end{vmatrix} = -(2-x)(x^2-1) = -(x-2)(x-1)(x+1).$$

Thus **A** has three distinct eigenvalues, $2, 1, -1$, and $m_A(x) = (x-2)(x-1)(x+1)$.

Example 11.19 Let **A** be the matrix

$$\begin{pmatrix} 3 & -4 & 2 \\ 1 & -1 & 1 \\ 1 & -2 & 2 \end{pmatrix}.$$

We compute $(\mathbf{A} - \mathbf{I})(\mathbf{A} - 2\mathbf{I})$.

$$(\mathbf{A} - \mathbf{I})(\mathbf{A} - 2\mathbf{I}) = \begin{pmatrix} 2 & -4 & 2 \\ 1 & -2 & 1 \\ 1 & -2 & 1 \end{pmatrix} \begin{pmatrix} 1 & -4 & 2 \\ 1 & -3 & 1 \\ 1 & -2 & 0 \end{pmatrix} = \begin{pmatrix} 0 & 0 & 0 \\ 0 & 0 & 0 \\ 0 & 0 & 0 \end{pmatrix}.$$

Since $\mathbf{A} - \mathbf{I} \neq \mathbf{0}$ and $\mathbf{A} - 2\mathbf{I} \neq \mathbf{0}$, we deduce that $m_{\mathbf{A}}(x) = (x - 1)(x - 2)$ and that 1 and 2 are the only eigenvalues of **A**.

The following terminology is often useful.

Definition 11.20 *The **algebraic multiplicity** of an eigenvalue λ of a linear transformation f or a matrix **A** is the multiplicity of λ as a root of $\chi_f(x)$ or $\chi_{\mathbf{A}}(x)$.*

In Example 11.17, 4 is an eigenvalue of **A** of algebraic multiplicity 4, and the eigenvalues in Example 11.18 all have algebraic multiplicity 1. One of the eigenvalues $1, 2$ in Example 11.19 has algebraic multiplicity 1 and the other has algebraic multiplicity 2—it is not possible to determine which is which without further calculation.

We conclude this chapter with another proof of the Cayley–Hamilton theorem, this time without using the notion of algebraic closure of a field. This is the usual proof found in most textbooks. It is rather striking in the way the cancellations appear at the end, and perhaps for this reason only it is well worth reading, but to us it is not as clear an explanation of *why* the theorem is true as the proof given earlier.

Second proof of Theorem 11.14   Let $\mathbf{B} = \mathbf{A} - x\mathbf{I}$, so that $\chi_{\mathbf{A}}(x) = \det(\mathbf{B})$. Since this is a polynomial of degree n we can write

$$\chi_{\mathbf{A}}(x) = b_0 + b_1 x + \cdots + b_n x^n.$$

The crux of this proof is to consider the adjugate matrix, adj(**B**), of **B**, which has the property that $\mathbf{B}\,\text{adj}(\mathbf{B}) = \det(\mathbf{B})\mathbf{I}$. This matrix, adj(**B**), is the transpose of the matrix of cofactors of **B**, an $n \times n$ matrix whose entries are plus or minus determinants of certain $(n - 1) \times (n - 1)$ submatrices of **B**. Each of these determinants is therefore a polynomial of degree at most $n - 1$, so

$$\text{adj}(\mathbf{B}) = \begin{pmatrix} p_{11}(x) & \cdots & p_{1n}(x) \\ \vdots & \ddots & \vdots \\ p_{n1}(x) & \cdots & p_{nn}(x) \end{pmatrix}$$

where each $p_{ij}(x)$ is a polynomial of degree at most $n - 1$. Now we can separate out the constant terms of the entries of adj(**B**) into a constant matrix, and the x terms into x times a constant matrix, and so on, to give

$$\mathrm{adj}(\mathbf{B}) = \mathbf{B}_0 + \mathbf{B}_1 x + \cdots + \mathbf{B}_{n-1} x^{n-1} \qquad (2)$$

where $\mathbf{B}_0, \dots, \mathbf{B}_{n-1}$ are $n \times n$ matrices.

We know that $\mathbf{B}\,\mathrm{adj}(\mathbf{B}) = \det(\mathbf{B})\mathbf{I}$, so

$$
\begin{aligned}
\det(\mathbf{B})\mathbf{I} &= \mathbf{B}\,\mathrm{adj}(\mathbf{B}) \\
&= (\mathbf{A} - x\mathbf{I})\,\mathrm{adj}(\mathbf{B}) \\
&= \mathbf{A}\,\mathrm{adj}(\mathbf{B}) - x\,\mathrm{adj}(\mathbf{B}),
\end{aligned}
$$

and both sides are polynomials in x with matrix coefficients. Now

$$\det(\mathbf{B}) = \chi_{\mathbf{A}}(x) = b_0 + b_1 x + \cdots + b_n x^n$$

so we have

$$\det(\mathbf{B})\mathbf{I} = b_0\mathbf{I} + b_1 x\mathbf{I} + \cdots + b_n x^n \mathbf{I}.$$

Also, using the expression (2) for $\mathrm{adj}(\mathbf{B})$ above, we have

$$
\begin{aligned}
\mathbf{A}\,\mathrm{adj}(\mathbf{B}) - x\,\mathrm{adj}(\mathbf{B}) = {} & \mathbf{A}\mathbf{B}_0 + \mathbf{A}\mathbf{B}_1 x + \cdots + \mathbf{A}\mathbf{B}_{n-1} x^{n-1} \\
& - \mathbf{B}_0 x - \mathbf{B}_1 x^2 - \cdots - \mathbf{B}_{n-1} x^n.
\end{aligned}
$$

Equating coefficients of powers of x in these two expressions gives

$$
\begin{aligned}
b_0 \mathbf{I} &= \mathbf{A}\mathbf{B}_0 \\
b_1 \mathbf{I} &= -\mathbf{B}_0 + \mathbf{A}\mathbf{B}_1 \\
b_2 \mathbf{I} &= -\mathbf{B}_1 + \mathbf{A}\mathbf{B}_2 \\
&\ \ \vdots \\
b_{n-1} \mathbf{I} &= -\mathbf{B}_{n-2} + \mathbf{A}\mathbf{B}_{n-1} \\
b_n \mathbf{I} &= -\mathbf{B}_{n-1}.
\end{aligned}
$$

If we multiply these $n+1$ equations on the left by $\mathbf{I}, \mathbf{A}, \mathbf{A}^2, \dots, \mathbf{A}^n$ respectively, we obtain

$$
\begin{aligned}
b_0 \mathbf{I} &= \mathbf{A}\mathbf{B}_0 \\
b_1 \mathbf{A} &= -\mathbf{A}\mathbf{B}_0 + \mathbf{A}^2 \mathbf{B}_1 \\
b_2 \mathbf{A}^2 &= -\mathbf{A}^2 \mathbf{B}_1 + \mathbf{A}^3 \mathbf{B}_2 \\
&\ \ \vdots \\
b_{n-1} \mathbf{A}^{n-1} &= -\mathbf{A}^{n-1} \mathbf{B}_{n-2} + \mathbf{A}^n \mathbf{B}_{n-1} \\
b_n \mathbf{A}^n &= -\mathbf{A}^n \mathbf{B}_{n-1}.
\end{aligned}
$$

Adding up these equations, all the terms on the right-hand side cancel out, and we are left with

$$b_0\mathbf{I} + b_1\mathbf{A} + \cdots + b_n\mathbf{A}^n = 0.$$

In other words, $\chi_{\mathbf{A}}(\mathbf{A}) = 0$. $\qquad\qquad\square$

Exercises

Exercise 11.1 Calculate the eigenvalues of the following matrices.

(a) $\begin{pmatrix} 1 & 2 & 0 \\ 2 & 1 & 2 \\ 0 & 2 & 1 \end{pmatrix}$ (b) $\begin{pmatrix} 1 & 0 & 2 \\ 0 & 1 & 0 \\ 2 & 0 & 1 \end{pmatrix}$ (c) $\begin{pmatrix} 1 & i & 0 \\ -i & 1 & 1 \\ 0 & 1 & -1/3 \end{pmatrix}$.

Exercise 11.2 Calculate the minimum polynomial and the characteristic polynomial of each of the following matrices.

(a) $\begin{pmatrix} 1 & 1 & 0 \\ -9 & -4 & 1 \\ -3 & 3 & 2 \end{pmatrix}$ (b) $\begin{pmatrix} -2 & -3 & -3 \\ -1 & 0 & -1 \\ 0 & 1 & -1 \end{pmatrix}$ (c) $\begin{pmatrix} -1 & -3 & 6 \\ -1 & 1 & -7 \\ 0 & 1 & -3 \end{pmatrix}$.

Exercise 11.3 Let $\mathbf{A}$ be the matrix

$$\begin{pmatrix} 3 & -2 & -3 & -2 \\ -1 & 4 & 3 & 2 \\ 1 & 2 & 3 & -2 \\ 0 & -1 & -1 & 4 \end{pmatrix}.$$

Compute $(\mathbf{A} - 4\mathbf{I})(\mathbf{A} - 6\mathbf{I})(\mathbf{A} - 2\mathbf{I})$ and $(\mathbf{A} - 4\mathbf{I})(\mathbf{A} - 6\mathbf{I})(\mathbf{A} - 2\mathbf{I})^2$.

Say what you can deduce from your answers.

Exercise 11.4 Which of the eigenvectors 1, 2 in Example 11.19 has algebraic multiplicity 2?

Exercise 11.5 Let $\mathbf{A} = \begin{pmatrix} 1 & 2 & 2 \\ 2 & 1 & 2 \\ 2 & 2 & 1 \end{pmatrix}$. Verify that 5 is an eigenvalue and find all

other eigenvalues and their algebraic multiplicities.

For each eigenvalue λ, find a basis of the subspace

$$\ker(\mathbf{A} - \lambda\mathbf{I}) = \{\mathbf{v} \in \mathbb{R}^3 : (\mathbf{A} - \lambda\mathbf{I})\mathbf{v} = 0\}$$

of the real vector space $\mathbb{R}^3$.

Exercise 11.6 Find the roots and their multiplicities of the characteristic equation of the 4×4 real symmetric matrix $\mathbf{A}$ where

$$\mathbf{A} = \begin{pmatrix} 0 & 1 & 0 & 1 \\ 1 & 0 & 1 & 0 \\ 0 & 1 & 0 & 1 \\ 1 & 0 & 1 & 0 \end{pmatrix}.$$

(You may find it easier to spot roots from the determinant using row operations rather than calculating the characteristic polynomial.)

Hence find a basis $\{\mathbf{v}_1, \mathbf{v}_2, \mathbf{v}_3, \mathbf{v}_4\}$ of $\mathbb{R}^4$ consisting of eigenvectors of $\mathbf{A}$. Write down the matrix of the linear transformation $T: \mathbb{R}^4 \to \mathbb{R}^4$ defined by $T(\mathbf{v}) = \mathbf{A}\mathbf{v}$ with respect to your basis.

Exercise 11.7 Compute

$$\begin{pmatrix} 0 & 1 & 0 & \ldots & 0 \\ 0 & 0 & 1 & \ldots & 0 \\ \vdots & & \ddots & \ddots & \vdots \\ \vdots & & & \ddots & 1 \\ 0 & \ldots & & 0 & 0 \end{pmatrix} \begin{pmatrix} a_{11} & \ldots & a_{1n} \\ a_{21} & \ldots & a_{2n} \\ \vdots & & \vdots \\ \vdots & & \vdots \\ a_{n1} & \ldots & a_{nn} \end{pmatrix}.$$

Hence find the characteristic and minimum polynomials of

$$\begin{pmatrix} \lambda & 1 & 0 & \ldots & 0 \\ 0 & \lambda & 1 & \ldots & 0 \\ \vdots & & \ddots & \ddots & \vdots \\ \vdots & & & \ddots & 1 \\ 0 & \ldots & & 0 & \lambda \end{pmatrix}.$$

Exercise 11.8 Define $f: \mathbb{C}^3 \to \mathbb{C}^3$ by $f(a,b,c)^T = (a+3b-c, 2a-b, b+2c)^T$. Find the characteristic polynomial $\chi_f(x)$ of f, and verify that $\chi_f(f)$ is the zero map. Show that $\chi_f(x)$ has exactly one real root, which is between -2 and -3, and deduce that the minimum polynomial $m_f(x)$ of f is the negative of the characteristic polynomial.

Exercise 11.9 Let $\mathbf{A}$ be an $n \times n$ matrix over $\mathbb{C}$. Use the invariance of trace to prove that the trace of $\mathbf{A}$ is equal to the sum of the eigenvalues of $\mathbf{A}$, counting each eigenvalue m times where m is its algebraic multiplicity.

Similarly, show that the determinant of $\mathbf{A}$ is the product of the eigenvalues counting algebraic multiplicity.

Exercise 11.10 Using the definition of the (j,k)th entry in a matrix product, show that $\mathrm{tr}(\mathbf{AB}) = \mathrm{tr}(\mathbf{BA})$ for matrices $\mathbf{A} = (a_{ij})$ and $\mathbf{B} = (b_{ij})$. Hence deduce $\mathrm{tr}(\mathbf{P}^{-1}\mathbf{AP}) = \mathrm{tr}(\mathbf{A})$ for any invertible matrix $\mathbf{P}$.

12

Diagonalization

Our basic aim is to find as nice a matrix as possible representing a given linear transformation. That is, given a vector space V and a linear transformation $f: V \to V$, we want to find a basis for V with respect to which the matrix of f is 'nice'. In this chapter, 'nice' will mean *diagonal*, if possible.

12.1 Diagonal matrices

The process of diagonalization is where one tries to transform a matrix $\mathbf{A}$ to a matrix $\mathbf{P}^{-1}\mathbf{A}\mathbf{P}$ which is diagonal, or to find a basis with respect to which the matrix representation of a given linear transformation f is diagonal. We have already seen in examples that this process is often useful or necessary for solving certain kinds of simultaneous equations, such as the example in Section 10.1.

Unfortunately, some matrices or linear transformations cannot be diagonalized in this way, and there are two main restrictions preventing diagonalization.

The first restriction is illustrated by Example 10.14. There, we saw that the matrix

$$\begin{pmatrix} 0 & 1 \\ -1 & 0 \end{pmatrix}$$

cannot even be put in upper triangular form over the reals. The problem here is that this matrix does not have real eigenvalues, and as pointed out in Proposition 10.15, the diagonal entries in any upper triangular form for a matrix $\mathbf{A}$ are precisely the eigenvalues of $\mathbf{A}$.

The eigenvalues of the matrix $\mathbf{A}$ can be described as being the roots of a particular polynomial $p(x)$. (Here, $p(x)$ may be taken to be the minimum polynomial of $\mathbf{A}$ or the characteristic polynomial of $\mathbf{A}$, it doesn't matter.) So, for $\mathbf{A}$ to be diagonalizable, it is necessary that $p(x)$ has the maximum number of roots in the field of scalars. For this reason, when we are investigating whether or not $\mathbf{A}$ can be diagonalized, we will usually assume these roots always exist; for example, by working over the field of complex numbers or some other algebraically closed field, or by simply assuming that the characteristic polynomial has its maximum number of roots in the field.

It is still not always possible to get a basis with respect to which the matrix is diagonal, even if the matrix in question is upper triangular, as the following example shows.

Example 12.1 Let $V = \mathbb{C}^2$, and define $f: V \to V$ by

$$f\begin{pmatrix} x \\ y \end{pmatrix} = \begin{pmatrix} x + y \\ y \end{pmatrix}.$$

With respect to the standard basis of V, the matrix of f is $\begin{pmatrix} 1 & 1 \\ 0 & 1 \end{pmatrix}$. So if we apply the base-change matrix $\begin{pmatrix} a & b \\ c & d \end{pmatrix}$ (assuming this *is* a base-change matrix, i.e. is nonsingular, so has determinant $ad - bc \neq 0$) we obtain the new matrix for f as follows.

$$\begin{pmatrix} a & b \\ c & d \end{pmatrix}^{-1} \begin{pmatrix} 1 & 1 \\ 0 & 1 \end{pmatrix} \begin{pmatrix} a & b \\ c & d \end{pmatrix} = \frac{1}{ad - bc} \begin{pmatrix} d & -b \\ -c & a \end{pmatrix} \begin{pmatrix} a + c & b + d \\ c & d \end{pmatrix}$$

$$= \frac{1}{ad - bc} \begin{pmatrix} ad + cd - bc & d^2 \\ -c^2 & -bc - cd + ad \end{pmatrix}.$$

If this is diagonal, then $c = d = 0$, which contradicts the fact that $\begin{pmatrix} a & b \\ c & d \end{pmatrix}$ is invertible.

The purpose of this chapter is to investigate this second restriction in more detail, and to show how to find diagonal representations when these exist. We start with a definition of the concept to be studied.

Definition 12.2 *A linear transformation* $f: V \to V$ *is called* **diagonalizable** *if there exists a basis of V with respect to which the matrix of f is diagonal.*

Similarly, a matrix *which represents a linear transformation* is called *diagonalizable* if there exists a matrix $\mathbf{P}$ such that $\mathbf{P}^{-1}\mathbf{A}\mathbf{P}$ is diagonal.

This concept is intimately connected with eigenvectors.

Proposition 12.3 *A linear transformation* $f: V \to V$ *is diagonalizable if and only if there is a basis of V consisting of eigenvectors of f.*

Proof If $\{v_1, \dots, v_n\}$ is a basis of V consisting of eigenvectors of f, then we have $f(v_i) = \lambda_i v_i$ for some scalars λ_i, and the matrix of f is by definition

$$\begin{pmatrix} \lambda_1 & 0 & \dots & 0 \\ 0 & \lambda_2 & & \vdots \\ \vdots & & \ddots & 0 \\ 0 & \dots & 0 & \lambda_n \end{pmatrix}.$$

Conversely, if the matrix of f with respect to $\{v_1, \dots, v_n\}$ is as above, then $f(v_i) = \lambda_i v_i$, so the basis vectors v_i are eigenvectors of f. $\qquad\square$

12.2   A criterion for diagonalizability

This section uses the ideas from the previous chapter to give a precise criterion for when a matrix or linear transformation is diagonalizable.

Recall that in the last chapter we identified two polynomials whose roots are precisely the eigenvalues of a linear transformation $f: V \to V$, namely the minimum polynomial $m_f(x)$ and the characteristic polynomial $\chi_f(x)$. If f is diagonalizable, then the minimum polynomial of f takes a particularly simple form.

Theorem 12.4 *Let V be a vector space, and let $f: V \to V$ be a linear transformation. If f is diagonalizable, then the minimum polynomial $m_f(x)$ of f has distinct roots.*

Proof Assume that f is diagonalizable, so there is a basis of eigenvectors $\{v_1, \ldots, v_n\}$ of V. Let $\lambda_1, \ldots, \lambda_r$ be the distinct eigenvalues of f, and define the polynomial

$$p(x) = (x - \lambda_1)(x - \lambda_2) \ldots (x - \lambda_r).$$

Then $p(f) = (f - \lambda_1 I)(f - \lambda_2 I) \ldots (f - \lambda_r I)$, and the factors $(f - \lambda_i I)$ all commute with one another by Proposition 9.5. Now each basis vector v_i is an eigenvector, with eigenvalue λ_j for some j, and therefore we have $(f - \lambda_j I)(v_i) = 0$. It follows that

$$
\begin{aligned}
p(f)(v_i) &= (f - \lambda_1 I)(f - \lambda_2 I) \ldots (f - \lambda_r I)v_i \\
&= (f - \lambda_1 I) \ldots (f - \lambda_{j-1} I)(f - \lambda_{j+1} I) \ldots (f - \lambda_r I)(f - \lambda_j I)v_i \\
&= 0
\end{aligned}
$$

for each i, and so $p(f)$ is the zero transformation. Therefore $m_f(x)$ divides $p(x)$, by Proposition 11.3. But $p(x)$ has distinct roots, and therefore $m_f(x)$ has distinct roots. $\qquad\square$

In fact, the converse of this result is also true, and we will prove it in due course (see Corollary 12.13). Notice that when we have proved this, we will have a criterion which we can use to determine if a particular linear transformation is diagonalizable. Specifically: f is diagonalizable if and only if $m_f(x)$ has distinct roots.

Recall from Theorem 10.2 that if $f: V \to V$, λ is a scalar, and $v \in V$, then $v \in \ker(f - \lambda I)$ if and only if v is either zero or an eigenvector of f with eigenvalue λ.

Definition 12.5 *If λ is an eigenvalue of a linear transformation $f: V \to V$, then the subspace $\ker(f - \lambda I)$ is called the λ-eigenspace of f. Its dimension is called the **geometric multiplicity** of the eigenvalue λ.*

We will prove that if the minimum polynomial of f has distinct roots, then V is the 'direct sum' of these eigenspaces. In other words, if we take whatever basis B_λ we like for the λ-eigenspace $\ker(f - \lambda I)$, and do this for all eigenvalues $\lambda_1, \lambda_2, \ldots, \lambda_k$, then

$$B_{\lambda_1} \cup B_{\lambda_2} \cup \cdots \cup B_{\lambda_k}$$

is automatically a basis of V, provided $m_f(x)$ has distinct roots.

Definition 12.6 V *is the* **direct sum** *of subspaces* $V_1, \ldots, V_r$, *if every vector* $v \in V$ *can be written uniquely as a sum* $v = v_1 + \cdots + v_r$, *where* $v_i \in V_i$. *We write* $V = V_1 \oplus \cdots \oplus V_r$ *or* $V = \bigoplus_{i=1}^r V_i$.

A useful way to think of direct sums is that V is a direct sum $V = \bigoplus_{i=1}^r V_i$ if and only if whenever we have bases B_i of V_i, then their union $\bigcup_i B_i$ is a basis of V, as the following proposition shows.

Proposition 12.7 *If* V_i *is a subspace of a vector space* V, *where* $V = \bigoplus_{i=1}^r V_i$, *and if* B_i *is a basis for* V_i *for each* i, *then* $\bigcup_{i=1}^r B_i$ *is a basis for* V.

Proof First, every vector $v \in V$ can be written as $v = v_1 + \cdots + v_r$, where $v_i \in V_i$. Therefore for each i, v_i is a linear combination of the vectors in B_i, and so v is a linear combination of the vectors in $\bigcup_{i=1}^r B_i$. Thus $\bigcup_{i=1}^r B_i$ spans the space V.

Now suppose that there is a linear dependence among the vectors of $\bigcup_{i=1}^r B_i$. Write $B_1 = \{a_1, \ldots, a_k\}$, $B_2 = \{b_1, \ldots, b_l\}$, $\ldots$, $B_r = \{z_1, \ldots, z_t\}$, and suppose the linear dependence is

$$(\lambda_1 a_1 + \cdots + \lambda_k a_k) + (\mu_1 b_1 + \cdots + \mu_l b_l) + \cdots + (\xi_1 z_1 + \cdots + \xi_t z_t) = 0.$$

Then $v_1 = \lambda_1 a_1 + \cdots + \lambda_k a_k \in V_1$, $v_2 = \mu_1 b_1 + \cdots + \mu_l b_l \in V_2$, and so on. Thus we have written the zero vector as

$$0 = v_1 + v_2 + \cdots + v_r,$$

where $v_i \in V_i$. But since $0 = 0 + \cdots + 0$ is the *unique* way of writing 0 as a sum of vectors in the V_i, we must have $v_i = 0$ for all i. This implies that $\lambda_1 a_1 + \cdots + \lambda_k a_k = 0$, so all the $\lambda_i = 0$, since $\{a_1, \ldots, a_k\}$ is a basis for V_1. Similarly, $\mu_i = 0, \ldots, \xi_i = 0$. In other words $\bigcup_{i=1}^r B_i$ is a linearly independent subset. $\square$

Corollary 12.8 *If* $V = \bigoplus_{i=1}^r V_i$, *then* $\dim(V) = \sum_{i=1}^r \dim(V_i)$.

Proof Immediate from Proposition 12.7. $\square$

The main theorem of this section is the following, from which we can easily deduce the converse to Theorem 12.4.

Theorem 12.9 *Let* V *be a complex vector space, and let* $f : V \to V$ *be a linear transformation. Suppose that*

$$m_f(x) = (x - \lambda_1)(x - \lambda_2) \ldots (x - \lambda_r)$$

with $\lambda_1, \lambda_2, \ldots, \lambda_r$ *distinct, and let* V_i *be the* λ_i-*eigenspace of* f. *Then*

$$V = V_1 \oplus V_2 \oplus \cdots \oplus V_r.$$

According to Definition 12.6, there are two parts to proving V is a direct sum of subspaces: uniqueness (proved below in Corollary 12.11) and existence (Proposition 12.12). The first part uses the fact that eigenvectors with distinct eigenvalues are linearly independent. More formally,

Proposition 12.10 *Let $f: V \to V$ be a linear transformation, and suppose that $v_1, \ldots, v_r$ are eigenvectors of f with distinct eigenvalues $\lambda_1, \ldots, \lambda_r$ respectively. Then $\{v_1, \ldots, v_r\}$ is a linearly independent set.*

Proof Suppose not, and let k be the smallest integer such that $\{v_1, \ldots, v_k\}$ is linearly dependent. In particular, $\{v_1, \ldots, v_{k-1}\}$ is linearly independent, and there exists a linear dependence

$$\alpha_1 v_1 + \cdots + \alpha_k v_k = 0$$

with $\alpha_k \neq 0$. Moreover, $v_k \neq 0$ so at least one other α_i is nonzero ($i < k$). Applying f to both sides of this equation we obtain

$$
\begin{aligned}
0 &= f(0) \\
&= f(\alpha_1 v_1 + \cdots + \alpha_k v_k) \\
&= \alpha_1 f(v_1) + \cdots + \alpha_k f(v_k) \\
&= \alpha_1 \lambda_1 v_1 + \cdots + \alpha_k \lambda_k v_k.
\end{aligned}
$$

Now subtract λ_k times the first equation from the second, to obtain

$$\alpha_1(\lambda_1 - \lambda_k)v_1 + \alpha_2(\lambda_2 - \lambda_k)v_2 + \cdots + \alpha_{k-1}(\lambda_{k-1} - \lambda_k)v_{k-1} = 0.$$

But $\lambda_i - \lambda_k \neq 0$ since $\lambda_1, \ldots, \lambda_k$ are distinct, so this is a nontrivial linear dependence among $\{v_1, \ldots, v_{k-1}\}$, contradicting the fact that these vectors are linearly independent. $\qquad\square$

Corollary 12.11 *Let $f: V \to V$ be a linear transformation, and suppose that $v \in V$ can be written as $v = v_1 + \cdots + v_r$ where v_i is an eigenvector of f with eigenvalue λ_i, and $\lambda_1, \ldots, \lambda_r$ are distinct. If also $v = w_1 + \cdots + w_r$ with each w_i an eigenvector of f with eigenvalue λ_i, then $v_i = w_i$ for each i.*

Proof Otherwise

$$(v_1 - w_1) + \cdots + (v_r - w_r) = 0$$

is a nontrivial linear dependence of eigenvectors with distinct eigenvalues, contradicting Proposition 12.10. $\qquad\square$

This proves the uniqueness part of Theorem 12.9. The existence part is a little harder.

Proposition 12.12 *Suppose that $f: V \to V$ is a linear transformation with minimum polynomial*

$$m_f(x) = (x - \lambda_1) \ldots (x - \lambda_r)$$

where $\lambda_1, \ldots, \lambda_r$ are distinct, and suppose that $v \in V$. Then there exist eigenvectors $v_1, \ldots, v_r$ of f such that $v = v_1 + \cdots + v_r$.

Proof For each j in turn consider the polynomial $p_j(x)$ defined by

$$p_j(x) = \frac{(x - \lambda_1)\ldots(x - \lambda_{j-1})(x - \lambda_{j+1})\ldots(x - \lambda_r)}{(\lambda_j - \lambda_1)\ldots(\lambda_j - \lambda_{j-1})(\lambda_j - \lambda_{j+1})\ldots(\lambda_j - \lambda_r)}$$
$$= \prod_{i \neq j} \frac{x - \lambda_i}{\lambda_j - \lambda_i}.$$

Note that $p_j(x)$ is well-defined (since all the λ_i are distinct) and that $p_j(\lambda_j) = 1$, while $p_j(\lambda_k) = 0$ if $k \neq j$.

Now consider the polynomial $p(x)$ defined by

$$p(x) = \sum_{j=1}^{r} p_j(x).$$

This has the property that $p(\lambda_i) = 1$ for each i since $p_j(\lambda_i) = 0$ if $i \neq j$ and $p_i(\lambda_i) = 1$. Thus the polynomial $p(x) - 1$ has roots $\lambda_1, \ldots, \lambda_r$. But $p(x) - 1$ has degree at most $r - 1$, since each $p_j(x)$ has degree $r - 1$, so $p(x) - 1$ has at most $r - 1$ roots. Since all the λ_i are distinct, the only way this can happen is if $p(x) - 1$ is identically 0. Thus $p(x)$ is identically 1. Hence $p(f)$ is the identity linear transformation, and so $p(f)(v) = v$. But

$$p(f) = \sum_{j=1}^{r} p_j(f)$$

so

$$v = p(f)(v) = \left(\sum_{j=1}^{r} p_j(f)\right)(v)$$
$$= \sum_{j=1}^{r} p_j(f)(v) = \sum_{j=1}^{r} v_j,$$

where

$$v_j = p_j(f)(v)$$
$$= \frac{(f - \lambda_1 \mathrm{I})\ldots(f - \lambda_{j-1}\mathrm{I})(f - \lambda_{j+1}\mathrm{I})\ldots(f - \lambda_r \mathrm{I})}{(\lambda_j - \lambda_1)\ldots(\lambda_j - \lambda_{j-1})(\lambda_j - \lambda_{j+1})\ldots(\lambda_j - \lambda_r)}(v).$$

Applying $f - \lambda_j \mathrm{I}$ to both sides of this we conclude that $(f - \lambda_j \mathrm{I})(v_j)$ is a scalar multiple of $m_f(f)(v)$, which is 0. Therefore $f(v_j) = \lambda_j v_j$, so v_j is an eigenvector of f with eigenvalue λ_j, as required. $\qquad\square$

An alternative proof of this proposition can be given by Proposition 9.14 and induction. See also Proposition 12.17 below.

Theorem 12.9 now follows immediately from Corollary 12.11 and Proposition 12.12.

Corollary 12.13 *Suppose that* $f \colon V \to V$ *is a linear transformation with minimum polynomial*

$$m_f(x) = (x - \lambda_1) \dots (x - \lambda_r)$$

where $\lambda_1, \dots, \lambda_r$ *are distinct. Then* f *is diagonalizable.*

Proof Choose a basis B_i for each eigenspace V_i. Then by Proposition 12.7, $\bigcup_{i=1}^r B_i$ is a basis for V, and every element of this basis is by definition an eigenvector of f. The result follows from Proposition 12.3. $\qquad\square$

12.3 Examples

We shall start by discussing some matrices whose minimum polynomials were calculated in the previous chapter.

Example 12.14 In Example 11.17 we showed that the matrix

$$\mathbf{A} = \begin{pmatrix} 5 & 3 & 1 \\ -1 & 1 & -1 \\ 2 & 6 & 6 \end{pmatrix}$$

has minimum polynomial $m_{\mathbf{A}}(x) = x^2 - 8x + 16$. Since this polynomial factorizes completely into linear factors over $\mathbb{R}$, $\mathbf{A}$ is similar to an upper triangular matrix, but $m_{\mathbf{A}}(x) = (x - 4)^2$ has 4 as a repeated root, so Theorem 12.4 says that $\mathbf{A}$ cannot be diagonalized.

Example 12.15 The matrix

$$\mathbf{A} = \begin{pmatrix} 1 & -1 \\ 1 & 1 \end{pmatrix}$$

over $\mathbb{R}$ has minimum polynomial $x^2 - 2x + 2$. To see this, it suffices to check that $\mathbf{A}^2 - 2\mathbf{A} + 2\mathbf{I} = \mathbf{0}$, and that $x^2 - 2x + 2$ cannot be factorized over $\mathbb{R}$. However, $m_{\mathbf{A}}(x)$ has no real roots, so $\mathbf{A}$ is not similar to any upper triangular matrix, let alone a diagonal one.

Over $\mathbb{C}$, the situation is different as $x^2 - 2x + 2 = (x - (1 + i))(x - (1 - i))$ which has two distinct roots in $\mathbb{C}$, so $\mathbf{A}$ is diagonalizable over $\mathbb{C}$. In fact, we find that

$$\begin{pmatrix} i & -i \\ 1 & 1 \end{pmatrix}^{-1} \mathbf{A} \begin{pmatrix} i & -i \\ 1 & 1 \end{pmatrix} = \begin{pmatrix} 1 + i & 0 \\ 0 & 1 - i \end{pmatrix},$$

as you may check.

Example 12.16 Consider the matrix

$$\mathbf{A} = \begin{pmatrix} 3 & 0 & 2 \\ -4 & 2 & -5 \\ -4 & 0 & -3 \end{pmatrix}$$

of Example 11.18. This has

$$m_{\mathbf{A}}(x) = (x - 2)(x - 1)(x + 1)$$

so the eigenvalues of $\mathbf{A}$ are $2, 1, -1$. The minimum polynomial $m_{\mathbf{A}}(x)$ has its maximum number of roots (three) in $\mathbb{R}$ and all these roots are distinct, so $\mathbf{A}$ is diagonalizable. In other words, there is an invertible 3×3 matrix $\mathbf{P}$ with real entries so that

$$\mathbf{P}^{-1}\mathbf{A}\mathbf{P} = \begin{pmatrix} 2 & 0 & 0 \\ 0 & 1 & 0 \\ 0 & 0 & -1 \end{pmatrix}.$$

To find such a matrix $\mathbf{P}$, it suffices to find a basis of eigenvectors of $\mathbf{A}$, since $\mathbf{P}$ is just the base-change matrix from the usual basis to a basis of eigenvectors. These eigenvectors can be found as usual by solving simultaneous equations.

For eigenvalue 2, we need to solve

$$(\mathbf{A} - 2\mathbf{I}) \begin{pmatrix} x \\ y \\ z \end{pmatrix} = \begin{pmatrix} 1 & 0 & 2 \\ -4 & 0 & -5 \\ -4 & 0 & -5 \end{pmatrix} \begin{pmatrix} x \\ y \\ z \end{pmatrix} = \mathbf{0}.$$

The full solution is that $(x, y, z)^T$ is any scalar multiple of $(0, 1, 0)^T$, so we may take $(0, 1, 0)^T$ as our first basis vector. Similarly, $(1, -1, -1)^T$ and $(1, -2, -2)^T$ are eigenvectors with eigenvalues $1, -1$ respectively. Proposition 12.10 says that these three vectors form a basis, and so (taking them in the same order as we took the eigenvalues $2, 1, -1$) we see we may take

$$\mathbf{P} = \begin{pmatrix} 0 & 1 & 1 \\ 1 & -1 & -2 \\ 0 & -1 & -2 \end{pmatrix}.$$

Note, however, that any three eigenvectors for the three eigenvalues form a basis with respect to which the matrix of the transformation $\mathbf{x} \mapsto \mathbf{A}\mathbf{x}$ is diagonal, so the choice of $\mathbf{P}$ above is by no means unique.

Looking ahead to Chapter 14 and the primary decomposition theorem (Theorem 14.3), we can point out an alternative method for finding eigenvectors other than solving the obvious simultaneous equations. With $\mathbf{A}$ as in the previous example, we already identified $m_{\mathbf{A}}(x) = (x - 2)(x - 1)(x + 1)$, so it is clear that $\operatorname{im} \mathbf{B} \subseteq \ker \mathbf{C}$ where $\mathbf{B} = (\mathbf{A} - \mathbf{I})(\mathbf{A} + \mathbf{I})$ and $\mathbf{C} = (\mathbf{A} - 2\mathbf{I})$, since the product $\mathbf{C}\mathbf{B}$ of these two matrices is zero. It turns out in fact that these subspaces are actually equal, $\operatorname{im} \mathbf{B} = \ker \mathbf{C}$, so to find the eigenvectors with eigenvalue 2 it suffices to compute $\mathbf{B}$ and find a basis of its image:

$$\mathbf{B} = (\mathbf{A} - \mathbf{I})(\mathbf{A} + \mathbf{I}) = \begin{pmatrix} 2 & 0 & 2 \\ -4 & 1 & -5 \\ -4 & 0 & -4 \end{pmatrix} \begin{pmatrix} 4 & 0 & 2 \\ -4 & 3 & -5 \\ -4 & 0 & -2 \end{pmatrix} = \begin{pmatrix} 0 & 0 & 0 \\ 0 & 3 & -3 \\ 0 & 0 & 0 \end{pmatrix},$$

so the image is spanned by $(0, 1, 0)^T$. The other two eigenspaces can be computed in a similar way.

In practice, this method seems useful for more simple matrices, especially when the work in identifying the minimum polynomial has already been done. But, in general, calculating bases of the image of a matrix still involves computing echelon forms, so there may not be any real saving in effort for more complicated examples. For the interested reader, we give the result that states that this method works as follows.

Proposition 12.17 *For any $n \times n$ matrix $\mathbf{A}$ over a field F, if $\mathbf{A}$ has minimum polynomial $m_{\mathbf{A}}(x) = p(x)q(x)$, where $p(x)$ and $q(x)$ do not have any nonconstant factor in common, then $\operatorname{im}(p(\mathbf{A})) = \ker(q(\mathbf{A}))$.*

The proof uses the version of Euclid's algorithm given as Proposition 9.14, and is left as an exercise for the reader.

Diagonalization is frequently applied in the solution of *simultaneous linear difference equations* and *simultaneous linear differential equations*. For example, if sequences x_n, y_n are defined by

$$\begin{pmatrix} x_{n+1} \\ y_{n+1} \end{pmatrix} = \begin{pmatrix} a_{11} & a_{12} \\ a_{21} & a_{22} \end{pmatrix} \begin{pmatrix} x_n \\ y_n \end{pmatrix}, \qquad \begin{pmatrix} x_0 \\ y_0 \end{pmatrix} = \begin{pmatrix} r \\ s \end{pmatrix},$$

we would like to find formulae for x_n and y_n in terms of the known quantities $a_{11}, a_{12}, a_{21}, a_{22}, r, s$. Of course, we may write

$$\begin{pmatrix} x_n \\ y_n \end{pmatrix} = \begin{pmatrix} a_{11} & a_{12} \\ a_{21} & a_{22} \end{pmatrix}^n \begin{pmatrix} r \\ s \end{pmatrix},$$

but this just begs the question of determining a formula for the nth power of a square matrix. Diagonalization helps here, since if

$$\mathbf{A} = \begin{pmatrix} a_{11} & a_{12} \\ a_{21} & a_{22} \end{pmatrix} \quad \text{and} \quad \mathbf{P}^{-1}\mathbf{A}\mathbf{P} = \begin{pmatrix} \lambda & 0 \\ 0 & \mu \end{pmatrix}$$

then

$$(\mathbf{P}^{-1}\mathbf{A}\mathbf{P})^n = \mathbf{P}^{-1}\mathbf{A}\mathbf{P}\mathbf{P}^{-1}\mathbf{A}\mathbf{P}\dots\mathbf{P}^{-1}\mathbf{A}\mathbf{P} = \mathbf{P}^{-1}\mathbf{A}^n\mathbf{P}$$

and

$$\begin{pmatrix} \lambda & 0 \\ 0 & \mu \end{pmatrix}^n = \begin{pmatrix} \lambda^n & 0 \\ 0 & \mu^n \end{pmatrix}.$$

So

$$\mathbf{A}^n = \mathbf{P}(\mathbf{P}^{-1}\mathbf{A}\mathbf{P})^n\mathbf{P}^{-1} = \mathbf{P}\begin{pmatrix} \lambda^n & 0 \\ 0 & \mu^n \end{pmatrix}\mathbf{P}^{-1},$$

which gives $\mathbf{A}^n$ in terms of the eigenvalues λ, μ of $\mathbf{A}$ and a basis of eigenvectors given by $\mathbf{P}$. This is what is going on in the example in Section 10.1, and of course there is nothing special about 2×2 matrices here.

Example 12.18 We solve the system of difference equations

$$x_{n+1} = 3x_n - 4y_n + 2z_n, \qquad x_0 = 1,$$
$$y_{n+1} = x_n - y_n + z_n, \qquad y_0 = 2,$$
$$z_{n+1} = x_n - 2y_n + 2z_n, \qquad z_0 = -1.$$

The solution is

$$\begin{pmatrix} x_n \\ y_n \\ z_n \end{pmatrix} = \mathbf{A}^n \begin{pmatrix} 1 \\ 2 \\ -1 \end{pmatrix}$$

where

$$\mathbf{A} = \begin{pmatrix} 3 & -4 & 2 \\ 1 & -1 & 1 \\ 1 & -2 & 2 \end{pmatrix}.$$

Now, the minimum polynomial of $\mathbf{A}$ was computed in Example 11.19 and found to be $(x - 1)(x - 2)$, which is a product of distinct linear factors, so $\mathbf{A}$ is diagonalizable. To find a basis of $\mathbb{R}^3$ of eigenvectors we must solve the simultaneous equations $(\mathbf{A} - \mathbf{I})(x, y, z)^T = \mathbf{0}$ and $(\mathbf{A} - 2\mathbf{I})(x, y, z)^T = \mathbf{0}$. The first of these is

$$\begin{pmatrix} 2 & -4 & 2 \\ 1 & -2 & 1 \\ 1 & -2 & 1 \end{pmatrix} \begin{pmatrix} x \\ y \\ z \end{pmatrix} = \mathbf{0}$$

which has solution space

$$\ker(\mathbf{A} - \mathbf{I}) = \mathrm{span}\big((1, 1, 1)^T, (2, 1, 0)^T\big)$$

as you may check. The second eigenspace is the set of solutions of

$$\begin{pmatrix} 1 & -4 & 2 \\ 1 & -3 & 1 \\ 1 & -2 & 0 \end{pmatrix} \begin{pmatrix} x \\ y \\ z \end{pmatrix} = \mathbf{0}$$

which is

$$\ker(\mathbf{A} - 2\mathbf{I}) = \mathrm{span}\big((2, 1, 1)^T\big).$$

Therefore,

$$\mathbf{P}^{-1}\mathbf{A}\mathbf{P} = \begin{pmatrix} 1 & 0 & 0 \\ 0 & 1 & 0 \\ 0 & 0 & 2 \end{pmatrix} \quad \text{where} \quad \mathbf{P} = \begin{pmatrix} 1 & 2 & 2 \\ 1 & 1 & 1 \\ 1 & 0 & 1 \end{pmatrix}.$$

(Of course, there are many other bases for the eigenspaces, and so many other suitable base-change matrices one might take.) This gives

$$\mathbf{A}^n = \mathbf{P} \begin{pmatrix} 1 & 0 & 0 \\ 0 & 1 & 0 \\ 0 & 0 & 2^n \end{pmatrix} \mathbf{P}^{-1} = \begin{pmatrix} 1 & 2 & 2^{n+1} \\ 1 & 1 & 2^n \\ 1 & 0 & 2^n \end{pmatrix} \begin{pmatrix} -1 & 2 & 0 \\ 0 & 1 & -1 \\ 1 & -2 & 1 \end{pmatrix}$$

$$= \begin{pmatrix} -1 + 2^{n+1} & 4 - 2^{n+2} & -2 + 2^{n+1} \\ -1 + 2^n & 3 - 2^{n+1} & -1 + 2^n \\ -1 + 2^n & 2 - 2^{n+1} & 2^n \end{pmatrix}$$

so

$$\begin{pmatrix} x_n \\ y_n \\ z_n \end{pmatrix} = \mathbf{A}^n \begin{pmatrix} 1 \\ 2 \\ -1 \end{pmatrix} = \begin{pmatrix} 9 - 2^{n+3} \\ 6 - 2^{n+2} \\ 3 - 2^{n+2} \end{pmatrix}.$$

Similar methods can be used to solve simultaneous differential equations.

Example 12.19 Suppose variables u, v depend on time, t, according to the equations

$$\frac{du}{dt} = u + 3v \qquad \frac{dv}{dt} = u - v,$$

which can be written as

$$\frac{d}{dt} \begin{pmatrix} u \\ v \end{pmatrix} = \begin{pmatrix} 1 & 3 \\ 1 & -1 \end{pmatrix} \begin{pmatrix} u \\ v \end{pmatrix}.$$

It turns out that this 2×2 matrix has eigenvalues 2 and -2, so can be diagonalized. In fact

$$\mathbf{P}^{-1} \begin{pmatrix} 1 & 3 \\ 1 & -1 \end{pmatrix} \mathbf{P} = \begin{pmatrix} 2 & 0 \\ 0 & -2 \end{pmatrix} \qquad \text{where} \quad \mathbf{P} = \begin{pmatrix} 3 & -1 \\ 1 & -1 \end{pmatrix}.$$

This suggests introducing variables x, y with $(x, y)^T = \mathbf{P}^{-1}(u, v)^T$, or

$$x = (u + v)/4, \qquad y = (-u + 3v)/4,$$

for then

$$\frac{d}{dt} \begin{pmatrix} u \\ v \end{pmatrix} = \frac{d}{dt} \mathbf{P} \begin{pmatrix} x \\ y \end{pmatrix} = \begin{pmatrix} 1 & 3 \\ 1 & -1 \end{pmatrix} \mathbf{P} \begin{pmatrix} x \\ y \end{pmatrix},$$

or

$$\frac{d}{dt} \begin{pmatrix} x \\ y \end{pmatrix} = \mathbf{P}^{-1} \begin{pmatrix} 1 & 3 \\ 1 & -1 \end{pmatrix} \mathbf{P} \begin{pmatrix} x \\ y \end{pmatrix}.$$

This gives $dx/dt = 2x$, $dy/dt = -2y$ so $x = Ae^{2t}$, $y = Be^{-2t}$ for some positive constants A, B, so the solution is $(u, v)^T = \mathbf{P}(x, y)^T$, or

$$u = 3Ae^{2t} - Be^{-2t} \qquad v = Ae^{2t} + Be^{-2t}.$$

The constants A, B can be found as usual from boundary conditions. For example, if we are given that $u = u_0$ and $v = v_0$ at time $t = 0$, then $A = (u_0 + v_0)/4$ and $B = (3v_0 - u_0)/4$.

The exercises following provide more examples, and one or two hints on some useful tricks that can be applied in similar cases. Not all matrices can be diagonalized, though, and when you meet such an example the methods of this chapter cannot be used. Instead, it may be necessary to put the matrix in *Jordan normal form*, and use the ideas from Chapter 14 below.

Exercises

Exercise 12.1 Calculate the characteristic polynomials and minimum polynomials of the following matrices.

$$\text{(a)} \begin{pmatrix} 1 & 2 & 3 \\ 0 & 1 & 2 \\ 0 & 0 & 1 \end{pmatrix} \quad \text{(b)} \begin{pmatrix} 1 & 0 & 2 \\ 0 & 1 & 0 \\ 0 & 0 & 1 \end{pmatrix}$$

$$\text{(c)} \begin{pmatrix} 1 & 0 & 2 & 1 \\ 0 & 1 & 0 & 1 \\ 0 & 0 & 2 & 3 \\ 0 & 0 & 0 & 2 \end{pmatrix} \quad \text{(d)} \begin{pmatrix} 1 & 0 & 2 & 1 \\ 0 & 1 & 0 & 1 \\ 0 & 0 & 2 & 3 \\ 0 & 0 & 3 & 2 \end{pmatrix}.$$

Which of these (if any) is diagonalizable?

Exercise 12.2 Show that, regarded as 2 × 2 matrices over $\mathbb{C}$,

$$\begin{pmatrix} \cos\theta & -\sin\theta \\ \sin\theta & \cos\theta \end{pmatrix} \quad \text{and} \quad \begin{pmatrix} e^{i\theta} & 0 \\ 0 & e^{-i\theta} \end{pmatrix}$$

are similar.

Exercise 12.3 Which of the following are diagonalizable? Explain your answers, but try to do as little work as possible, using results from this and previous chapters where applicable.

$$\text{(a)} \begin{pmatrix} 1 & 2 & 3 \\ 4 & 5 & 6 \\ 7 & 8 & 9 \end{pmatrix} \quad \text{(b)} \begin{pmatrix} 0 & 0 & -2 \\ 1 & 1 & 2 \\ 1 & 0 & 3 \end{pmatrix} \quad \text{(c)} \begin{pmatrix} 0 & 4 & -6 \\ 1 & -1 & 4 \\ 1 & -2 & 5 \end{pmatrix}$$

$$\text{(d)} \begin{pmatrix} 5 & -1 & -3 \\ -2 & 1 & 2 \\ 6 & -2 & -4 \end{pmatrix} \text{ over } \mathbb{R} \quad \text{(e)} \begin{pmatrix} 5 & -1 & -3 \\ -2 & 1 & 2 \\ 6 & -2 & -4 \end{pmatrix} \text{ over } \mathbb{C}.$$

Exercise 12.4 (a) Find two 2 × 2 matrices over $\mathbb{R}$ which have the same characteristic polynomial but which are not similar.

(b) Find two 3 × 3 matrices over $\mathbb{R}$ which have the same minimum polynomial but which are not similar.

(c) Find two 4 × 4 matrices over $\mathbb{R}$ which have the same minimum polynomial and the same characteristic polynomial, but which are not similar.

Exercise 12.5 The *Fibonacci numbers* x_n are defined by $x_{n+2} = x_{n+1} + x_n$, $x_0 = x_1 = 1$. Let $u_n = x_{2n}$, $v_n = x_{2n+1}$ and find a matrix $\mathbf{A}$ so that

$$\begin{pmatrix} u_{n+1} \\ v_{n+1} \end{pmatrix} = \mathbf{A} \begin{pmatrix} u_n \\ v_n \end{pmatrix}.$$

Diagonalize $\mathbf{A}$ and hence find a formula for x_n in terms of n.

Exercise 12.6 Solve $x_{n+1} = -y_n + z_n$; $y_{n+1} = -y_n$; $z_{n+1} = 2x_n - 2y_n + z_n$; with initial values $x_0 = y_0 = 1$, $z_0 = 2$.

Exercise 12.7 Solve

(a) $x_{n+1} = x_n + 2y_n$; $y_{n+1} = 2x_n + y_n + 1$; where $x_0 = y_0 = 1$.
   [Hint: introduce z_n with $z_{n+1} = z_n$ and $z_0 = 1$.]
(b) $x_{n+1} = 2x_n + 3y_n$; $y_{n+1} = 3x_n + 2y_n + 2^n$; where $x_0 = 1$, $y_0 = 2$.
   [Hint: introduce some suitable z_n.]

Exercise 12.8 Solve

(a) $x_{n+1} = x_n + 4x_n + 1$; $y_{n+1} = x_n + y_n$; $x_0 = y_0 = 1$.
(b) $x_{n+1} = 2x_n + y_n + 1$; $y_{n+1} = x_n + 2y_n$; $x_0 = y_0 = 1$.

[Hint: in each case, introduce $u_n = x_n + an + b$, $v_n = y_n + cn + d$ for certain constants a, b, c, d.]

Exercise 12.9 Solve the following systems of differential equations for functions $x(t)$, $y(t)$, and $z(t)$, where a dot denotes differentiation with respect to t.

(a) $\dot{x} = -y + z$; $\dot{y} = -y$; $\dot{z} = 2x - 2y + z$; with boundary conditions $x(0) = y(0) = 1$, $z(0) = 2$.
(b) $\dot{x} = x + 2y$; $\dot{y} = 2x + y + 1$; where $x(0) = y(0) = 1$.
(c) $\dot{x} = 2x + 3y$; $\dot{y} = 3x + 2y + e^{2t}$; where $x(0) = 1$, $y(0) = 2$.
(d) $\dot{x} = x + 4x + 1$; $\dot{y} = x + y$; $x(0) = y(0) = 1$.
(e) $\dot{x} = 2x + y + 1$; $\dot{y} = x + 2y$; $x(0) = y(0) = 1$.

Exercise 12.10 Show that V is the direct sum of subspaces U, W if and only if (a) every $v \in V$ is equal to $u + w$ for some $u \in U$ and $w \in W$ and (b) $U \cap W = \varnothing$. That is, show that Definition 5.28 and Definition 12.6 agree.

Exercise 12.11 Prove Proposition 12.17. Hence, using induction on dimension, give an alternative proof of Proposition 12.12.

13
Self-adjoint transformations

This chapter combines material concerning quadratic forms with material from the previous chapter on diagonalization. Throughout, V is a finite dimensional vector space over $\mathbb{R}$ or $\mathbb{C}$ *with an inner product* $\langle\,|\,\rangle$.

The main goal in this chapter is to understand the nature of quadratic forms, symmetric bilinear forms and conjugate-symmetric sesquilinear forms on a finite dimensional *inner product* space V—in particular, how the form relates to the inner product on V. It turns out that the key to describing such a form is an associated linear transformation on V. The linear transformations here are of interest in their own right, and have the property of being *self-adjoint* (as defined below). They can be diagonalized using methods in the last chapter, and this diagonalization provides a complete description of the bilinear or sesquilinear form we are interested in.

13.1 Orthogonal and unitary transformations

In earlier chapters we studied the behaviour of quadratic forms (or equivalently, symmetric bilinear forms) on arbitrary vector spaces over $\mathbb{R}$. The goal was to find a change of basis that diagonalizes the form, or at least makes it look as 'nice' as possible. In many ways, particularly so for applications to geometry, quantum mechanics, etc., it is much more interesting to study forms on *inner product spaces*. What this means is that a base-change transformation f must preserve the inner product, i.e. must send a vector v to another vector $f(v)$ of the same length as v, and send an orthogonal pair of vectors v, w to another orthogonal pair $f(v), f(w)$. An equivalent view is that our base-change transformations should send orthonormal bases to orthonormal bases, so instead of allowing ourselves to use arbitrary bases, we only allow *orthonormal* bases.

Example 13.1 Let us suppose that we are working in $\mathbb{R}^2$ with the standard inner product, and we are considering the quadratic form $Q(x, y) = (x/a)^2 + (y/b)^2$. You should recognise the equation $Q(x, y) = 1$ as the equation of an ellipse. By scaling the coordinates, changing the basis to $(a, 0)^T, (0, b)^T$ (which is an orthogonal basis, but *not* orthonormal), we can write Q as $Q(u, v) = u^2 + v^2$, and the ellipse turns into a unit circle. If on the other hand we only allowed ourselves to use *orthonormal* bases, then our ellipse would keep its shape, but it might be rotated. For example, changing basis to the orthonormal basis

$(1/\sqrt{2}, 1/\sqrt{2})^T, (-1/\sqrt{2}, 1/\sqrt{2})^T$ represents a rotation by $\pi/4$ about the origin, and hence preserves orthogonality and length.

From now to the end of this chapter we will deal with the real case and the complex case at the same time by writing complex-conjugate signs where they are required in the complex case. (In the real case, these complex-conjugate signs can always be ignored since the number in question is real.)

Definition 13.2 *Let V be a finite dimensional inner product space, and suppose f is a linear transformation $V \to V$. We say f is **orthogonal** (when V is a vector space over $\mathbb{R}$) or **unitary** (when V is a vector space over $\mathbb{C}$) if it preserves the inner product $\langle v|w \rangle$ on V, i.e. if for all $v, w \in V$ we have*

$$\langle v|w \rangle = \langle f(v)|f(w) \rangle.$$

Proposition 13.3 *Let V be a finite dimensional inner product space, and suppose f is a linear transformation $V \to V$. Suppose that $e_1, e_2, \ldots, e_n$ is an orthonormal basis of V. Then f preserves the inner product $\langle v|w \rangle$ on V if and only if $f(e_1), f(e_2), \ldots, f(e_n)$ is orthonormal.*

Proof   One direction is easy. For the other, suppose $\{f(e_1), f(e_2), \ldots, f(e_n)\}$ is orthonormal, and consider two arbitrary vectors $u, v \in V$. Writing u, v in terms of the basis $e_1, \ldots, e_n$ we have scalars λ_i and μ_j with

$$u = \sum_{i=1}^{n} \lambda_i e_i, \qquad v = \sum_{j=1}^{n} \mu_j e_j,$$

so that, using linearity of f and the orthonormality of $\{f(e_1), \ldots, f(e_n)\}$,

$$\begin{aligned}
\langle f(u)|f(v) \rangle &= \left\langle f\left(\sum_{i=1}^{n} \lambda_i e_i\right) \middle| f\left(\sum_{j=1}^{n} \mu_j e_j\right) \right\rangle \\
&= \left\langle \sum_{i=1}^{n} \lambda_i f(e_i) \middle| \sum_{j=1}^{n} \mu_j f(e_j) \right\rangle \\
&= \sum_{i=1}^{n} \sum_{j=1}^{n} \overline{\lambda_i} \mu_j \langle f(e_i)|f(e_j) \rangle \\
&= \sum_{i=1}^{n} \sum_{j=1}^{n} \overline{\lambda_i} \mu_j \langle e_i|e_j \rangle \\
&= \left\langle \sum_{i=1}^{n} \lambda_i e_i \middle| \sum_{j=1}^{n} \mu_j e_j \right\rangle \\
&= \langle u|v \rangle,
\end{aligned}$$

as required. $\square$

Example 13.4 Let V be the real vector space $\mathbb{R}^3$ with the usual inner product and the usual basis $\mathbf{e}_1 = (1, 0, 0)^T, \mathbf{e}_2 = (0, 1, 0)^T, \mathbf{e}_3 = (0, 0, 1)^T$ and change basis to the orthonormal basis

$$\mathbf{f}_1 = (1/\sqrt{2}, -1/\sqrt{2}, 0)^T,$$
$$\mathbf{f}_2 = (1/\sqrt{6}, 1/\sqrt{6}, -2/\sqrt{6})^T,$$
$$\mathbf{f}_3 = (1/\sqrt{3}, 1/\sqrt{3}, 1/\sqrt{3})^T.$$

The base-change matrix from $\mathbf{e}_1, \mathbf{e}_2, \mathbf{e}_3$ to $\mathbf{f}_1, \mathbf{f}_2, \mathbf{f}_3$ is

$$\mathbf{P} = \begin{pmatrix} 1/\sqrt{2} & 1/\sqrt{6} & 1/\sqrt{3} \\ -1/\sqrt{2} & 1/\sqrt{6} & 1/\sqrt{3} \\ 0 & -2/\sqrt{6} & 1/\sqrt{3} \end{pmatrix} = \frac{1}{\sqrt{6}} \begin{pmatrix} \sqrt{3} & 1 & \sqrt{2} \\ -\sqrt{3} & 1 & \sqrt{2} \\ 0 & -2 & \sqrt{2} \end{pmatrix}.$$

We now calculate $\mathbf{P}^T \mathbf{P}$:

$$\mathbf{P}^T \mathbf{P} = \frac{1}{6} \begin{pmatrix} \sqrt{3} & -\sqrt{3} & 0 \\ 1 & 1 & -2 \\ \sqrt{2} & \sqrt{2} & \sqrt{2} \end{pmatrix} \begin{pmatrix} \sqrt{3} & 1 & \sqrt{2} \\ -\sqrt{3} & 1 & \sqrt{2} \\ 0 & -2 & \sqrt{2} \end{pmatrix} = \frac{1}{6} \begin{pmatrix} 6 & 0 & 0 \\ 0 & 6 & 0 \\ 0 & 0 & 6 \end{pmatrix} = \mathbf{I}.$$

So $\mathbf{P}^{-1} = \mathbf{P}^T$. You should verify the matrix multiplication here, and notice as you do so that each entry in $\mathbf{P}^T \mathbf{P}$ is precisely the inner product of the vector given by a row of $\mathbf{P}^T$ with a column of $\mathbf{P}$. Since the columns of $\mathbf{P}$ are precisely the elements of the orthonormal basis $\mathbf{f}_1, \mathbf{f}_2, \mathbf{f}_3$ we then see why the inverse of $\mathbf{P}$ is $\mathbf{P}^T$.

This example was over the real numbers. The same works in general over $\mathbb{C}$ provided we add a complex-conjugate sign.

Proposition 13.5 *Let V be a finite dimensional inner product space and let $e_1, \dots, e_n$ be an orthonormal basis for V. Suppose $f: V \to V$ is a linear transformation with matrix $\mathbf{P}$ with respect to $e_1, \dots, e_n$. Then f preserves the inner product $\langle | \rangle$ (i.e. is orthogonal or unitary) if and only if $\mathbf{P}^{-1}$ exists and equals $\overline{\mathbf{P}}^T$.*

Proof The matrix $\mathbf{P}$ of f with respect to $e_1, \dots, e_n$ is precisely the matrix formed by columns equal to the column vector representations of $f(e_1), \dots, f(e_n)$ with respect to the same basis. Since $e_1, \dots, e_n$ is orthonormal, the (i, j)th entry in $\overline{\mathbf{P}}^T \mathbf{P}$ is just $\langle f(e_i)|f(e_j)\rangle$; hence $\overline{\mathbf{P}}^T \mathbf{P} = \mathbf{I}$ if and only if $\{f(e_1), \dots, f(e_n)\}$ is orthonormal. The proposition now follows from Proposition 13.3. $\square$

Example 13.6 We consider a slightly less trivial example than that in Example 13.1, on $\mathbb{R}^2$ with the usual inner product. Consider the quadratic form $Q(x, y) = 8x^2 + 24xy + y^2$ on $\mathbb{R}^2$ with the usual inner product. Completing the square gives us $Q(x, y) = 2(2x + 3y)^2 - 17y^2$, so this form has rank 2 and signature 0. However, our new basis is not even orthogonal, let alone orthonormal, so this tells us nothing about the shape of the hyperbola $Q(x, y) = 1$. On the other hand, it turns out that there is an orthogonal base-change matrix,

$$\mathbf{P} = \begin{pmatrix} -3/5 & 4/5 \\ 4/5 & 3/5 \end{pmatrix},$$

such that

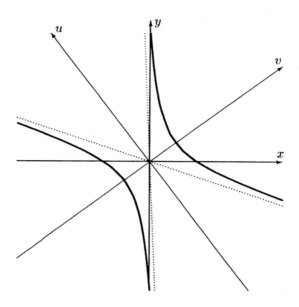

Fig. 13.1 The hyperbola $8x^2 + 24xy + y^2 = 1$

$$\mathbf{P}^T\mathbf{AP} = \begin{pmatrix} -3/5 & 4/5 \\ 4/5 & 3/5 \end{pmatrix} \begin{pmatrix} 8 & 12 \\ 12 & 1 \end{pmatrix} \begin{pmatrix} -3/5 & 4/5 \\ 4/5 & 3/5 \end{pmatrix} = \begin{pmatrix} -8 & 0 \\ 0 & 17 \end{pmatrix}.$$

Thus there is an orthonormal basis given by vectors $\mathbf{f}_1 = (-3/5, 4/5)^T$ and $\mathbf{f}_2 = (4/5, 3/5)^T$ with respect to which Q has the form $-8u^2 + 17v^2$.

From this we can calculate the slope of the asymptotes of the hyperbola $Q = 1$, as well as the points where it cuts the v-axis. In u, v coordinates, the equations of the asymptotes of $-8u^2 + 17v^2 = 1$ are $u = \pm\sqrt{17/8}\,v$, which on substituting $u = (-3x + 4y)/5$ and $v = (4x + 3y)/5$ become

$$(4\sqrt{17} + 6\sqrt{2})x = (-3\sqrt{17} + 8\sqrt{2})y$$

and

$$(4\sqrt{17} - 6\sqrt{2})x = (-3\sqrt{17} - 8\sqrt{2})y.$$

The hyperbola passes through the v-axis at $u = 0$, $v = \pm 1/\sqrt{17}$, or in other words at the points (x, y) given by

$$(4/5\sqrt{17}, 3/5\sqrt{17}) \qquad (-4/5\sqrt{17}, -3/5\sqrt{17}).$$

(See Figure 13.1.)

13.2   From forms to transformations

Suppose V is a finite dimensional inner product space V over $\mathbb{R}$ or $\mathbb{C}$. We saw in the last section that the linear transformations of V that preserve the inner product are precisely the orthogonal transformations. If $\mathbf{e}_1, \ldots, \mathbf{e}_n$ is an *orthonormal basis* of V, we saw that these transformations have matrices $\mathbf{P}$ which satisfy $\mathbf{P}^{-1} = \overline{\mathbf{P}}^T$. Now if $\mathbf{A}$ is an $n \times n$ matrix representing a linear transformation f and we consider $\mathbf{P}$ as a base-change matrix, then the matrix $\mathbf{A}$ is transformed to $\mathbf{P}^{-1}\mathbf{AP}$ by this base change. On the other hand, if $\mathbf{A}$ is an $n \times n$ matrix representing a bilinear form F and we consider $\mathbf{P}$ as a base-change matrix, then $\mathbf{A}$ should be transformed to $\overline{\mathbf{P}}^T\mathbf{AP}$. But $\overline{\mathbf{P}}^T = \mathbf{P}^{-1}$, so these two transformed matrices are the same. In other words, *provided we restrict ourselves to orthogonal base changes* the matrix $\mathbf{A}$ can be thought of as either the matrix of a linear transformation or the matrix of a bilinear form.

All this suggests the idea of switching from looking at a bilinear form with matrix $\mathbf{A}$ to considering the linear transformation with the same matrix, and back again.

Given a bilinear form $F(u,v)$ on a Euclidean space V with orthonormal basis $e_1, e_2, \ldots, e_n$, or a sesquilinear form $F(u,v)$ on a unitary space V with orthonormal basis $e_1, e_2, \ldots, e_n$, we can define an associated map $f \colon V \to V$ by

$$f(v) = \sum_{i=1}^{n} F(e_i, v)e_i. \tag{1}$$

This is a linear transformation by linearity of F in the second argument, since

$$f(\lambda v + \mu w) = \sum_{i=1}^{n} F(e_i, \lambda v + \mu w)e_i$$
$$= \sum_{i=1}^{n} (\lambda F(e_i, v) + \mu F(e_i, w))e_i$$
$$= \lambda \sum_{i=1}^{n} F(e_i, v)e_i + \mu \sum_{i=1}^{n} F(e_i, w)e_i$$
$$= \lambda f(v) + \mu f(w).$$

At this stage, it is conceivable that the map f depends on the basis $e_1, \ldots, e_n$ used in (1). It follows from the calculations below, however, that it is actually independent of the choice of basis.

Working in the other direction, suppose that a linear transformation $f \colon V \to V$ is given. Then we can define a form $F(v, w)$ by

$$F(v, w) = \langle v | f(w) \rangle \tag{2}$$

for all $v, w \in V$. It is easy to see that F is bilinear (or sesquilinear in the complex case) since

$$F(u, \lambda v + \mu w) = \langle u | f(\lambda v + \mu w) \rangle$$
$$= \langle u | \lambda f(v) + \mu f(w) \rangle$$
$$= \lambda \langle u | f(v) \rangle + \mu \langle u | f(w) \rangle$$
$$= \lambda F(u, v) + \mu F(u, w)$$

and

$$F(\lambda u + \mu v, w) = \langle \lambda u + \mu v | f(w) \rangle$$
$$= \overline{\lambda} \langle u | f(w) \rangle + \overline{\mu} \langle v | f(w) \rangle$$
$$= \overline{\lambda} F(u, w) + \overline{\mu} F(v, w).$$

These two processes, of going from F to f, and of going from f to F, are inverse to each other; that is, if we start with F, say, pass to the associated linear transformation f, and then to its associated form, we get

$$\left\langle u \,\Big|\, \sum_{i=1}^{n} F(e_i, v) e_i \right\rangle = \sum_{i=1}^{n} F(e_i, v) \langle u | e_i \rangle$$
$$= \sum_{i=1}^{n} \langle u | e_i \rangle F(e_i, v)$$
$$= \sum_{i=1}^{n} \overline{\langle e_i | u \rangle} F(e_i, v)$$
$$= F\left(\sum_{i=1}^{n} \langle e_i | u \rangle e_i, v \right) = F(u, v),$$

by the Fourier expansion formula $u = \sum_i \langle e_i | u \rangle e_i$. In the other direction, if we start with f, compute its associated form F, and the linear transformation associated with that, we get

$$\sum_{i=1}^{n} F(e_i, v) e_i = \sum_{i=1}^{n} \langle e_i | f(v) \rangle e_i = f(v),$$

by the Fourier expansion formula again. This, incidentally, also shows that the definition of f from the form F is independent of the choice of orthonormal basis $e_1, \ldots, e_n$ used.

We now return to the ideas mentioned at the beginning of this section, and show that f and F are represented by the same matrix.

Proposition 13.7 *Suppose that* $f \colon V \to V$ *is a linear transformation on an inner product space* V, *suppose that* F *is the corresponding conjugate-symmetric sesquilinear form, and that* $\{e_1, \ldots, e_n\}$ *is an orthonormal basis of* V. *Then* f *and* F *are represented by the same matrix with respect to the ordered basis* $e_1, \ldots, e_n$.

Proof   The matrix of F is $\mathbf{A} = (a_{ij})$, where $a_{ij} = F(e_i, e_j)$. On the other hand, the matrix of f is $\mathbf{B} = (b_{ij})$ where

$$f(e_j) = \sum_{i=1}^{n} b_{ij} e_i.$$

However, our formula for f gives

$$f(e_j) = \sum_{i=1}^{n} F(e_i, e_j)e_i;$$

thus $b_{ij} = F(e_i, e_j) = a_{ij}$ so $\mathbf{A} = \mathbf{B}$. □

In the real case, we are interested in symmetric bilinear forms F, i.e. forms satisfying $F(v, w) = F(w, v)$. If f is the corresponding linear transformation, this equation translates into

$$\langle v|f(w)\rangle = F(v, w) = F(w, v) = \langle w|f(v)\rangle = \langle f(v)|w\rangle$$

for all $v, w \in V$.

In the complex case, we are interested in conjugate-symmetric sesquilinear forms F, i.e. satisfying $F(v, w) = \overline{F(w, v)}$. Again, if f is the corresponding linear transformation, this is equivalent to

$$\langle v|f(w)\rangle = F(v, w) = \overline{F(w, v)} = \overline{\langle w|f(v)\rangle} = \langle f(v)|w\rangle.$$

In both the real and complex cases we obtain $\langle f(v)|w\rangle = \langle v|f(w)\rangle$. Linear transformations f with this very special property are called self-adjoint.

Definition 13.8 *A linear transformation* $f : V \to V$ *of an inner product space* V *(over* $\mathbb{R}$ *or* $\mathbb{C}$*) is said to be* ***self-adjoint*** *if* $\langle f(v)|w\rangle = \langle v|f(w)\rangle$ *for all* $v, w \in V$.

We conclude this section by discussing the properties of matrices representing self-adjoint transformations.

Proposition 13.9 *If* f *is a self-adjoint transformation of an inner product space* V, *and if* $\{e_1, \ldots, e_n\}$ *is an orthonormal basis of* V, *then the matrix* $\mathbf{A}$ *of* f *with respect to the ordered basis* $e_1, \ldots, e_n$ *is conjugate-symmetric, i.e.* $\overline{\mathbf{A}}^T = \mathbf{A}$.

Proof Since f is self-adjoint, the corresponding form F is conjugate-symmetric,

$$F(u, v) = \langle u|f(v)\rangle = \overline{\langle f(u)|v\rangle} = \overline{\langle v|f(u)\rangle} = \overline{F(v, u)},$$

so the matrix of F is conjugate-symmetric. But with respect to $e_1, \ldots, e_n$, the matrices of f and F are the same by the last proposition, since $\{e_1, \ldots, e_n\}$ is orthonormal. □

There is a useful converse to this, characterizing self-adjoint transformations.

Proposition 13.10 *If* f *is a linear transformation of an inner product space* V, *and if* $e_1, \ldots, e_n$ *is an ordered orthonormal basis of* V *with respect to which the matrix* $\mathbf{A}$ *of* f *is conjugate-symmetric, then* f *is self-adjoint.*

Proof Let F be the corresponding form $\langle u|f(v)\rangle$. Then the matrix of F is the same as that of f, and is conjugate-symmetric. Therefore F is conjugate-symmetric, and hence f is self-adjoint. □

Note that we have proved slightly more than was stated: if f has conjugate-symmetric matrix with respect to *some* orthonormal basis, then f is self-adjoint and so has conjugate-symmetric matrix with respect to *every* orthonormal basis. This is *not* true for arbitrary bases, since for example the real 2×2 matrix

$$\mathbf{A} = \begin{pmatrix} 0 & 1 \\ 2 & 0 \end{pmatrix}$$

is not symmetric, but

$$\mathbf{P}^{-1}\mathbf{A}\mathbf{P} = \begin{pmatrix} -2 & 0 \\ 0 & 2 \end{pmatrix}$$

is symmetric, where

$$\mathbf{P} = \begin{pmatrix} 1 & 1 \\ -2 & 2 \end{pmatrix}.$$

13.3   Eigenvalues and diagonalization

The main aim of this section is to prove that given any conjugate-symmetric sesquilinear form (or real symmetric bilinear form) there is an *orthonormal* basis with respect to which its matrix is diagonal.

Recall that if a matrix (a_{ij}) is conjugate-symmetric, then its diagonal entries are real, as $a_{ii} = \overline{a_{ii}}$.

Theorem 13.11 *If f is a self-adjoint linear transformation of an inner product space V and λ is an eigenvalue of f, then λ is real.*

Proof   We have $f(v) = \lambda v$ for some nonzero vector v, and

$$\begin{aligned}
\lambda\langle v|v\rangle &= \langle v|\lambda v\rangle \\
&= \langle v|f(v)\rangle \\
&= \langle f(v)|v\rangle \\
&= \langle \lambda v|v\rangle \\
&= \overline{\lambda}\langle v|v\rangle,
\end{aligned}$$

so $(\lambda - \overline{\lambda})\langle v|v\rangle = 0$. But $\langle v|v\rangle \neq 0$ since $v \neq 0$ and hence $\lambda = \overline{\lambda}$.   □

Corollary 13.12 *For any symmetric real $n \times n$ matrix $\mathbf{A}$ or any conjugate-symmetric complex $n \times n$ matrix $\mathbf{A}$, the characteristic polynomial $\chi_{\mathbf{A}}(x)$ has n real roots (counting multiplicities).*

Proof $\chi_{\mathbf{A}}(x)$ has n complex roots, but each of these roots is real by the preceding theorem.   □

Theorem 13.13 *The minimum polynomial $m_f(x)$ of a self-adjoint linear transformation $f: V \to V$ of a finite dimensional inner product space V has no repeated roots.*

Proof If not, suppose $m_f(x) = (x - \lambda)^2 p(x)$ for some polynomial $p(x)$. Then $(f - \lambda)p(f) \neq 0$, so there is $v \in V$ with $(f - \lambda)p(f)(v) \neq 0$ and $(f - \lambda)^2 p(f)(v) = 0$. But then

$$0 \neq \langle (f - \lambda)p(f)(v)|(f - \lambda)p(f)(v)\rangle = \langle p(f)(v)|(f - \lambda)(f - \lambda)p(f)(v)\rangle = 0$$

since $(f - \lambda)$ is self-adjoint. This is a contradiction. □

Corollary 13.14 *Any self-adjoint* $f: V \to V$ *of a finite dimensional inner product space V is diagonalizable.*

Proof By the previous theorem and Corollary 12.13. □

In Proposition 12.7 we proved that if $v_1, v_2, \ldots, v_k$ are eigenvectors of a linear transformation f, where $f(v_i) = \lambda_i v_i$ and the λ_i are all distinct, then $\{v_1, v_2, \ldots, v_k\}$ is linearly independent. For self-adjoint f we can make the stronger statement that the v_i are *orthogonal*.

Theorem 13.15 *Let f be a self-adjoint linear transformation* $f: V \to V$*, and suppose v_1, v_2 are eigenvectors of f with corresponding eigenvalues λ_1, λ_2. If $\lambda_1 \neq \lambda_2$ then v_1 and v_2 are orthogonal.*

Proof We have

$$\begin{aligned}
\lambda_1 \langle v_1 | v_2 \rangle &= \overline{\lambda_1} \langle v_1 | v_2 \rangle \\
&= \langle \lambda_1 v_1 | v_2 \rangle \\
&= \langle f(v_1) | v_2 \rangle \\
&= \langle v_1 | f(v_2) \rangle \\
&= \langle v_1 | \lambda_2 v_2 \rangle \\
&= \lambda_2 \langle v_1 | v_2 \rangle
\end{aligned}$$

since λ_1 is real, by Theorem 13.11. But then $(\lambda_1 - \lambda_2)\langle v_1 | v_2 \rangle = 0$, and hence $\langle v_1 | v_2 \rangle = 0$, as $\lambda_1 \neq \lambda_2$. □

What this means is that, for a self-adjoint linear transformation f, we can always find an *orthogonal* basis of eigenvectors. In fact, we can do even better: by normalizing in the usual way there is an *orthonormal* basis of eigenvectors. To see this, we first find any basis of eigenvectors. Then for each eigenvalue λ, we take the set of basis vectors which have that eigenvalue, and apply the Gram–Schmidt algorithm to it. The result will be an orthonormal basis for the eigenspace, since any nonzero linear combination of eigenvectors with eigenvalue λ is itself an eigenvector with eigenvalue λ.

The base-change matrix $\mathbf{P}$ from the usual orthonormal basis to this orthonormal basis of eigenvectors will be unitary, i.e. $\overline{\mathbf{P}}^T = \mathbf{P}^{-1}$, since the new basis is orthonormal. It follows that, given a real symmetric matrix $\mathbf{A}$, or a complex conjugate-symmetric matrix **A**, we can find an orthonormal basis of $\mathbb{R}^n$ (respectively $\mathbb{C}^n$) for which both the linear transformation

$$f(\mathbf{v}) = \mathbf{A}\mathbf{v}$$

and the (symmetric bilinear, or sesquilinear) form

$$F(\mathbf{v}, \mathbf{w}) = \overline{\mathbf{v}}^T\mathbf{A}\mathbf{w}$$

are represented by the same diagonal matrix.

The proof just given of Corollary 13.14 and Theorem 13.15 is somewhat indirect. Using the notion of orthogonal complement from the 'optional' Section 5.4, it is possible to give a direct proof of these results. We do this now for the benefit of readers who have read the material on orthogonal complements.

Theorem 13.16 *Suppose* $f: V \to V$ *is a self-adjoint linear transformation of a finite dimensional inner product space* V *over* $\mathbb{R}$ *or* $\mathbb{C}$. *Then there is an orthonormal basis* $\{v_1, v_2, \dots, v_n\}$ *of* V *such that each* v_i *is an eigenvector of* f.

Proof We use induction on the dimension n of V. If $n = 0$ there is nothing to prove as the empty set $\varnothing$ is a suitable basis of V.

Since f has a real eigenvalue λ_1, there is a nonzero $v_1 \in V$ with $f(v_1) = \lambda_1 v_1$ and $\|v_1\| = 1$. Let $U = \mathrm{span}(v_1)$ and $W = U^{\perp} = \{w \in V : \langle u|w \rangle = 0\}$. Then for $w \in W$ we have $\langle f(w)|u \rangle = \langle w|f(u) \rangle = \langle w|\lambda_1 u \rangle = \lambda_1 \langle w|u \rangle = 0$, so $f(w) \in W$; thus we may regard f as a self-adjoint linear transformation of W. Also, $U \oplus W = V$ and U has dimension 1, so W has dimension $n - 1$. By our induction hypothesis, there is an orthonormal basis $v_2, \dots, v_n$ of W consisting of eigenvectors of f, and clearly $v_1, v_2, \dots, v_n$ is the required basis of V. $\qquad\square$

13.4 Applications

We shall indicate some of the applications of the results in the previous section here by way of some examples, all of which concern real vector spaces.

One obvious place in which real symmetric matrices arise is as the representing matrix of the symmetric bilinear form corresponding to a quadratic form Q on $\mathbb{R}^n$. For example, given a quadratic form $Q(x, y, z)$ on $\mathbb{R}^3$, the equation $Q(x, y, z) = a$ represents a surface which we might want to describe. Simply completing the square as we did in the last part to find the rank and signature of Q gives some information, but we lose the additional structure given by the usual inner product on $\mathbb{R}^3$ in the process. Somehow, we need to diagonalize the form Q and the usual inner product simultaneously to get a full picture.

This is illustrated by the following example.

Example 13.17 Consider the surface in $\mathbb{R}^3$ defined by the equation

$$5x^2 + 5y^2 + 5z^2 - 2xy - 2yz - 2zx = 3.$$

This is $Q(x, y, z) = 3$ where Q is the quadratic form

$$Q(x,y,z) = (x,y,z) \begin{pmatrix} 5 & -1 & -1 \\ -1 & 5 & -1 \\ -1 & -1 & 5 \end{pmatrix} \begin{pmatrix} x \\ y \\ z \end{pmatrix}.$$

We now diagonalize the matrix

$$\mathbf{A} = \begin{pmatrix} 5 & -1 & -1 \\ -1 & 5 & -1 \\ -1 & -1 & 5 \end{pmatrix}$$

as if it represented a linear transformation not a bilinear form. This is legitimate, provided we keep to orthonormal bases, since we know that Q corresponds to a symmetric bilinear form, which in turn corresponds to a self-adjoint linear transformation. It is symmetric, and hence the corresponding linear transformation $f_{\mathbf{A}}$ is self-adjoint, and is diagonalizable. In fact, it turns out that this matrix has a basis of eigenvectors $(1,1,1)^T, (1,-1,0)^T, (1,0,-1)^T$ with corresponding eigenvalues $3, 6, 6$, as you can check. However, in this example we want an *orthonormal* basis of eigenvectors. Using the Gram–Schmidt process to orthogonalize $\mathbf{u}_1 = (1,-1,0)^T, \mathbf{u}_2 = (1,0,-1)^T$ we set $\mathbf{v}_1 = \mathbf{u}_1$ and

$$\mathbf{v}_2 = \mathbf{u}_2 - \frac{\langle \mathbf{u}_1 | \mathbf{u}_2 \rangle}{\langle \mathbf{u}_1 | \mathbf{u}_1 \rangle} \mathbf{u}_1 = \begin{pmatrix} 1 \\ 0 \\ -1 \end{pmatrix} - \frac{1}{2} \begin{pmatrix} 1 \\ -1 \\ 0 \end{pmatrix} = \begin{pmatrix} 1/2 \\ 1/2 \\ -1 \end{pmatrix}$$

This gives the following *orthogonal* basis of eigenvectors of $\mathbf{A}$

$$\mathbf{v}_1 = \begin{pmatrix} 1 \\ -1 \\ 0 \end{pmatrix}, \quad \mathbf{v}_2 = \begin{pmatrix} 1/2 \\ 1/2 \\ -1 \end{pmatrix}, \quad \mathbf{v}_3 = \begin{pmatrix} 1 \\ 1 \\ 1 \end{pmatrix}.$$

Now normalize:

$$\mathbf{w}_1 = \begin{pmatrix} 1/\sqrt{2} \\ -1/\sqrt{2} \\ 0 \end{pmatrix}, \quad \mathbf{w}_2 = \begin{pmatrix} 1/\sqrt{6} \\ 1/\sqrt{6} \\ -2/\sqrt{6} \end{pmatrix}, \quad \mathbf{w}_3 = \begin{pmatrix} 1/\sqrt{3} \\ 1/\sqrt{3} \\ 1/\sqrt{3} \end{pmatrix}.$$

This gives the base-change matrix

$$\mathbf{P} = \begin{pmatrix} 1/\sqrt{2} & 1/\sqrt{6} & 1/\sqrt{3} \\ -1/\sqrt{2} & 1/\sqrt{6} & 1/\sqrt{3} \\ 0 & -2/\sqrt{6} & 1/\sqrt{3} \end{pmatrix}$$

of Example 13.4. The point of that example was to show that $\mathbf{P}^T = \mathbf{P}^{-1}$ and so

$$\mathbf{P}^T \mathbf{A} \mathbf{P} = \mathbf{P}^{-1} \mathbf{A} \mathbf{P} = \begin{pmatrix} 6 & 0 & 0 \\ 0 & 6 & 0 \\ 0 & 0 & 3 \end{pmatrix}.$$

This matrix is diagonal, so *the base-change matrix* $\mathbf{P}$ *diagonalizes the quadratic form* Q *as well as the linear transformation* $f_{\mathbf{A}}$.

Now introduce 'new coordinates' a, b, c by the rule

$$\begin{pmatrix} a \\ b \\ c \end{pmatrix} = \mathbf{P}^{-1} \begin{pmatrix} x \\ y \\ z \end{pmatrix}$$

so

$$\begin{aligned} Q(x, y, z) &= (x, y, z)\mathbf{P}(\mathbf{P}^{-1}\mathbf{A}\mathbf{P})\mathbf{P}^{-1} \begin{pmatrix} x \\ y \\ z \end{pmatrix} \\ &= (a, b, c) \begin{pmatrix} 6 & 0 & 0 \\ 0 & 6 & 0 \\ 0 & 0 & 3 \end{pmatrix} \begin{pmatrix} a \\ b \\ c \end{pmatrix} \\ &= 6a^2 + 6b^2 + 3c^2, \end{aligned}$$

since $(x, y, z)\mathbf{P} = \mathbf{P}^T(x, y, z)^T = \mathbf{P}^{-1}(x, y, z)^T$. Thus the surface $Q(x, y, z) = 3$ is given by the equation

$$2a^2 + 2b^2 + c^2 = 1$$

with respect to the new coordinates a, b, c in the directions given by the vectors $\mathbf{w}_1, \mathbf{w}_2, \mathbf{w}_3$.

This surface is an ellipsoid, with centre at the origin, and elongated in the c or $\mathbf{w}_3$ direction with radius 1 in this direction, and with radius $1/\sqrt{2}$ in the directions orthogonal to this.

A quadratic form of rank 3 over $\mathbb{R}^3$ can always be diagonalized as

$$Q(x, y, z) = \pm(x/a)^2 \pm (y/b)^2 \pm (z/c)^2.$$

The surfaces given by the equation $Q(x, y, z) = 1$ have different shapes according to the signs (i.e. the signature). The surface

$$Q(x, y, z) = (x/a)^2 + (y/b)^2 + (z/c)^2 = 1$$

is an ellipsoid, with semi-major axes a, b, c. The surface

$$Q(x, y, z) = (x/a)^2 + (y/b)^2 - (z/c)^2 = 1$$

is a one-sheet hyperboloid, something like a cooling tower extending to infinity in both directions. The surface

$$Q(x, y, z) = (x/a)^2 - (y/b)^2 - (z/c)^2 = 1$$

is a two-sheet hyperboloid, like a hill reflected in the sky. The final equation,

$$Q(x, y, z) = -(x/a)^2 - (y/b)^2 - (z/c)^2 = 1,$$

clearly has no real solutions. (See Figure 13.2.)

(a) Ellipsoid

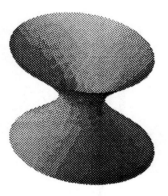

(b) One-sheet hyperboloid

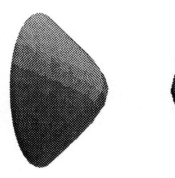

(c) Two-sheet hyperboloid

Fig. 13.2 Surfaces defined by quadratic forms of rank 3

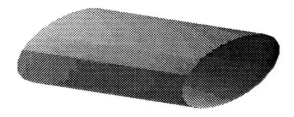

(a) Elliptical cylinder

(b) Hyperbolic cylinder

Fig. 13.3 Surfaces defined by quadratic forms of rank 2

The degenerate cases, when the quadratic form has smaller rank, are also worth noting. The surface $Q(x, y, z) = (x/a)^2 + (y/b)^2 = 1$ is an elliptical cylinder, while the surface $Q(x, y, z) = (x/a)^2 - (y/b)^2 = 1$ is a hyperbolic cylinder (if that makes sense!). (See Figure 13.3.)

We may also consider surfaces of the form $Q(x, y, z) = 0$. These are degenerate cases of $Q(x, y, z) = \epsilon$ when $\epsilon \to 0$. In particular, the surface $(x/a)^2 - (y/b)^2 - (z/b)^2 = 0$ is that of two cones joined together at their apexes, whereas the more general case $(x/a)^2 - (y/b)^2 - (z/c)^2 = 0$ is similar except the cross sections of the 'cones' are ellipses. The other degenerate case of this type is exemplified by $(x/a)^2 - (y/b)^2 = 0$ which is a pair of planes meeting at the line $x = y = 0$.

Exercises

Exercise 13.1 For each case, sketch the graph of the curve in question and describe all eigenvectors of the matrix **A** geometrically.

(a) $x^2 + 4xy + y^2 = 7$, and $\mathbf{A} = \begin{pmatrix} 1 & 2 \\ 2 & 1 \end{pmatrix}$.

(b) $x^2 + 2xy + y^2 = 7$, and $\mathbf{A} = \begin{pmatrix} 1 & 1 \\ 1 & 1 \end{pmatrix}$.

(c) $5x^2 + 4xy + 5y^2 = 7$, and $\mathbf{A} = \begin{pmatrix} 5 & 2 \\ 2 & 5 \end{pmatrix}$.

(d) $x^2 + y^2 = 7$, and $\mathbf{A} = \begin{pmatrix} 1 & 0 \\ 0 & 1 \end{pmatrix}$.

Exercise 13.2 Let $T \colon \mathbb{C}^2 \to \mathbb{C}^2$ be defined by $T((x, y)^T) = (2ix + y, x)^T$.

(a) Write down the matrix $\mathbf{A}$ of T with respect to the usual basis of $\mathbb{C}^2$.
(b) Is $\mathbf{A}$ symmetric?
(c) Is $\mathbf{A}$ conjugate-symmetric?
(d) What are the eigenvalues of $\mathbf{A}$?
(e) Is $\mathbf{A}$ diagonalizable?

Exercise 13.3 A matrix $\mathbf{A}$ is of the form $\begin{pmatrix} a & c \\ c & b \end{pmatrix}$, where $a, b, c \in \mathbb{R}$. Suppose that $\mathbf{A}$ has an eigenvalue λ of algebraic multiplicity 2. Prove that $a = b$ and calculate the value of c.

Exercise 13.4 Sketch the graph of each of the following.

(a) $5x^2 - 8xy + 5y^2 = 9$.
(b) $11x^2 - 24xy + 4y^2 + 6x + 8y = -15$.
(c) $16x^2 - 24xy + 9y^2 - 30x + 40y = 5$.

[Hint: for (b) and (c) diagonalize the matrix for the quadratic form first, then transform the whole equation including the nonquadratic parts.]

Exercise 13.5 Describe the set of points

$$\{(x, y, z) \in \mathbb{R}^3 : x^2 - y^2/3 + z^2 - 2xy - 2yz + 2xz = 1\}$$

mentioning any rotational or translational symmetries that you can find. [Translational symmetry is when a figure looks the same after it has been shifted by a translation vector $\mathbf{v}$. Rotational symmetry is similar, but the figure is rotated through an angle θ about a given axis.]

Exercise 13.6 The form Q is defined on $\mathbb{R}^3$ by

$$Q((x, y, z)^T) = x^2 + y^2 + 4z^2 + 14xy + 8xz + 8yz.$$

By finding a suitable orthogonal matrix $\mathbf{P}$ and defining 'new coordinates' a, b, c by

$$(a, b, c)^T = \mathbf{P}^{-1}(x, y, z)^T,$$

write $Q(x, y, z)^T$ as

$$\lambda a^2 + \mu b^2 + \nu c^2$$

for some real constants λ, μ, and ν. Hence describe the following surfaces.

(a) $Q(x, y, z) = 1$.

(b) $Q(x, y, z) = 0$.

(c) $Q(x, y, z) + x + y - 2z = 1$.

(d) $Q(x, y, z) + x + y + z = 1$.

(e) $Q(x, y, z) + x + y + z = 0$.

(f) $Q(x, y, z) + x = 1$.

[Hint: for some of these, you may find it helpful to change the origin.]

Exercise 13.7 (This exercise is for students wondering why the terminology 'self-adjoint' is used.) Let V be a finite dimensional inner product space, and let $e_1, \ldots, e_n$ be an ordered orthonormal basis of V. Suppose $f : V \to V$ is a linear transformation with matrix $\mathbf{A}$ with respect to this ordered basis, and define $f^\dagger : V \to V$ by

$$f^\dagger(v) = \sum_{i=1}^{n} \langle v | f(e_i) \rangle e_i.$$

$f^\dagger$ is called the *adjoint* of f.

(a) Show that $f^\dagger$ is a linear transformation of V.

(b) Show that

$$\langle e_i | f(e_j) \rangle = \sum_{k=1}^{n} \langle e_k | f(e_j) \rangle \langle e_i | e_j \rangle = \sum_{k=1}^{n} \langle e_k | f(e_i) \rangle \langle e_i | e_j \rangle = \langle f^\dagger(e_i) | e_j \rangle$$

for all i, j.

(c) Using the previous part and linearity, show that

$$\langle u | f(v) \rangle = \langle f^\dagger(u) | v \rangle$$

for all $u, v \in V$.

(d) Deduce that the matrix of $f^\dagger$ with respect to the basis $e_1, \ldots, e_n$ is $\overline{\mathbf{A}}^T$, and that the definition of $f^\dagger$ is independent of the orthonormal basis $e_1, \ldots, e_n$ taken.

Exercise 13.8 Let V be an inner product space over $\mathbb{R}$ or $\mathbb{C}$, and suppose that f is a self-adjoint linear transformation from V to V. Given $p(x)$, a polynomial with coefficients from the field of scalars for V, show that $p(f)$ is a self-adjoint linear transformation $V \to V$.

Exercise 13.9 Let $\alpha, \beta \in \mathscr{L}(V, V)$ be self-adjoint, where V is an inner product space over $\mathbb{R}$ or $\mathbb{C}$. Write $\alpha\beta$ as

$$\tfrac{1}{2}(\alpha\beta - \beta\alpha) + \tfrac{1}{2}(\alpha\beta + \beta\alpha),$$

to show that

$$|\langle v | \alpha\beta(v) \rangle|^2 \geqslant \tfrac{1}{4} |\langle v | (\alpha\beta - \beta\alpha)(v) \rangle|^2$$

for all vectors $v \in V$.

14

The Jordan normal form

If a linear transformation $f: V \to V$ is not diagonalizable, we may still ask for a basis with respect to which the matrix of f is as 'nice as possible'. It turns out that we can always obtain such a basis where this matrix is in *Jordan normal form*; that is, a special upper triangular form where the only nonzero entries off the diagonal are entries equal to 1 just above repeated eigenvalues. Such forms will enable us to solve a much greater variety of simultaneous difference and differential equations.

14.1 Jordan normal form

We have proved that if $f: V \to V$ is a linear transformation whose minimum polynomial has distinct roots, then f is diagonalizable. We now consider the general case, when $m_f(x)$ may have repeated roots.

First we need to generalize the concept of eigenspaces.

Definition 14.1 *Suppose* $f : V \to V$ *has minimum polynomial*

$$m_f(x) = (x - \lambda_1)^{e_1} \ldots (x - \lambda_r)^{e_r},$$

where $\lambda_1, \ldots, \lambda_r$ *are the distinct eigenvalues of* f. *Then the subspaces*

$$\ker((f - \lambda_i \mathrm{I})^{e_i})$$

of V *are called the* **generalized eigenspaces** *of* f.

Notice that if $e_i = 1$, i.e. λ_i is an eigenvalue which occurs with multiplicity 1 as a root of the minimum polynomial, then this is just the usual eigenspace.

The most important result for our purposes is that V is the direct sum of these generalized eigenspaces. This is a generalization of Theorem 12.9, and is stated below as Theorem 14.3 and will be proved in Section 14.4. But before we give this result formally, let us consider an example by way of illustration.

Example 14.2 Let $V = \mathbb{C}^3$, and define the linear map $f : V \to V$ by

$$f \begin{pmatrix} x \\ y \\ z \end{pmatrix} = \begin{pmatrix} 2x \\ -x - 3y - z \\ -x + 4y + z \end{pmatrix}.$$

Then f is represented with respect to the standard basis by the matrix

$$\mathbf{A} = \begin{pmatrix} 2 & 0 & 0 \\ -1 & -3 & -1 \\ -1 & 4 & 1 \end{pmatrix},$$

so we can calculate its characteristic polynomial in the usual way, as

$$\chi_f(x) = \chi_{\mathbf{A}}(x) = (2-x)(x^2 + 2x + 1) = (2-x)(x+1)^2.$$

Thus $m_f(x)$ is either $(x-2)(x+1)$ or $(x-2)(x+1)^2$. But

$$(\mathbf{A} - 2\mathbf{I})(\mathbf{A} + \mathbf{I}) = \begin{pmatrix} 0 & 0 & 0 \\ -1 & -5 & -1 \\ -1 & 4 & -1 \end{pmatrix} \begin{pmatrix} 3 & 0 & 0 \\ -1 & -2 & -1 \\ -1 & 4 & 2 \end{pmatrix} = \begin{pmatrix} 0 & 0 & 0 \\ 3 & 6 & 3 \\ -6 & -12 & -6 \end{pmatrix},$$

which is not the zero matrix, so $m_f(x) = m_{\mathbf{A}}(x) = (x-2)(x+1)^2$.

For the eigenvalue 2, the generalized eigenspace is the same as the ordinary eigenspace, and is just the kernel of the linear map $g = f - 2\mathbf{I}$. Now

$$g \begin{pmatrix} x \\ y \\ z \end{pmatrix} = \begin{pmatrix} 0 \\ -x - 5y - z \\ -x + 4y - z \end{pmatrix}$$

so the kernel of g is the set of all vectors $(x, y, z)^T$ satisfying the equations $-x - 5y - z = 0$ and $-x + 4y - z = 0$, equivalently, $y = 0$ and $x = -z$. So $\ker(g) = \{(-z, 0, z)^T : z \in \mathbb{C}\}$.

For the eigenvalue -1, we work out the ordinary eigenspace in the same way. Writing $h = f + \mathbf{I}$ we have

$$h \begin{pmatrix} x \\ y \\ z \end{pmatrix} = \begin{pmatrix} 3x \\ -x - 2y - z \\ -x + 4y + 2z \end{pmatrix}$$

and

$$\ker(h) = \{(0, y, z)^T : y, z \in \mathbb{C}, -2y - z = 0, 4y + 2z = 0\}$$
$$= \{(0, y, -2y)^T : y \in \mathbb{C}\}.$$

Now

$$h^2 \begin{pmatrix} x \\ y \\ z \end{pmatrix} = h \begin{pmatrix} 3x \\ -x - 2y - z \\ -x + 4y + 2z \end{pmatrix}$$
$$= \begin{pmatrix} 9x \\ -3x - 2(-x - 2y - z) - (-x + 4y + 2z) \\ -3x + 4(-x - 2y - z) + 2(-x + 4y + 2z) \end{pmatrix}$$
$$= \begin{pmatrix} 9x \\ 0 \\ -9x \end{pmatrix}$$

so

$$\ker(h^2) = \{(x, y, z)^T : 9x = 0\} = \{(0, y, z)^T : y, z \in \mathbb{C}\},$$

which has dimension 2.

Note that the image of h^2 is the set of all vectors of the form $(9x, 0, -9x)^T$, which is the same as the kernel of g. That is, $\mathrm{im}(h^2) = \ker(g)$. Similarly we have $\mathrm{im}(g) = \ker(h^2)$. (Compare Proposition 12.17 and the example preceding it.)

We now state our promised generalization of Theorem 12.9.

Theorem 14.3 (Primary decomposition) *Let $f : V \to V$ be a linear transformation with minimum polynomial*

$$m_f(x) = (x - \lambda_1)^{e_1} \ldots (x - \lambda_r)^{e_r},$$

where $\lambda_1, \ldots, \lambda_r$ are the distinct eigenvalues of f, and $e_1, \ldots, e_r$ are positive integers. Let $V_1, \ldots, V_r$ denote the corresponding generalized eigenspaces, i.e. $V_i = \ker((f - \lambda_i \mathrm{I})^{e_i})$. Then

$$V = V_1 \oplus V_2 \oplus \cdots \oplus V_r.$$

Proof See Section 14.4, Theorem 14.15. □

This is sometimes called the *primary decomposition* of V with respect to f, and $V_1, \ldots, V_r$ are the *primary components*. We have already seen an example of this in Example 14.2. In that example, the generalized eigenspaces are

$$V_1 = \ker(g) = \{(x, 0, -x)^T : x \in \mathbb{C}\}, \text{ and}$$

$$V_2 = \ker(h^2) = \{(0, y, z)^T : y, z \in \mathbb{C}\},$$

and it is easy to check that $V = V_1 \oplus V_2$ in this case. If we now choose a basis B_1 for V_1 and a basis B_2 for V_2, then $B = B_1 \cup B_2$ is a basis for V, since $V = V_1 \oplus V_2$ is a direct sum, and then we can write f with respect to the basis B. For example, take $B_1 = (1, 0, -1)^T$ and $B_2 = (0, 1, 0)^T, (0, 0, 1)^T$, and calculate

$$
\begin{array}{rcll}
f(1, 0, -1)^T & = & (2, 0, -2)^T & = & 2(1, 0, -1)^T \\
f(0, 1, 0)^T & = & (0, -3, 4)^T & = & -3(0, 1, 0)^T + 4(0, 0, 1)^T \\
f(0, 0, 1)^T & = & (0, -1, 1)^T & = & -(0, 1, 0)^T + (0, 0, 1)^T
\end{array}
$$

so the matrix of f with respect to the basis B is

$$
\begin{pmatrix}
2 & 0 & 0 \\
0 & -3 & -1 \\
0 & 4 & 1
\end{pmatrix}.
$$

Observe that this matrix is in *block diagonal form*, with the blocks corresponding to the different generalized eigenspaces of f.

Proposition 14.4 *With the notation of Theorem 14.3, if B_i is a basis for V_i, then $\bigcup_{i=1}^{r} B_i$ is a basis for V, and with respect to this basis the matrix of f has block diagonal form, i.e.*

$$\begin{pmatrix} \mathbf{A}_1 & \mathbf{0} & \cdots & \mathbf{0} \\ \mathbf{0} & \mathbf{A}_2 & \cdots & \mathbf{0} \\ \vdots & & \ddots & \vdots \\ \mathbf{0} & \mathbf{0} & \cdots & \mathbf{A}_r \end{pmatrix},$$

where $\mathbf{A}_1, \ldots, \mathbf{A}_r$ are square matrices giving the action of f on $V_1, \ldots, V_r$.

Proof We use the fact that f and $(f - \lambda_i \mathrm{I})^{e_i}$ commute with each other, for all i.

If $v_i \in V_i = \ker((f - \lambda_i \mathrm{I})^{e_i})$, then

$$(f - \lambda_i \mathrm{I})^{e_i}(v_i) = 0$$

so

$$f((f - \lambda_i \mathrm{I})^{e_i}(v_i)) = 0,$$

and

$$(f - \lambda_i \mathrm{I})^{e_i}(f(v_i)) = 0;$$

hence $f(v_i) \in V_i$.   □

The primary decomposition uses the factorization of the minimum polynomial of f to give us a block diagonal form for the matrix of f. Each block now has a single eigenvalue: the minimum polynomial of an $n \times n$ block is, say, $(x - \lambda)^k$, and the characteristic polynomial is $(x - \lambda)^n$. Our next task is to simplify the shape of these blocks. In other words, we try to find as nice a basis as possible for each generalized eigenspace. We have already seen in Example 12.1 that in general we cannot find a basis with respect to which the matrix is diagonal. However, we can get close. To see the kind of thing that we can do, let us consider an example.

Example 14.5 Let $f \colon \mathbb{R}^3 \to \mathbb{R}^3$ be defined by

$$f\begin{pmatrix} x \\ y \\ z \end{pmatrix} = \begin{pmatrix} -x - y + z \\ 14x + 8y - 7z \\ 10x + 5y - 4z \end{pmatrix},$$

so that f is represented with respect to the standard basis by the matrix

$$\mathbf{B} = \begin{pmatrix} -1 & -1 & 1 \\ 14 & 8 & -7 \\ 10 & 5 & -4 \end{pmatrix}.$$

First we calculate the characteristic polynomial $\chi_\mathbf{B}(x) = (1 - x)^3$ and minimum polynomial $m_\mathbf{B}(x) = (x - 1)^2$. In particular there is a single eigenvalue, namely

1, and its algebraic multiplicity is 3. Next we work out the eigenspace, $\ker(\mathbf{B} - \mathbf{I})$, which consists of all vectors (x, y, z) satisfying

$$\begin{pmatrix} -1 & -1 & 1 \\ 14 & 8 & -7 \\ 10 & 5 & -4 \end{pmatrix} \begin{pmatrix} x \\ y \\ z \end{pmatrix} = \begin{pmatrix} x \\ y \\ z \end{pmatrix}.$$

Solving these equations in the usual way, we obtain a two-dimensional space of solutions, spanned by eigenvectors such as $(1, 0, 2)^T$ and $(0, 1, 1)^T$, for example. (Thus the geometric multiplicity of the eigenvalue is 2.) Moreover, as $(\mathbf{B} - \mathbf{I})^2$ is the zero matrix, $\ker(\mathbf{B} - \mathbf{I})^2$ is the whole space. To get a nice basis for the space, we first take a basis for $\ker(\mathbf{B} - \mathbf{I})$ and then extend to a basis for $\ker(\mathbf{B} - \mathbf{I})^2$. For example, we could take the ordered basis

$$(1, 0, 2)^T, (0, 1, 1)^T, (1, 0, 0)^T.$$

Applying the corresponding base-change matrix

$$\mathbf{Q} = \begin{pmatrix} 1 & 0 & 1 \\ 0 & 1 & 0 \\ 2 & 1 & 0 \end{pmatrix}$$

we obtain the new matrix

$$\mathbf{Q}^{-1}\mathbf{B}\mathbf{Q} = \begin{pmatrix} 1 & 0 & -2 \\ 0 & 1 & 14 \\ 0 & 0 & 1 \end{pmatrix},$$

which is now an upper triangular matrix, with the eigenvalues on the diagonal.

But we can do better than this. If we apply $\mathbf{B} - \mathbf{I}$ to the basis vector $(1, 0, 0)^T$, then the image vector $(-2, 14, 10)^T$ is in $\ker(\mathbf{B} - \mathbf{I})$ since $\mathbf{w} = (\mathbf{B} - \mathbf{I})(1, 0, 0)^T$ has $(\mathbf{B} - \mathbf{I})\mathbf{w} = (\mathbf{B} - \mathbf{I})^2(1, 0, 0)^T = \mathbf{0}$ as $\ker(\mathbf{B} - \mathbf{I})^2 = \mathbb{R}^3$. So let us change our basis of $\ker(\mathbf{B} - \mathbf{I})$ to include this vector. For example, we could take our new basis for the whole space to be

$$(1, 0, 0)^T, (-2, 14, 10)^T, (0, 1, 1)^T$$

which would give a base-change matrix

$$\mathbf{R} = \begin{pmatrix} 1 & -2 & 0 \\ 0 & 14 & 1 \\ 0 & 10 & 1 \end{pmatrix}$$

and a new matrix

$$\mathbf{R}^{-1}\mathbf{B}\mathbf{R} = \begin{pmatrix} 1 & 1 & 0 \\ 0 & 1 & 0 \\ 0 & 0 & 1 \end{pmatrix},$$

which is in so-called *Jordan normal form*. The only nonzero entries off the diagonal are entries equal to 1, one place immediately above the diagonal.

Theorem 14.6 *If* $f: W \to W$ *is a linear transformation with minimum polynomial* $m_f(x) = (x - \lambda)^k$, *then there is a basis of* W *with respect to which the matrix of* f *has* λ *on the diagonal, 1 or 0 in each entry immediately above the diagonal, and 0 elsewhere. That is, the matrix of* f *has the form*

$$\begin{pmatrix} \lambda & 1 & 0 & \cdots & & & & \\ 0 & \lambda & 1 & 0 & & & & \\ \vdots & & \ddots & 1 & & & & \\ 0 & \cdots & 0 & \lambda & & & & \\ & & & & \lambda & 1 & 0 & \cdots \\ & & & & 0 & \lambda & 1 & 0 \\ & & & & \vdots & & \ddots & 1 \\ & & & & 0 & \cdots & 0 & \lambda \\ & & & & & & & & \ddots \end{pmatrix}$$

with zeros everywhere except as indicated.

Proof See Section 14.2. □

This theorem tells you how each of the blocks $\mathbf{A}_i$ in Proposition 14.4 can be rewritten. Putting all the blocks together again, we get the Jordan normal form of an arbitrary matrix, which has blocks of the shape given in Theorem 14.6, for various values of λ.

The small blocks which make up this matrix, of the form

$$\mathbf{E} = \begin{pmatrix} \lambda & 1 & 0 & \cdots & 0 \\ 0 & \lambda & 1 & & 0 \\ 0 & & \ddots & & 0 \\ \vdots & & & & 1 \\ 0 & 0 & 0 & \cdots & \lambda \end{pmatrix},$$

are called *elementary Jordan matrices*. If $\mathbf{E}$ is a $k \times k$ matrix of this form, then it is easy to show that $(\mathbf{E} - \lambda \mathbf{I}_k)^k = 0$, but that $(\mathbf{E} - \lambda \mathbf{I}_k)^{k-1} \neq 0$. Thus if a Jordan matrix $\mathbf{J}$ has a $k \times k$ block $\mathbf{E}$ as above, then the minimum polynomial must be divisible by $(x - \lambda)^k$. Indeed we have the following result (see Exercise 11.7).

Proposition 14.7 *If* $f: V \to V$ *is a linear transformation and*

$$m_f(x) = (x - \lambda_1)^{e_1} \ldots (x - \lambda_r)^{e_r},$$

then, in a matrix representation of f *in Jordan normal form, the largest elementary Jordan matrix with eigenvalue* λ_i *is an* $e_i \times e_i$ *matrix.*

With the same matrix $\mathbf{E}$ as above, suppose that $\mathbf{v} = (v_1, \ldots, v_k)^T$ is an eigenvector of $\mathbf{E}$. Then $\mathbf{E}\mathbf{v} = \lambda \mathbf{v}$, i.e.

$$\mathbf{E}\mathbf{v} = (\lambda v_1 + v_2, \lambda v_2 + v_3, \dots, \lambda v_{k-1} + v_k, \lambda v_k)^T$$
$$= \lambda(v_1, v_2, \dots, v_k)^T,$$

so $v_1 = v_2 = \cdots = v_{k-1} = 0$. So up to a scalar multiple, we have $\mathbf{v} = (0, 0, \dots, 1)^T$. Thus each elementary Jordan matrix has a one-dimensional eigenspace. Putting all these together we obtain the following.

Proposition 14.8 *The dimension of the λ-eigenspace of f (i.e. the geometric multiplicity of λ) is equal to the number of elementary Jordan matrices for λ in the Jordan normal form for f.*

14.2 Obtaining the Jordan normal form

Here, we will be rather more precise on how the Jordan form of an arbitrary square matrix can be obtained.

Suppose $f \colon V \to V$ is a linear transformation. First, the primary decomposition theorem (Theorem 14.3 or Theorem 14.15) shows how we can get a block diagonal form for the matrix of f, by finding bases of the generalized eigenspaces. (All you need to know to be able to carry out this calculation is the definition of the generalized eigenspaces. In particular, you don't need to know the proof of the primary decomposition theorem.) This reduces the problem to finding a 'nice' representation for each block, i.e. finding a 'nice' basis for each generalized eigenspace.

Each block corresponding to one of these generalized eigenspaces has a single eigenvalue. The minimum polynomial of an $n \times n$ block is $(x - \lambda)^k$, say, and the characteristic polynomial is $(x - \lambda)^n$. We suppose, therefore, that we have a linear transformation $f \colon V \to V$ with minimum polynomial $m_f(x) = (x - \lambda)^k$, and for simplicity we consider the linear transformation $g = f - \lambda \mathbf{I}$ instead. Then we have $m_g(x) = x^k$ and $\chi_g(x) = x^n$, so the only eigenvalue of g is 0.

As $m_g(x) = x^k$, g^k is the zero map, so $V = \ker(g^k)$. Now clearly $\ker g^{r+1} \supseteq \ker g^r$ for all r, for if $v \in \ker(g^r)$ then $g^r(v) = 0$ and $g^{r+1}(v) = g(g^r(v)) = g(0) = 0$, so $v \in \ker(g^{r+1})$. This means we get a chain of subspaces

$$V = \ker g^k \supseteq \ker g^{k-1} \supseteq \cdots \supseteq \ker g^2 \supseteq \ker g \supseteq \ker g^0 = \{0\}.$$

The general method for finding a suitable basis of V is as follows. First take a basis $v_1, \dots, v_{r_1}$ of $\ker g$, extend this to a basis $v_1, \dots, v_{r_1}, v_{r_1+1}, \dots, v_{r_2}$ of $\ker g^2$, and so on, until we have a basis

$$v_1, \dots, v_{r_1}, v_{r_1+1}, \dots, v_{r_2}, v_{r_2+1}, \dots, v_{r_k}$$

of $V = \ker g^k$. We now modify this basis: first write down those basis elements $v_{r_{k-1}+1}, \dots, v_{r_k}$ of $\ker g^k$ not in $\ker g^{k-1}$ as $a_1 = v_{r_{k-1}+1}, \dots, a_{n_1} = v_{r_k}$, giving

$$\operatorname{span}(a_1, \dots, a_{n_1}) = \operatorname{span}(v_{r_{k-1}+1}, \dots, v_{r_k}).$$

Next calculate $b_1 = g(a_1), \dots, b_{n_1} = g(a_{n_1})$ and write these down underneath the a_i. These b_i are all elements of $\ker g^{k-1}$ since $g^{k-1}(b_i) = g^k(a_i) = 0$, and it will turn out that all the vectors a_i, b_j form a linearly independent set. Because of this, we can extend the list of the b_i to $b_{n_1+1}, \ldots, b_{n_2}$ so that

$$\text{span}(a_1, \ldots, a_{n_1}, b_1, \ldots, b_{n_2}) = \text{span}(v_{r_{k-2}+1}, \ldots, v_{r_{k-1}+1}, \ldots, v_{r_k}).$$

We then work out $c_i = g(b_i)$ for each i, write these underneath, and extend what we have got to a basis of $\text{span}(v_{r_{k-3}+1}, \ldots, v_{r_k})$. When this process is complete, we will have a basis of the whole space V written as a table of the form

$$
\begin{array}{llllllll}
a_1 & \cdots & a_{n_1} & & & & & \\
b_1 & \cdots & b_{n_1} & b_{n_1+1} & \cdots & b_{n_2} & & \\
c_1 & \cdots & c_{n_1} & c_{n_1+1} & \cdots & c_{n_2} & c_{n_2+1} & \cdots & c_{n_3} \\
\vdots & & & & & & & \\
z_1 & \cdots & z_{n_1} & z_{n_1+1} & \cdots & z_{n_2} & z_{n_2+1} & \cdots & z_{n_3} & \cdots & z_{n_k}.
\end{array}
$$

All that is required is to order this basis in a suitable way. To do this, note that $g(a_i) = b_i$, $g(b_i) = c_i$, etc., so we order the basis *reading up the columns first*, and then left to right, as

$$z_1, \ldots, c_1, b_1, a_1, z_2, \ldots, c_2, b_2, a_2, \ldots \,.$$

Because $g(a_i) = b_i$, $g(b_i) = c_i$, etc., the matrix of g will be in Jordan normal form, with an elementary Jordan matrix of the form

$$
\begin{pmatrix}
0 & 1 & 0 & \cdots & 0 \\
0 & 0 & 1 & & 0 \\
0 & & \ddots & & 0 \\
\vdots & & & & 1 \\
0 & 0 & 0 & \cdots & 0
\end{pmatrix}
$$

for each column of the table. The matrix of the original linear transformation $f = g + \lambda I$ is then formed of elementary Jordan matrices

$$
\begin{pmatrix}
\lambda & 1 & 0 & \cdots & 0 \\
0 & \lambda & 1 & & 0 \\
0 & & \ddots & & 0 \\
\vdots & & & & 1 \\
0 & 0 & 0 & \cdots & \lambda
\end{pmatrix}
$$

as required.

Clearly, the crucial point to this construction (and it is not immediately obvious) is that the basis modification actually does give a basis. The lemma that tells us that it really does work is the following.

Lemma 14.9 *If $\{u_1, \ldots, u_r\}$ is a basis for $\ker(g^j)$, is extended to a basis*

$$\{u_1, \ldots, u_r, v_1, \ldots, v_s\}$$

of $\ker(g^{j+1})$, and to a basis

$$\{u_1, \ldots, u_r, v_1, \ldots, v_s, w_1, \ldots, w_t\}$$

of $\ker(g^{j+2})$, then $\{u_1, \ldots, u_r, g(w_1), \ldots, g(w_t)\}$ is a linearly independent subset of $\ker(g^{j+1})$.

Proof First note that $g^{j+1}(g(w_i)) = g^{j+2}(w_i) = 0$ so $g(w_i) \in \ker(g^{j+1})$. To show linear independence, suppose we have a linear dependence

$$\sum_{i=1}^{r} \lambda_i u_i + \sum_{i=1}^{t} \mu_i g(w_i) = 0,$$

so that

$$\sum_{i=1}^{t} \mu_i g(w_i) = -\sum_{i=1}^{r} \lambda_i u_i \in \ker(g^j).$$

Therefore

$$0 = g^j\left(\sum_{i=1}^{t} \mu_i g(w_i)\right) = g^{j+1}\left(\sum_{i=1}^{t} \mu_i w_i\right)$$

so $\sum_{i=1}^{t} \mu_i w_i \in \ker(g^{j+1})$ which means that it can be written as a linear combination of $\{u_1, \ldots, u_r, v_1, \ldots, v_s\}$. But

$$\{u_1, \ldots, u_r, v_1, \ldots, v_s, w_1, \ldots, w_t\}$$

is a linearly independent set, so all $\mu_i = 0$. Therefore $\sum_{i=1}^{r} \lambda_i u_i = 0$, so all the $\lambda_i = 0$. $\qquad\square$

Example 14.10 Let **A** be the matrix

$$\mathbf{A} = \begin{pmatrix} 3 & 1 & 0 & 0 & 0 \\ -1 & 2 & 1 & 0 & 0 \\ 1 & 0 & 1 & 0 & 0 \\ -1 & 0 & 1 & 3 & 1 \\ 1 & 0 & -1 & -1 & 1 \end{pmatrix}$$

You can calculate that $\chi_\mathbf{A} = (2 - x)^5$ and $m_\mathbf{A}(x) = (x - 2)^3$. Now let $g(\mathbf{v}) = \mathbf{B}\mathbf{v}$ where

$$\mathbf{B} = \mathbf{A} - 2\mathbf{I} = \begin{pmatrix} 1 & 1 & 0 & 0 & 0 \\ -1 & 0 & 1 & 0 & 0 \\ 1 & 0 & -1 & 0 & 0 \\ -1 & 0 & 1 & 1 & 1 \\ 1 & 0 & -1 & -1 & -1 \end{pmatrix}.$$

Calculating kernels as usual, we find that

$$\ker g = \{(x_1, x_2, x_3, x_4, x_5)^T : x_1 = -x_2,\ x_1 = x_3,\ x_4 = -x_5\}$$

with basis $(1, -1, 1, 0, 0)^T, (0, 0, 0, 1, -1)^T$,

$$\ker g^2 = \{(x_1, x_2, x_3, x_4, x_5)^T : x_2 = -x_3\}$$

with basis

$$(1, -1, 1, 0, 0)^T, (0, 0, 0, 1, -1)^T, (1, 0, 0, 0, 0)^T, (0, 0, 0, 1, 0)^T,$$

and $\ker g^3 = \mathbb{R}^5$, with basis

$$(1, -1, 1, 0, 0)^T, (0, 0, 0, 1, -1)^T, (1, 0, 0, 0, 0)^T, (0, 0, 0, 1, 0)^T, (0, 1, 0, 0, 0)^T.$$

We now modify this basis according to the rules above. First, we set $\mathbf{a}_1 = (0, 1, 0, 0, 0)^T$. Next, take $\mathbf{b}_1 = g(\mathbf{a}_1) = (1, 0, 0, 0, 0)^T$, and extend by adding $\mathbf{b}_2 = (0, 0, 0, 1, 0)^T$. Finally, we set $\mathbf{c}_1 = g(\mathbf{b}_1) = (1, -1, 1, -1, 1)^T$ and $\mathbf{c}_2 = g(\mathbf{b}_2) = (0, 0, 0, 1, -1)^T$. These vectors are organized in the following way,

$$
\begin{array}{ll}
\mathbf{a}_1 & \\
\mathbf{b}_1 & \mathbf{b}_2 \\
\mathbf{c}_1 & \mathbf{c}_2,
\end{array}
$$

and we can order the basis we have just found reading up columns and across from left to right as $\mathbf{c}_1, \mathbf{b}_1, \mathbf{a}_1, \mathbf{c}_2, \mathbf{b}_2$. The corresponding base-change matrix $\mathbf{P}$ and its inverse are

$$
\mathbf{P} = \begin{pmatrix}
1 & 1 & 0 & 0 & 0 \\
-1 & 0 & 1 & 0 & 0 \\
1 & 0 & 0 & 0 & 0 \\
-1 & 0 & 0 & 1 & 1 \\
1 & 0 & 0 & -1 & 0
\end{pmatrix}, \qquad
\mathbf{P}^{-1} = \begin{pmatrix}
0 & 0 & 1 & 0 & 0 \\
1 & 0 & -1 & 0 & 0 \\
0 & 1 & 1 & 0 & 0 \\
0 & 0 & 1 & 0 & -1 \\
0 & 0 & 0 & 1 & 1
\end{pmatrix}
$$

and you can check that

$$
\mathbf{P}^{-1}\mathbf{A}\mathbf{P} = \begin{pmatrix}
2 & 1 & 0 & 0 & 0 \\
0 & 2 & 1 & 0 & 0 \\
0 & 0 & 2 & 0 & 0 \\
0 & 0 & 0 & 2 & 1 \\
0 & 0 & 0 & 0 & 2
\end{pmatrix},
$$

in Jordan normal form.

14.3   Applications

Just as with diagonalization, Jordan form gives us a useful method for solving many kinds of simultaneous difference and differential equations. We illustrate the method here with an example of simultaneous difference equations.

Example 14.11 Solve the equations

$$x_{n+1} = 3x_n + z_n$$
$$y_{n+1} = -x_n + y_n - z_n$$
$$z_{n+1} = y_n + 2z_n$$

subject to $x_0 = y_0 = z_0 = 1$.

First, in matrix form this becomes

$$\begin{pmatrix} x_{n+1} \\ y_{n+1} \\ z_{n+1} \end{pmatrix} = \mathbf{A} \begin{pmatrix} x_n \\ y_n \\ z_n \end{pmatrix}$$

where

$$\mathbf{A} = \begin{pmatrix} 3 & 0 & 1 \\ -1 & 1 & -1 \\ 0 & 1 & 2 \end{pmatrix}.$$

By the usual calculations, we find that

$$\chi_\mathbf{A}(x) = (2 - x)^3, \qquad m_\mathbf{A}(x) = (x - 2)^3.$$

Put $g(\mathbf{v}) = (\mathbf{A} - 2\mathbf{I})\mathbf{v}$, so $\ker g$ has basis $(1, 0, -1)^T$, $\ker g^2$ has basis

$$(1, 0, -1)^T, (1, -1, 0)^T,$$

and $\ker g^3$ has basis

$$(1, 0, -1)^T, (1, -1, 0)^T, (1, 0, 0)^T.$$

Set $\mathbf{a}_1 = (1, 0, 0)^T$, $\mathbf{b}_1 = g(\mathbf{a}_1) = (1, -1, 0)^T$, and $\mathbf{c}_1 = g(\mathbf{b}_1) = (1, 0, -1)^T$. This gives a base-change matrix $\mathbf{P}$ and its inverse given by

$$\mathbf{P} = \begin{pmatrix} 1 & 1 & 1 \\ 0 & -1 & 0 \\ -1 & 0 & 0 \end{pmatrix}, \qquad \mathbf{P}^{-1} = \begin{pmatrix} 0 & 0 & -1 \\ 0 & -1 & 0 \\ 1 & 1 & 1 \end{pmatrix},$$

and

$$\mathbf{P}^{-1}\mathbf{A}\mathbf{P} = \begin{pmatrix} 2 & 1 & 0 \\ 0 & 2 & 1 \\ 0 & 0 & 2 \end{pmatrix}.$$

As usual, we 'change coordinates' to

$$\begin{pmatrix} a_n \\ b_n \\ c_n \end{pmatrix} = \mathbf{P}^{-1} \begin{pmatrix} x_n \\ y_n \\ z_n \end{pmatrix}$$

giving

$$\begin{pmatrix} a_{n+1} \\ b_{n+1} \\ c_{n+1} \end{pmatrix} = \mathbf{P}^{-1}\mathbf{A}\mathbf{P} \begin{pmatrix} a_n \\ b_n \\ c_n \end{pmatrix} = \begin{pmatrix} 2 & 1 & 0 \\ 0 & 2 & 1 \\ 0 & 0 & 2 \end{pmatrix} \begin{pmatrix} a_n \\ b_n \\ c_n \end{pmatrix}.$$

Now the general solution to difference equations like this is

$$a_n = 2^n a_0 + n2^{n-1} b_0 + \frac{n(n-1)}{2} 2^{n-2} c_0$$
$$b_n = 2^n b_0 + n2^{n-1} c_0$$
$$c_n = 2^n c_0.$$

Substituting

$$\begin{pmatrix} a_0 \\ b_0 \\ c_0 \end{pmatrix} = \mathbf{P}^{-1} \begin{pmatrix} 1 \\ 1 \\ 1 \end{pmatrix} = \begin{pmatrix} -1 \\ -1 \\ 3 \end{pmatrix}$$

into this we get

$$a_n = 2^{n-2}(3n^2 - 7n/2 - 4)$$
$$b_n = 2^{n-1}(3n - 2)$$
$$c_n = 2^n \cdot 3,$$

which gives

$$x_n = 2^{n-2}(3n^2 + 5n/2 + 4)$$
$$y_n = 2^{n-2}(4 - 6n)$$
$$z_n = 2^{n-2}(4 + 7n/2 - 3n^2).$$

For solving equations like these, the following 'standard forms' are useful.

Theorem 14.12 *The general solution of the difference equations*

$$\begin{pmatrix} x_1(i+1) \\ x_2(i+1) \\ x_3(i+1) \\ \vdots \\ x_k(i+1) \end{pmatrix} = \begin{pmatrix} \lambda & 1 & 0 & \cdots & 0 \\ 0 & \lambda & 1 & & 0 \\ 0 & & \ddots & & 0 \\ \vdots & & & & 1 \\ 0 & 0 & 0 & \cdots & \lambda \end{pmatrix} \begin{pmatrix} x_1(i) \\ x_2(i) \\ x_3(i) \\ \vdots \\ x_k(i) \end{pmatrix}$$

is

$$x_{k-i}(n) = \lambda^n x_{k-i}(0) + {}^nC_1 \lambda^{n-1} x_{k-i+1}(0) + \cdots + {}^nC_i \lambda^{n-i} x_k(0)$$

where

$$^nC_i = n!/(i!(n-i)!)$$

is the coefficient of x^i in $(x+1)^n$, so $^nC_0 = 1$, $^nC_1 = n$, $^nC_2 = n(n-1)/2, \ldots,$ $^nC_n = 1$, *etc.*

Proof Use induction on n together with the familiar identity

$$^nC_i + {}^nC_{i+1} = {}^{n+1}C_{i+1}.$$

to obtain the theorem. □

The same ideas can be applied to differential equations too.

Example 14.13 Quantities $x(t), y(t), z(t)$ vary with time, t, and satisfy the equations

$$\frac{dx}{dt} = 3x + z$$

$$\frac{dy}{dt} = -x + y - z$$

$$\frac{dz}{dt} = y + 2z$$

and boundary conditions $x(0) = y(0) = z(0) = 1$. We find $x(t), y(t), z(t)$.

In matrix form, we have

$$\frac{d}{dt} \begin{pmatrix} x \\ y \\ z \end{pmatrix} = \mathbf{A} \begin{pmatrix} x \\ y \\ z \end{pmatrix}$$

where

$$\mathbf{A} = \begin{pmatrix} 3 & 0 & 1 \\ -1 & 1 & -1 \\ 0 & 1 & 2 \end{pmatrix}$$

is the matrix of the previous example. This suggests using the same base-change matrix $\mathbf{P}$, and defining new quantities $u(t), v(t), w(t)$ by

$$\begin{pmatrix} u \\ v \\ w \end{pmatrix} = \mathbf{P}^{-1} \begin{pmatrix} x \\ y \\ z \end{pmatrix}.$$

Then

$$\frac{d}{dt} \begin{pmatrix} u \\ v \\ w \end{pmatrix} = \mathbf{P}^{-1} \mathbf{A} \mathbf{P} \begin{pmatrix} u \\ v \\ w \end{pmatrix} = \begin{pmatrix} 2 & 1 & 0 \\ 0 & 2 & 1 \\ 0 & 0 & 2 \end{pmatrix} \begin{pmatrix} u \\ v \\ w \end{pmatrix}$$

or

$$\frac{dw}{dt} = 2w \qquad \frac{dv}{dt} = 2v + w \qquad \frac{du}{dt} = 2u + v.$$

The solution to this standard system of differential equations is

$$w = Ae^{2t} \qquad v = Ate^{2t} + Be^{2t} \qquad u = \tfrac{1}{2}At^2e^{2t} + Bte^{2t} + Ce^{2t},$$

for constants of integration A, B, C. These constants are found using the boundary conditions

$$\begin{pmatrix} u(0) \\ v(0) \\ w(0) \end{pmatrix} = \mathbf{P}^{-1} \begin{pmatrix} x(0) \\ y(0) \\ z(0) \end{pmatrix} = \mathbf{P}^{-1} \begin{pmatrix} 1 \\ 1 \\ 1 \end{pmatrix} = \begin{pmatrix} -1 \\ -1 \\ 3 \end{pmatrix}.$$

So $A = 3$, $B = -1$, and $C = -1$. This gives the required solution

$$x = \tfrac{3}{2}t^2e^{2t} + 2te^{2t} + e^{2t}$$
$$y = -3te^{2t} + e^{2t}$$
$$z = -\tfrac{3}{2}t^2e^{2t} + te^{2t} + e^{2t}$$

of the original differential equations.

14.4   Proof of the primary decomposition theorem

Here we prove the primary decomposition theorem. As will be clear, if the theorem is taken on trust, the proof is not required in calculations. However, the proof is given here for those readers who like to see the complete story.

Suppose that we have a linear map $f : V \to V$ with minimum polynomial $m_f(x) = (x - \lambda_1)^{e_1} \dots (x - \lambda_r)^{e_r}$, where $\lambda_1, \dots, \lambda_r$ are the distinct eigenvalues of f, and let $V_1, \dots, V_r$ be the generalized eigenspaces, defined by

$$V_i = \ker((f - \lambda_i I)^{e_i}).$$

We want to prove that V is the direct sum of these generalized eigenspaces.

Define $p(x) = (x - \lambda_1)^{e_1}$ and $q(x) = (x - \lambda_2)^{e_2} \dots (x - \lambda_r)^{e_r}$, so that $m_f(x) = p(x)q(x)$ and $V_1 = \ker(p(f))$. Also define $W_1 = \ker(q(f))$. Our plan is to show that $V_1 \oplus W_1 = V$, and then use induction on $\dim V$ to obtain a decomposition $W_1 = \bigoplus_{i \geqslant 2} V_i$.

We shall use the result in Proposition 9.9 (or Proposition 9.14) which says there are polynomials $t(x)$ and $s(x)$ such that

$$t(x)p(x) + s(x)q(x) = 1.$$

We shall also use the fact from Section 9.2 that $p(f)q(f) = q(f)p(f)$, etc., throughout.

Lemma 14.14 $V_1 \oplus W_1 = V$.

Proof Let $v \in V$. Then

$$v = Iv = (t(f)p(f) + s(f)q(f))v = t(f)p(f)v + s(f)q(f)v.$$

But $t(f)p(f)v \in W_1$ since $q(f)(t(f)p(f)v) = t(f)(p(f)q(f)v) = 0$ as $p(f)q(f) = 0$. Similarly, $s(f)q(f)v \in V_1$, so $V = V_1 + W_1$.

To show that this sum is direct, suppose $v \in V_1$, $w \in W_1$, and $v + w = 0$. We must show $v = w = 0$. But $v + w = 0$ implies $p(f)v + p(f)w = 0$ but $p(f)v = 0$ as $v \in \ker p(f)$, so $p(f)w = 0$. But this means

$$w = 1w = (t(f)p(f) + s(f)q(f))w = t(f)p(f)w + s(f)q(f)w = 0$$

as $q(f)w = 0$ since $w \in \ker q(f)$. So $w = 0$. We prove $v = 0$ in exactly the same way. $\qquad\square$

Theorem 14.15 *If* $f: V \to V$ *is a linear map with minimum polynomial*

$$m_f(x) = (x - \lambda_1)^{e_1} \dots (x - \lambda_r)^{e_r},$$

where $\lambda_1, \dots, \lambda_r$ *are the (distinct) eigenvalues of* f, *let* $V_1, \dots, V_r$ *be the generalized eigenspaces, defined by*

$$V_i = \ker((f - \lambda_i 1)^{e_i}).$$

Then $V = V_1 \oplus \cdots \oplus V_r$.

Proof By induction on the number r of distinct eigenvalues of f.

If $r = 1$ there is nothing to prove. If $r > 1$, we have $V = V_1 \oplus W_1$, and we can define $g: W_1 \to W_1$ by $g(w) = f(w)$. We only need to check that $g(w) \in W_1$. But if $w \in W_1 = \mathrm{im}(p(f))$ then $w = p(f)(v)$ for some $v \in V$, so $g(w) = f(w) = (f \circ p(f))(v) = (p(f) \circ f)(v) = p(f)(f(v)) \in \mathrm{im}(p(f)) = W_1$. Moreover, $W_1 = \ker(q(f))$, so $q(g)$ maps every vector in W_1 to 0. Therefore

$$q(x) = (x - \lambda_2)^{e_2} \dots (x - \lambda_r)^{e_r}$$

divides the minimal polynomial $m_g(x)$ of g. It follows that g restricted to W_1 has $r - 1$ eigenvalues, and we can use induction to say that $W_1 = V_2 \oplus \cdots \oplus V_r$, so that $V = V_1 \oplus W_1 = V_1 \oplus V_2 \oplus \cdots \oplus V_r$. $\qquad\square$

Let us look at an example in detail to see how this primary decomposition works. Let $V = \mathbb{R}^3$ and suppose that $f: V \to V$ is given by

$$f\begin{pmatrix} a \\ b \\ c \end{pmatrix} = \begin{pmatrix} -a - c \\ 4a - b + 7c \\ 4a + 3c \end{pmatrix},$$

so that

$$
\begin{aligned}
f^2\begin{pmatrix} a \\ b \\ c \end{pmatrix} &= f\begin{pmatrix} -a - c \\ 4a - b + 7c \\ 4a + 3c \end{pmatrix} \\
&= \begin{pmatrix} -(-a - c) - (4a + 3c) \\ 4(-a - c) - (4a - b + 7c) + 7(4a + 3c) \\ 4(-a - c) + 3(4a + 3c) \end{pmatrix} \\
&= \begin{pmatrix} -3a - 2c \\ 20a + b + 10c \\ 8a + 5c \end{pmatrix}.
\end{aligned}
$$

Then f is represented with respect to the standard basis by the matrix

$$\mathbf{A} = \begin{pmatrix} -1 & 4 & 4 \\ 0 & -1 & 0 \\ -1 & 7 & 3 \end{pmatrix}$$

and the characteristic polynomial of f is

$$\chi_f(x) = \chi_\mathbf{A}(x) = (x-1)^2(x+1).$$

Therefore the minimal polynomial of f, $m_f(x) = m_\mathbf{A}(x)$, is either $(x-1)(x+1)$ or $(x-1)^2(x+1)$. But $(\mathbf{A}-\mathbf{I})(\mathbf{A}+\mathbf{I}) = \mathbf{A}^2 - \mathbf{I} \neq 0$, so in fact $m_\mathbf{A}(x) = (x-1)^2(x+1)$. (Note: throughout these calculations you can work either with the matrix $\mathbf{A}$ or with the linear map f.)

Now we want to take out one linear factor of $m_\mathbf{A}(x)$, to the full power, say $p(x) = (x-1)^2$, and $q(x)$ is what is left, namely $q(x) = (x+1)$. By definition we have $m_f(x) = p(x)q(x)$. Substituting f into this identity, we obtain $p(f)q(f) = m_f(f) = 0$, so all vectors get mapped to 0 by $p(f) \circ q(f)$. Of course, some vectors go to 0 under $p(f)$ on its own (these are exactly the elements of $\ker(p(f))$), and some go to 0 under $q(f)$ on its own (these vectors form $\ker(q(f))$).

We have $q(x) = x + 1$, so $q(f) = f + I$ is defined by $(q(f))(v) = f(v) + v$, i.e.

$$q(f)\colon (a,b,c)^T \mapsto (-c, 4a + 7c, 4a + 4c)^T.$$

Therefore the image of $q(f)$ is spanned by the vectors $(0,4,4)^T$ and $(-1,7,4)^T$, or to take a simpler basis, $\{(0,1,1)^T, (1,0,3)^T\}$. Moreover, the kernel of $q(f)$ consists of all vectors $(a,b,c)^T$ such that $(-c, 4a+7c, 4a+4c) = (0,0,0)$, in other words $\ker(q(f)) = \langle (0,1,0)^T \rangle$. It is easy to see now that in this example $V = \mathrm{im}(q(f)) \oplus \ker(q(f))$.

Similarly we have $p(x) = (x-1)^2$, so $p(f) = f^2 - 2f + I$ and

$$(p(f)) \begin{pmatrix} a \\ b \\ c \end{pmatrix} = f^2 \begin{pmatrix} a \\ b \\ c \end{pmatrix} - 2f \begin{pmatrix} a \\ b \\ c \end{pmatrix} + \begin{pmatrix} a \\ b \\ c \end{pmatrix}$$

$$= \begin{pmatrix} -3a - 2c \\ 20a + b + 10c \\ 8a + 5c \end{pmatrix} - 2 \begin{pmatrix} -a - c \\ 4a - b + 7c \\ 4a + 3c \end{pmatrix} + \begin{pmatrix} a \\ b \\ c \end{pmatrix}$$

$$= \begin{pmatrix} 0 \\ 12a + 4b - 4c \\ 0 \end{pmatrix}.$$

Therefore the image of $p(f)$ is spanned by $(0,1,0)^T$. Also, the kernel of $p(f)$ consists of all vectors $(a,b,c)^T$ satisfying $12a + 4b - 4c = 0$. This is clearly a two-dimensional space, spanned by $(1,0,3)^T$ and $(0,1,1)^T$. Thus we see that the image of $p(f)$ is equal to the kernel of $q(f)$, and vice versa. In the notation above we have $V_1 = \mathrm{span}((1,0,3)^T, (0,1,1)^T)$, $W_1 = \mathrm{span}((0,1,0)^T)$, and $V = V_1 \oplus W_1$.

If we choose the new basis

$$v_1 = (1,0,3)^T, v_2 = (0,1,1)^T, v_3 = (0,1,0)^T$$

and write f with respect to this basis, we see that

$$
\begin{array}{rclcl}
f(v_1) &=& (-4,25,13)^T &=& -4v_1 + 25v_2 \\
f(v_2) &=& (-1,6,3)^T &=& -v_1 + 6v_2 \\
f(v_3) &=& (0,-1,0)^T &=& -v_3
\end{array}
$$

so the matrix of f with respect to the ordered basis v_1, v_2, v_3 is

$$\mathbf{B} = \begin{pmatrix} -4 & -1 & 0 \\ 25 & 6 & 0 \\ 0 & 0 & -1 \end{pmatrix}$$

which is in block diagonal form: we have separated out the different eigenvalues into different blocks.

Exercises

Exercise 14.1 Using matrices

$$\mathbf{P}^{-1} = \begin{pmatrix} -1 & 2 & 2 & -2 \\ 1 & -1 & -1 & 1 \\ 1 & -1 & -2 & 2 \\ -1 & 1 & 2 & -1 \end{pmatrix} \qquad \mathbf{P} = \begin{pmatrix} 1 & 2 & 0 & 0 \\ 1 & 0 & 1 & 0 \\ 0 & 1 & 0 & 1 \\ 0 & 0 & 1 & 1 \end{pmatrix}$$

compute $\mathbf{P}^{-1}\mathbf{AP}$, where

$$\mathbf{A} = \begin{pmatrix} 5 & -3 & -5 & 5 \\ 1 & 1 & -1 & 1 \\ 2 & -2 & -2 & 3 \\ 1 & -1 & -2 & 3 \end{pmatrix}.$$

What are the characteristic and minimum polynomials of $\mathbf{A}$? Give bases for each of the generalized eigenspaces $\ker(\mathbf{A} - \lambda \mathbf{I})^a$.

Exercise 14.2 For the matrix

$$\mathbf{A} = \begin{pmatrix} 0 & 2 & -1 \\ -2 & 3 & -2 \\ -3 & 2 & -2 \end{pmatrix},$$

compute a base-change matrix $\mathbf{P}$ such that $\mathbf{P}^{-1}\mathbf{AP}$ is in Jordan normal form, as follows.

(a) Compute $\chi_\mathbf{A}(x)$. Show that it is of the form $-(x - \lambda_1)(x - \lambda_2)^2$ for some distinct λ_1, λ_2.

(b) Find bases u of $\ker(\mathbf{A} - \lambda_1 \mathbf{I})$, v_1 of $\ker(\mathbf{A} - \lambda_1 \mathbf{I})$, and v_1, v_2 of $\ker(\mathbf{A} - \lambda_1 \mathbf{I})^2$.

(c) Working from first principles, explain why $(\mathbf{A} - \lambda_1\mathbf{I})v_2$ is a scalar multiple of v_1. [Hint: what is $(\mathbf{A} - \lambda_1\mathbf{I})(\mathbf{A} - \lambda_1\mathbf{I})v_2$?]

(d) Let $w_1 = u$, $w_2 = (\mathbf{A} - \lambda_1\mathbf{I})v_2$, and $w_3 = v_2$. What is the matrix of the linear transformation $\mathbf{A}$ with respect to this basis? Write down the base-change matrix $\mathbf{P}$.

Exercise 14.3 Repeat the last exercise (except for part (c)) for the matrix

$$\mathbf{A} = \begin{pmatrix} 0 & 2 & -1 \\ 0 & 1 & 0 \\ 1 & -2 & 2 \end{pmatrix}.$$

(This time $\chi_\mathbf{A}(x) = (\lambda - x)^3$ for some λ. Find bases for $\ker(\mathbf{A} - \lambda\mathbf{I})^n$ for $n = 1, 2, 3$.)

Exercise 14.4 Solve

(a) $x_{n+1} = 2y_n - z_n$
$y_{n+1} = y_n$
$z_{n+1} = x_n - 2y_n + 2z_n$
$x_0 = y_0 = z_0 = 1$

(b) $x_{n+1} = 5x_n - 3y_n - 5z_n + 5w_n$
$y_{n+1} = x_n + y_n - z_n + w_n$
$z_{n+1} = 2x_n - 2y_n - 2z_n + 3w_n$
$w_{n+1} = x_n - y_n - 2z_n + 3w_n$
$x_0 = y_0 = z_0 = w_0 = 1.$

Exercise 14.5 Find bases for the primary components of the linear map represented by the matrix

$$\mathbf{A} = \begin{pmatrix} -1 & 0 & -4 & 0 \\ 1 & -2 & -7 & 1 \\ 1 & 0 & 3 & 0 \\ 3 & 0 & 15 & -2 \end{pmatrix},$$

and hence find a matrix $\mathbf{P}$ such that $\mathbf{P}^{-1}\mathbf{A}\mathbf{P}$ is in block diagonal form.

Exercise 14.6 (a) Let

$$\mathbf{A} = \begin{pmatrix} 3 & 0 & 1 \\ -1 & 1 & -1 \\ 0 & 1 & 2 \end{pmatrix}.$$

Find the characteristic polynomial, eigenvalues, and the minimum polynomial of $\mathbf{A}$, and find the algebraic and geometric multiplicities of each eigenvalue. Write down the Jordan normal form $\mathbf{J}$ for $\mathbf{A}$.

(b) Do the same for the matrix $\mathbf{B} = \begin{pmatrix} i & -1 \\ 1 & i-2 \end{pmatrix}$.

Exercise 14.7 Write down all possible Jordan normal forms for matrices with characteristic polynomial $(x - \lambda)^5$. In each case, calculate the minimum polynomial and the geometric multiplicity of the eigenvalue λ. Verify that this information determines the Jordan normal form.

Exercise 14.8 Do the same for $(x - \lambda)^6$.

Exercise 14.9 Show that there are two 7×7 matrices in Jordan normal form which have the same minimum polynomial and for which λ has the same geometric multiplicity, but which are not similar.

Exercise 14.10 Show that the Jordan normal form of a 2×2 matrix is determined by its minimum polynomial, but that this is not true for 3×3 matrices.

Exercise 14.11 Show that the Jordan normal form of a 3×3 matrix is determined by its minimum polynomial and its characteristic polynomial, but that this is not true for 4×4 matrices.

Appendix A
A theorem of analysis

If I is an interval of the reals (e.g. $[a, b]$, (a, b), $[a, b)$ where $a < b$ and a, b are either real or $\pm\infty$), a function $f\colon I \to \mathbb{R}$ or $f\colon I \to \mathbb{C}$ is said to be *continuous* at a point $c \in I$ if for all $\epsilon > 0$ from $\mathbb{R}$ there is $\delta > 0$ in $\mathbb{R}$ with

$$|f(x) - f(c)| < \epsilon \text{ for all } x \in I \text{ with } 0 < |x - c| < \delta$$

where $|\,|$ is the usual absolute value in $\mathbb{R}$ or $\mathbb{C}$. The function $f\colon I \to \mathbb{R}$ or $\mathbb{C}$ is continuous if it is continuous at all points $c \in I$.

Continuous functions $f\colon I \to \mathbb{R}$ or $\mathbb{C}$ can always be integrated; that is, if $r \leqslant s$ are real numbers in I then the integral

$$\int_r^s f(x)\, dx$$

can be defined by some limiting process as 'the area under the line $y = f(x)$'. (Two methods of defining integrals are commonly used. The first is the *Riemann integral*, the second is the more general and more powerful *Lebesgue integral*; we don't go into the details, but either method is suitable for the discussion here.) If r or s is required to be infinite, we define the integral as a limit as follows,

$$\int_s^\infty f(x)\, dx = \lim_{r \to \infty} \int_s^r f(x)\, dx,$$

$$\int_{-\infty}^r f(x)\, dx = \lim_{s \to -\infty} \int_s^r f(x)\, dx,$$

$$\int_{-\infty}^\infty f(x)\, dx = \lim_{\substack{r \to \infty \\ s \to \infty}} \int_{-s}^r f(x)\, dx,$$

when these limits exist.

Here is a proof that $\langle f|g \rangle = \int_a^b f(x)g(x)\, dx$ is an inner product on $\mathscr{C}[a, b]$. (See Example 3.4.)

Theorem A.1 *Let V be the vector space $\mathscr{C}[a, b]$ of all continuous functions from the closed interval $[a, b]$ to $\mathbb{R}$, under pointwise addition and scalar multiplication. For any $f, g \in V$, define $\langle f|g \rangle = \int_a^b f(x)g(x)\, dx$. Then $\langle | \rangle$ is an inner product on V.*

The symmetry and bilinearity follow immediately from

$$\int_a^b f(x)g(x)\,dx = \int_a^b g(x)f(x)\,dx$$

and

$$\int_a^b (\alpha f(x) + \beta g(x))h(x)\,dx = \alpha \int_a^b f(x)h(x)\,dx + \beta \int_a^b g(x)h(x)\,dx.$$

Moreover, $\langle f | f \rangle = \int_a^b (f(x))^2\,dx \geqslant 0$ for any $f \in V$. Finally, we have to show that if $\int_a^b (f(x))^2\,dx = 0$ then $f = 0$ (i.e. $f(x) = 0$ for all $x \in [a, b]$). This follows immediately from the following result.

Lemma A.2 *If* $h : [a, b] \to \mathbb{R}$ *is a continuous function, with* $h(x) \geqslant 0$ *for all* $x \in [a, b]$, *and* h *is not identically zero, then* $\int_a^b h(x)\,dx > 0$.

Proof If $h(x) = 0$ for all $x \in (a, b)$, then by continuity $h(a) = h(b) = 0$ so h is identically zero, which is a contradiction. Therefore $\exists c \in (a, b)$ such that $h(c) \neq 0$. Since h is continuous, there is $\delta > 0$ such that $(c - \delta, c + \delta) \subseteq [a, b]$ and if $|x - c| < \delta$ then $|h(x) - h(c)| < \frac{1}{2}h(c)$ (in particular, $h(x) > \frac{1}{2}h(c)$). Thus there is a step function k (see diagram)

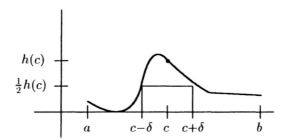

defined by $k(x) = \frac{1}{2}h(c)$ if $x \in (c - \delta, c + \delta)$ and $k(x) = 0$ otherwise, which is bounded by h—that is, $0 \leqslant k(x) \leqslant h(x)$ for all $x \in [a, b]$. So by the definition of the Riemann integral, $\int_a^b h(x)\,dx \geqslant \int_a^b k(x)\,dx = 2\delta\frac{1}{2}h(c) > 0$ as required. $\quad\square$

The proof of this result can be modified to give a proof of the corresponding result for infinite integrals, provided the functions are sufficiently well-behaved for the integrals to exist.

Appendix B

Applications to quantum mechanics

Many of the ideas in this book are applied in quantum mechanics. In particular the notions of an inner product and of self-adjoint linear transformations of certain infinite dimensional vector spaces over the complex numbers are essential for the theory of quantum mechanics. We include a very brief discussion of some of the ideas here, since they give useful motivation of many of the topics discussed in the book—especially the importance of inner product spaces over $\mathbb{C}$.

In quantum mechanics, a particle moving along the x-axis at a particular moment in time t is represented by a continuous complex-valued function $\psi\colon \mathbb{R} \to \mathbb{C}$ in the space V of Example 3.24, where the argument x of $\psi(x)$ denotes position along the axis. We further stipulate that for such a function representing the state of a particle we have

$$\|\psi\|^2 = \langle \psi | \psi \rangle = \int_{-\infty}^{\infty} \overline{\psi(x)} \psi(x)\, dx = 1. \tag{1}$$

The real-valued function of x, $|\psi(x)|^2$, is interpreted as a *probability density*— that is, for all $a < b$ in $\mathbb{R}$ the integral

$$\int_a^b |\psi(x)|^2\, dx$$

gives the probability that the particle will be found in the interval $[a, b]$. (The condition $\|\psi\|^2 = 1$ of (1) can be seen as just saying that the probability that the particle is *somewhere* on the x-axis is exactly 1.)

Many other expressions using the inner product have physical interpretations too. For example,

$$\int_{-\infty}^{\infty} \overline{\psi(x)} x \psi(x)\, dx = \langle \psi | \phi \rangle,$$

where ϕ is the function $x \mapsto x\psi(x)$, is the *expected* or *average position* of the particle, and is defined whenever the function $\phi\colon x \mapsto x\psi(x)$ is in V. Also, assuming $\psi(x)$ is differentiable with respect to x with continuous derivatives in V,

$$\int_{-\infty}^{\infty} \overline{\psi(x)}\, \frac{-ih}{2\pi}\frac{d\psi}{dx}\, dx = \frac{-ih}{2\pi}\langle\psi|\psi'\rangle$$

is the expected or average *momentum* of the particle at time t, where the physical constant $h \approx 6.63 \times 10^{-34}$ J s is *Planck's constant* and the prime in ψ' denotes differentiation.

The general form for these 'expected values' is that for a measurable quantity q, there is a self-adjoint linear transformation $\mathscr{O}_q \colon V \to V$ (usually called an 'operator') such that $\langle\psi|\mathscr{O}_q(\psi)\rangle$ gives the expected value for measurements of q. Thus the position operator $\mathscr{O}_x$ is pointwise multiplication by the position x, taking ψ to the function $x \mapsto x\psi(x)$, and the momentum operator $\mathscr{O}_p$ is $(-ih/2\pi)(d/dx)$ taking ψ to $(-ih/2\pi)\psi'$.

Exercise B.1 Use integration by parts to show that $\langle\psi|i\phi'\rangle = \langle i\psi'|\phi\rangle$ for continuously differentiable $\psi, \phi \colon \mathbb{R} \to \mathbb{C}$ in the vector space V of Example 3.24. Use $\langle\psi|\phi\rangle = \overline{\langle\phi|\psi\rangle}$ to deduce that $\langle\psi|i\psi'\rangle$ is real. (Note that the imaginary number i is required to make the signs turn out right.)

This last exercise shows that the operator $i(d/dx)$ (which is clearly a linear transformation $V \to V$) is indeed self-adjoint. In quantum mechanics, all 'measurable' quantities are represented by such operators.

We now consider a particle moving along the x-axis and described at time t by a function $\psi \in V$ which is sufficiently smooth (differentiable, with continuous derivatives, etc.) for the operators discussed here to make sense.

If q is a measurable quantity with corresponding operator $\mathscr{O}_q$, then $e_q = \langle\psi|\mathscr{O}_q(\psi)\rangle$ is the mean or expected value for q, and the 'uncertainty' of measurement of q is given by Δ_q^2, the mean square deviation from the mean, i.e. the expected value of $(q - e_q)^2$. In terms of operators, this is given by

$$\begin{aligned}
\Delta_q^2 &= \langle\psi|(\mathscr{O}_q - e_q)^2(\psi)\rangle \\
&= \langle\psi|\mathscr{O}_q^2(\psi)\rangle - 2e_q\langle\psi|\mathscr{O}_q(\psi)\rangle + \langle\psi|e_q^2\psi\rangle \\
&= \langle\psi|\mathscr{O}_q^2(\psi)\rangle - e_q^2,
\end{aligned}$$

recalling that $e_q = \langle\psi|\mathscr{O}_q(\psi)\rangle$ and $\|\psi\|^2 = 1$.

In the special cases of position x and momentum p, the operators for Δ_x^2 and Δ_p^2 are

$$\alpha = (x - e_x)^2$$

and

$$\beta = \left(-(ih/2\pi)(d/dx) - e_p\right)^2,$$

so

$$\Delta_x^2 \Delta_p^2 = \langle \psi | \alpha^2(\psi) \rangle \langle \psi | \beta^2(\psi) \rangle$$
$$= \langle \alpha(\psi) | \alpha(\psi) \rangle \langle \beta(\psi) | \beta(\psi) \rangle$$
$$\geqslant |\langle \alpha(\psi) | \beta(\psi) \rangle|^2$$
$$= |\langle \psi | \alpha\beta(\psi) \rangle|^2,$$

using the Cauchy–Schwarz inequality and the facts that α and β are self-adjoint. (This is obvious for α; for β it follows from Exercise B.1.)

The expression $|\langle \psi | \alpha\beta(\psi) \rangle|^2$ can be estimated, and it turns out that

$$|\langle \psi | \alpha\beta(\psi) \rangle|^2 \geqslant \tfrac{1}{4} |\langle \psi | (\alpha\beta - \beta\alpha)(\psi) \rangle|^2.$$

(See Exercise 13.9 in Chapter 13.) Also,

$$(\alpha\beta - \beta\alpha)(\psi) = \frac{-ih}{2\pi} \left[x \frac{d\psi}{dx} - \frac{d}{dx}(x\psi) \right] = \frac{-ih}{2\pi} \psi,$$

so (using $\|\psi\|^2 = 1$ again) we get

$$\Delta_x^2 \Delta_p^2 = |\langle \psi | \alpha\beta(\psi) \rangle|^2$$
$$\geqslant \tfrac{1}{4} |\langle \psi | (\alpha\beta - \beta\alpha)(\psi) \rangle|^2$$
$$\geqslant \tfrac{1}{4} |-ih/2\pi|^2 \|\psi\|^2$$
$$\geqslant h^2/16\pi^2$$

or

$$\Delta_x \Delta_p \geqslant h/4\pi,$$

which is the famous *Heisenberg uncertainty principle* that describes the theoretical limit of any attempt on the simultaneous measurement of both the momentum and position of a particle.

Index

Printed in the United States
35825LVS00003B/145-162